Mathematics for Physical Chemistry

Mathematics for Physical Chemistry

ROBERT G. MORTIMER

DEPARTMENT OF CHEMISTRY

Southwestern at Memphis

Macmillan Publishing Co., Inc.
NEW YORK

Collier Macmillan Publishers
LONDON

Macmillan Publishing Co., Inc.
866 Third Avenue, New York, New York 10022
Collier Macmillan Canada, Ltd.

Library of Congress Cataloging in Publication Data

Mortimer, Robert G
  Mathematics for physical chemistry.
  Includes bibliographies and index.
  1. Chemistry, Physical and theoretical—
Mathematics. I. Title.
QD455.3.M3M67   510'.24541   80-16615
ISBN 0-02-384000-5

Printing: 1 2 3 4 5 6 7 8   Year: 1 2 3 4 5 6 7

Preface

The purpose of this book is to provide a survey of the mathematics needed for physical chemistry courses at the undergraduate level. Although several kinds of elementary physical chemistry courses exist, all have some mathematics as a prerequisite. However, in nearly two decades of teaching physical chemistry, I have found that only a small minority of students have been introduced to all the mathematical topics needed in physical chemistry, and that most need some practice in applying their mathematical knowledge to the problems of physical chemistry.

Physical chemistry was defined by my first teacher of the subject as "anything that physical chemists are interested in," but a better definition is that physical chemistry is the branch of chemistry which applies the methods and theories of physics to chemical problems. As such, it involves more theory and more mathematics than other branches of chemistry, and in fact contains the theoretical foundation underlying all branches of chemistry.

Students arrive at their first course in physical chemistry with a variety of backgrounds, so I have not tried to write part of the book as a review of familiar topics and part of it as an introduction to new material. I have consciously tried to write all parts of the book so that they can be used for selfstudy by someone not familiar with the material, although of course any book such as this cannot be a substitute for the traditional training in mathematics required by students of chemistry. Solved examples and problems for the reader are interspersed throughout the presentations, and these form an important part of the presentations. As you study any topic in the book, you should follow the solution to each example, and work each problem as you come to it.

The first nine chapters of the book are constructed around a sequence of mathematical topics. Chapter 10 is a discussion of mathematical topics needed in the analysis of experimental data, and Chapter 11 is a brief introduction to computer programming in the BASIC language. Most of the material in at least the first four chapters should be a review for nearly all readers of the book. I have tried to write all of the chapters after the first four so that they can be studied in any sequence, or piecemeal as the need arises. The first section in each chapter (except Chapter 1) contains a list of instructional objectives, which can be used by a reader to determine whether he needs to study that chapter or not.

I hope that this book will serve three functions: (1) as a review or introduction to new topics for those preparing for a course in physical chemistry, (2) as a supplementary text to be used during a physical chemistry course, and (3) as a reference book for graduate students and practicing chemists.

It is a pleasure to acknowledge the help which I have received from James Smith and Gregory Payne of the Macmillan Publishing Co., and from the following persons, who have reviewed all or part of the manuscript:

Sydney Bluestone, Allen Denio, Gerald E. Doeden, Fred H. Dorer, Vojtech Fried, Michael E. Green, Clifford W. Hand, William Z. Plachy, and Daniel Ira Sverdlik. I also wish to express thanks to my teachers of physical chemistry. I have benefited from the instruction of Lowell Tensmeyer, the late Norman Bauer, Robert Mazo, Norman Davidson, Joseph E. Mayer, and Henry Eyring. Thanks are due the Research and Creative Activities Committee of Southwestern College for a summer research grant under which I began this book. Finally, I want to thank my family for exercising a considerable amount of patience while I was working on this book.

Robert G. Mortimer

Contents

CHAPTER ONE

Numbers, Mathematics, and Science—An Introduction to the Book

Section 1-1. GENERAL COMMENTS FROM THE AUTHOR TO THE READER

Mathematics and science have grown up together during the past three or four centuries, and some people, such as Isaac Newton, have made great contributions to both fields. However, there seems to be a gulf between pure mathematics, which is of interest to mathematicians, and applied mathematics, which is an indispensable tool for physicists, chemists, and biologists.

As you can see from the table of contents, this book is organized by mathematical topics. It is, however, not a mathematics book. The rigorous point of view of a trained mathematician is of great value, but I have not taken this point of view. The book presents mathematical concepts and facts without much derivation or rigor, and concentrates on the application of these concepts and facts to problems such as those arising in physical chemistry. Our approach is of course no substitute for a standard course of study in calculus and other areas of mathematics.

Physical chemistry is often the first course, outside of mathematics courses, in which a student encounters applications of mathematics beyond simple algebra and statistics. In my own first course in physical chemistry, it was like a revelation to me to see that chemists actually use calculus in their work, and also to learn that physical theory in a mathematical form underlies all of chemistry. I soon became glad that I was studying both chemistry and mathematics. I hope that you will gain the same feeling in your physical chemistry course, and that this book will assist you.

I have tried to write the book so that the further you go into the book, the more likely you will be to encounter material that is new to you. I have also tried to write the book so that it is not necessary to work through the book from front to back, and you may wish to study the chapters in a sequence of your own design.

Section 1-2. NUMBERS AND MEASUREMENTS

There are several sets into which we can classify numbers. Those which can represent actual measured physical quantities are called *real numbers*. They can range from positive numbers of indefinitely large magnitude to negative numbers of indefinitely large magnitude. Among the real numbers are the integers, 0, ± 1, ± 2, ± 3, etc. Other numbers are said to be *rational numbers*. They are quotients of two integers, such as $\frac{2}{3}$, $\frac{1}{2}$, $\frac{37}{53}$, etc. Of course, the integers are rational numbers. Other numbers are called *algebraic irrational numbers*. They include square roots of rational numbers, cube roots of rational numbers, etc., which are not themselves rational numbers. All of the rest of the real numbers are called *transcendental irrational numbers*. Two commonly encountered transcendental irrational numbers are the ratio of the circumference of a circle to its diameter, called π and given by 3.141592654 ..., and the base of natural logarithms, called e and given by 2.718281828 .... The three dots that follow the given digits indicate that more digits follow. Irrational numbers have the property that if you have some means of finding what the correct digits are, you will never reach a point beyond which all of the remaining digits are zero, or beyond which the digits form a repeating pattern. However, with a rational number, one or the other of these two things will always happen.[1]

Problem 1-1

Take a few simple fractions, such as $\frac{2}{3}$, $\frac{4}{9}$, or $\frac{3}{7}$, and express them as decimal numbers, finding either all the nonzero digits or the repeating pattern of digits. •

The most common use that chemists make of numbers is to report values for measured quantities. A measured quantity can almost never be known with complete accuracy, unless the number is required to be a fairly small integer. It is therefore a good idea to communicate the probable accuracy of a reported measurement.

For example, let us consider the specification of the length of an object, say a piece of glass tubing, which we have measured with a meter stick. We will discuss experimental error in Chapter 10, but let us now assume that we believe our experimental error to be no greater than 0.6 millimeter (mm) and that our measured value is 387.8 mm.

The best way to specify what we believe the length of the glass tubing to be is to say:

$$\text{length} = 387.8 \pm 0.6 \text{ mm}$$

---

[1] I have been told that early in the twentieth century, the legislature of the state of Indiana, in an effort to simplify things, passed a resolution that henceforth in that state, π should be equal to exactly 3.

If for some reason we cannot include a statement of the probable error, we must at least avoid including digits that are likely to be wrong. That is, we must *round off* the number. Since our error is somewhat less than 1 mm, the correct number is probably closer to 388 mm than to either 387 mm or 389 mm, so we report the length as 388 mm and assert that the three digits given are significant. All this means is that we think the given digits are correct. If we had reported the length as 387.8 mm, the last digit is insignificant. That is, it is probably wrong.

You should always avoid reporting digits that are not significant. When you carry out calculations involving measured quantities, you should always determine how many significant digits your answer can have and round off your result to that number of digits.

When you are multiplying a number of factors together, a good rule of thumb is that your answer will have the same number of significant digits as the factor with the *least* number. The same rule holds for division.

Example 1-1

What is the area of a rectangle whose length is given as 7.78 m and whose width is given as 3.486 m?

Solution

When we use a calculator to find the product of 7.78 m and 3.486 m, we get 27.12108 m^2. We must round this to 27.1 m^2. •

Example 1-2

Compute the smallest and largest values that the area of the rectangle might have, and determine whether the answer given in Example 1-1 is correctly stated.

Solution

The smallest value that the length might have, assuming the given value to have only significant digits, is 7.775 m, and the largest value that it might have is 7.785 m. The smallest possible value for the width is 3.4855 m and the largest value is 3.4865 m. The minimum value for the area is

$$A(\text{minimum}) = (7.775 \text{ m})(3.4855 \text{ m}) = 27.0997625 \text{ m}^2$$

The maximum value for the area is

$$A(\text{maximum}) = (7.785 \text{ m})(3.4865 \text{ m}) = 27.1424025 \text{ m}^2$$

To get an agreement between these numbers, we must round both of them to 27.1 m^2. The result in Example 1-1 was correctly stated. •

The rule of thumb for significant digits in addition or subtraction is slightly different: For a digit to be significant, it must arise from a significant digit in every term of the sum or difference.

Example 1-3

Determine the combined length of two objects, one of length 0.783 m and one of length 17.3184 m.

Solution

We add:

$$
\begin{array}{ll}
0.783 & \text{m} \\
17.3184 & \text{m} \\
\hline
18.1014 & \text{m}
\end{array}
$$

The final 4 is not a significant digit, because the fourth digit after the decimal point in the top number is unknown, and the 4 could be significant only if the fourth digit in the first number were a zero. We must round the answer to 18.101 m.   •

Some people round off the insignificant digits at each step of a calculation. However, this can lead to "round-off" error if a number of steps are required in a calculation. A reasonable policy is to carry along at least one insignificant digit during the calculation, and then to round off the insignificant digits at the final answer.

If your only insignificant digit is a 5, you must decide whether to round up to the next larger significant digit or to round down to the next lower significant digit. This problem can usually be avoided by carrying along two or more insignificant digits, but it is possible that a 50 or 500 will occur that must be rounded. It is best to be consistent, and either round up in all such cases, or always round down.

If you are carrying out operations other than additions, subtractions, multiplications, and divisions, you may have trouble in determining which digits are significant. If you must take sines, cosines, logarithms, etc., it may be necessary to do the operation with the smallest and the largest values that the number on which you must operate can have. However, rules of thumb can be found.[2]

Problem 1-2

Calculate the following to the proper numbers of significant digits.
a. $(37.815 + 0.00435)(17.01 + 3.713)$
b. $625[e^{12.1} + \sin(30°)]$
c. 65.718×14.3
d. $17.13 + 14.7651 + 3.123 + 7.654 - 8.123$ •

---

[2] Donald E. Jones, "Significant Digits in Logarithm–Antilogarithm Interconversions," *J. Chem. Educ.* **49**, 753 (1972).

Scientific Notation. A problem in communication arises if the last significant digit or digits in a number happen to be zeros. For example, say that a certain distance on the surface of the earth happens to be 6300 kilometers (km). That is, the distance is closer to 6300 km than it is to 6299 km or 6301 km. If you report the distance as 6300 km, most people will not assume that the two zeros are significant digits, but that they are included only to show where the decimal point belongs. In other words, they will assume that the distance is only known to lie between 6250 and 6350 km. However, if some leading zeros are necessary to locate a decimal point, everybody recognizes that there are not significant digits. Thus, 0.0000149 cm is a distance expressed with only three significant digits.

The communication difficulty just mentioned can be avoided by the use of *scientific notation*, in which a number is expressed as the product of two factors, one 10 raised to some power, and the other a number lying between 1 and 10. The distance mentioned above would thus be written as 6.300×10^3 km, and there are clearly four significant digits indicated, since the trailing zeros are not required to locate a decimal point. If the number were known only to two significant digits, it would be written as 6.3×10^3.

Scientific notation is convenient if extremely small or extremely large numbers must be written. For example, Avogadro's number, the number of molecules per mole, is much easier to write as 6.0220×10^{23} molecules mol^{-1} than as 602,200,000,000,000,000,000,000, and the charge on an electron is easier to write as 1.602×10^{-19} coulomb than as 0.0000000000000000001602 coulomb.

Problem 1-3

Convert the following numbers to scientific notation (assume that trailing zeros on the right are not significant).
a.   0.000645
b.   67,342,000
c.   0.000002
d.   6432   •

Section 1-3.   UNITS OF MEASUREMENT

In Section 1-2, we mentioned measuring the length of an object with a meter stick. Such a measurement would be impossible without a standard definition of the meter (or other unit of length), and for many years science and commerce were hampered by the lack of accurately defined units of measurement. This problem has been largely overcome by accurate measurements and international agreements.

The internationally accepted system of units of measurements is called the *Système International d' Unités*, abbreviated SI. This is an MKS system, which means that length is measured in meters, mass in kilograms, and time

Table 1-1. SI Units*

SI Base Units (quantities with independent definitions)			
Physical Quantity	*Name of Unit*	*Symbol*	*Definition*
length	meter	m	1, 650, 763.73 wavelengths in vacuum for a certain spectral line of krypton-86.
mass	kilogram	kg	The mass of a platinum-iridium cylinder kept at the Internation Bureau of Weights and Measures
time	second	s	The duration of 9, 192, 631, 770 cycles of the radiation of a certain emission of the cesium atom
electric current	ampere	A	The magnitude of current which, when flowing in each of two long parallel wires 1 m apart in free space, results in a force of 2×10^{-7} N per meter of length
temperature	kelvin	K	Absolute zero is 0 K, triple point of water is 273.16 K
luminous intensity	candela	cd	The luminous intensity, in prependicular direction, of surface of 1/600,000 sq m of a black body at temperature of freezing platinum at a pressure of 101325 N/m^2
amount of substance	mole	mol	Amount of substance which contains as many elementary units as there are carbon atoms in 0.012 kg of carbon-12.

Other SI Units				
Physical Quantity	*Name of Unit*	*Physical Dimensions*	*Symbol*	*Definition*
force	newton	kg m s^{-2}	N	$1\ N = 1\ kg\ m\ s^{-2}$
energy	joule	kg m^2 s^{-2}	J	$1\ J = 1\ kg\ m^2\ s^{-2}$
electrical charge	coulomb	A s	C	$1\ C = 1\ A\ s$
pressure	pascal	N m^{-2}	Pa	$1\ Pa = 1\ N\ m^{-2}$
magnetic field	tesla	kg s^{-2}A^{-1}	T	$1\ T = 1\ kg\ s^{-2}\ A^{-1} = 1$ weber m^{-2}
luminous flux	lumen	cd sr	lm	$1\ lm = 1\ cd\ sr$ (sr = steradian)

* Robert A. Alberty and Farrington Daniels, Physical Chemistry, fifth edition, p. 662, John Wiley and Sons, New York, 1979.

in seconds. These and the four other base units given in Table 1-1 form the heart of the system. Included in the table are also some "derived" units, which owe their definitions to the definitions of the seven base units. In 1960 the international chemical community agreed to use SI units, which had been in use by physicists for some time.[3]

Some non-SI units continue to be used, such as the atmosphere, which is a pressure equal to 101,325 N m^{-2}, and the torr, which is a pressure such that 760 torr equals 1 atmosphere. The Celsius temperature scale also remains in common use among chemists.

Multiples and submultiples of SI units are commonly used.[4] Examples are the millimeter and kilometer. These quantities are denoted by standard prefixes attached to the name of the unit, as listed in Table 1-2. The abbreviation for a multiple or submultiple is obtained by attaching the prefix abbreviation to the unit abbreviation, as in Gm (gigameter), or ns (nanosecond). Note that since the base unit of length is the kilogram, the table would imply the use of things such as the megakilogram. This is not done. We use gigagram instead of megakilogram.

Table 1-2. Prefixes for Multiple and Submultiple Units

Multiple	Prefix	Abbreviation
10^{12}	tera-	T
10^{9}	giga-	G
10^{6}	mega-	M
10^{3}	kilo-	k
1	—	—
10^{-1}	deci-*	d
10^{-2}	centi-*	c
10^{-3}	milli-	m
10^{-6}	micro-	μ
10^{-9}	nano-	n
10^{-12}	pico-	p
10^{-15}	femto-	f
10^{-18}	atto-	a

* The use of the prefixes for 10^{-1} and 10^{-2} is being discouraged, but centimeters will probably not be abandoned for many years to come.

[3] See "Policy for NBS Usage of SI Units," *J. Chem. Educ.* **48**, 569 (1971).

[4] There is a somewhat apocryphal story about Robert A. Millikan, a Nobel-prize-winning physicist who was not noted for false modesty. A rival is supposed to have told Millikan that he had defined a new unit for the quantitative measure of conceit, and had named the new unit the kan. However, 1 kan was an exceedingly large amount of conceit, so that for most purposes the practical unit was to be the millikan.

Any measured quantity is not completely specified until its units are given. If a is a length, one must say

$$a = 10.345 \text{ m} \tag{1.1}$$

not just

$$a = 10.345 \quad (not\ correct)$$

It is permissible to write

$$a/\text{m} = 10.345$$

which means that the length a divided by 1 meter (m) is 10.345, a dimensionless number.

When you make numerical calculations, you should always make certain that you use consistent units for all quantities. Otherwise, you will almost certainly get the wrong answer. This means that (1) you must convert all multiple and submultiple units to the base unit, and (2) you cannot mix different systems of units. For example, you cannot substitute a length in inches into a formula in which the other quantities are in SI units without converting. It is a good idea to write the unit as well as the number, as in Eq. (1.1), even for scratch calculations. This will help you avoid some kinds of mistakes, since you can inspect any equation and see that both sides are measured in the same units.

The Factor-Label Method. This is an elementary method for the routine conversion of a quantity measured in one unit to the same quantity measured in another unit. The method consists of multiplying the quantity by a fraction that is equal to unity in a physical sense, with the numerator and denominator equal to the same quantity expressed in different units. This does not change the quantity physically, but numerically expresses it in another unit, and so changes the number expressing the value of the quantity.

For example, to express 3.00 km in terms of meters, one writes

$$(3.00 \text{ km})\left(\frac{1000 \text{ m}}{1 \text{ km}}\right) = 3000 \text{ m} = 3.00 \times 10^3 \text{ m} \tag{1.2}$$

You can check the units by considering the same unit to "cancel" if it occurs in both the numerator and denominator. Thus, the left-hand side of Eq. (1.2) has units of meters, because the km on the top cancels the km on the bottom.

Example 1-4

Convert the speed of light, $2.9979 \times 10^8 \text{ m s}^{-1}$, to the same quantity in miles per hour. Use the definition of the inch, 1 in. = 0.0254 m (exactly).

Solution

$$2.9979 \times 10^8 \text{ m s}^{-1} \times \frac{1 \text{ in.}}{0.0254 \text{ m}} \times \frac{1 \text{ ft}}{12 \text{ in.}} \times \frac{1 \text{ mi}}{5280 \text{ ft}} \times \frac{60 \text{ s}}{1 \text{ min}} \times \frac{60 \text{ min}}{1 \text{ h}}$$

$$= 6.7061 \times 10^8 \text{ mi h}^{-1}$$

●

Problem 1-4

Express the following in terms of SI base units.
a. 26.17 mi   b. 55 mi h^{-1}
c. 7.5 nm ps^{-1} d. 13.6 eV (electron volts)
Note: The electron volt, a unit of energy, is defined in Appendix 1.   ●

Section 1-4. HOW TO SOLVE PROBLEMS

As you study physical chemistry, you must solve a large number of problems. In doing so, you will probably learn more than you do from reading your textbook or listening to lectures.

A typical physical chemistry problem is similar to what was once called a "story problem" in elementary school. You are given some factual information (or asked to find some), together with a verbal statement of what answer is required, but you must find your own method of obtaining the answer from the given information. In a simple problem, this may consist only of substituting numerical values into a formula, but in a more complicated problem you might have to derive your own mathematical formula, or you might have to carry out a complicated procedure such as drawing a graph and making conclusions from it or using statistical techniques. However, the method, or algorithm, must be developed for each problem.

Example 1-5

Calculate the volume occupied by 1.278 moles (mol) of an ideal gas if the pressure is 6.341 atmospheres (atm) and the temperature is 298.15 K.

Solution

The ideal gas equation of state gives

$$V = \frac{nRT}{P} \tag{1.3}$$

where V is the volume, n the number of moles of gas, T the temperature, P the pressure, and R the gas constant equal to 8.3144 J K^{-1} mol^{-1}. We calculate the volume by substitution into the formula

$$V = \frac{(1.278 \text{ mol})(8.3144 \text{ J K}^{-1} \text{ mol}^{-1})(298.15 \text{ K})}{6.341 \text{ atm}} \times \frac{1 \text{ atm}}{101{,}325 \text{ N m}^{-2}}$$

$$= 4.931 \times 10^{-3} \text{ J N}^{-1} \text{ m}^2 = 4.931 \times 10^{-3} \frac{\text{kg m}^2 \text{ s}^{-2}}{\text{kg m s}^{-2}} \text{ m}^2$$

$$= 4.931 \times 10^{-3} \text{ m}^3$$

Notice that in order to complete the "cancellation" of the units and determine the units of the result, it was necessary to express joules and newtons in terms of SI base units.   ●

The example just completed illustrates the fact that consistent units have to be used. If the conversion factor to change atmospheres to N m^{-2} had not been used, the units of the result would have been m^3 N m^{-2} atm^{-1}, which is not a simple unit of volume that we can use.

Let's summarize the procedure of Example 1-5. We first determined that enough information was given, and that the formula expressing the ideal gas equation of state was sufficient to work the problem. The given values of quantities were substituted into the formula, the necessary unit conversions were made, and the multiplications and divisions were carried out. This was a simple problem. In working a more complicated problem, it might even help to map out on a piece of paper how you are going to get from the given information to the desired answer.[5]

Problem 1-5

Calculate the temperature of 10.6 mol of an ideal gas if the volume is 37.6 liters and the pressure is 5476 torr. The quantity 1000 liters is exactly 1 cubic meter. ●

In the remaining chapters of this book, you will see a number of examples worked out, and you will be asked to work a number of problems. In most of these, a method must be found and applied that will lead from the given information to the desired answer.

In some problems there will be a choice of methods. Perhaps you must choose between a graphical procedure and a numerical procedure, or between an approximate formula and an exact formula. In some of these cases, it would be foolish to carry out a more difficult solution, because an approximate solution will give you an answer that will be sufficient for the purpose at hand. In other cases, you will need to carry out a more nearly exact solution. You will need to learn how to distinguish between these two cases.

ADDITIONAL PROBLEMS

1.6

The inch has been redefined to be exactly 0.0254 m in length.
a. Find the number of inches in a meter. How many significant digits could be given?
b. Find the number of meters in 1 mile, and the number of miles in 1 kilometer. How many significant digits could be given?

1.7

A furlong is one-eighth of a mile, and a fortnight is 2 weeks. Find the speed of light in furlongs per fortnight, specifying the correct number of significant digits.

---

[5] Some unkind soul has defined a mathematician as a person capable of traversing a mathematically precise path from an unwarranted assumption to a foregone conclusion.

1.8

The Rankine temperature scale is defined so that the Rankine degree is the same size as the Fahrenheit degree, and $0\,°R$ is the same as 0 K.

a. Find the Rankine temperature at $0\,°C$.

b. Find the Rankine temperature at $0\,°F$.

1.9

The volume of a right circular cylinder is given by

$$V = \pi r^2 h$$

where V is the volume, r the radius, and h the height. If a certain right circular cylinder has a radius given as 0.134 m and a height given as 0.318 m, find its volume, specifying it with the correct number of significant digits. Calculate the smallest and largest volumes that the cylinder might have, and check your first answer for the volume.

1.10

A certain angle is given as $31°$. Find the smallest and largest values that its sine and cosine might have, and specify the sine and cosine to the appropriate number of significant digits. Use a table or a calculator.

ADDITIONAL READING

M. L. McGlashan, *Physico-chemical Quantities and Units*, Royal Inst. Chem. Publ. No. 15, London, 1968.

This is a description of the SI.

M. A. Paul, "The International System of Units (SI). Development and Progress," *J. Chem. Doc.* **11**, 3(1971).

This is another description and explanation of the SI.

G. Polya, *How to Solve It—A New Aspect of Mathematical Method*, Princeton University Press, Princeton, N.J., 1945.

This small book, which should be in every college or university library, contains a detailed discussion of general methods of solving problems. The techniques apply to physical chemistry problems just as well as to mathematical problems.

Arthur W. Adamson, *Understanding Physical Chemistry*, W. A. Benjamin, New York, 1969.

This is a book of problems in physical chemistry, designed to supplement any physical chemistry textbook. There is a brief section in each chapter that summarizes the relevant theory. Solutions are given to the problems.

Leonard C. Labowitz and John S. Arents, *Physical Chemistry Problems and Solutions*, Academic Press, Inc., New York, 1969.

This is another book of problems, with solutions given to all problems. In each section, there are three categories of problems, arranged by difficulty.

CHAPTER TWO

Mathematical Variables and Operations

Section 2-1. INTRODUCTION

In this chapter, we discuss elementary operations on mathematical variables. Included are the simple algebraic operations, trigonometric functions, logarithms and exponentials of real scalar variables, algebraic operations on real vector variables, and algebraic operations on complex scalar variables.

After you have studied the chapter, you should be able to do all of the following:

1. Manipulate variables algebraically well enough to get the correct answer to moderately complicated problems.
2. Manipulate trigonometric functions correctly.
3. Work correctly with logarithms and exponentials.
4. Calculate correctly the sum, difference, scalar product, and vector product of any two vectors, whether constant or variable.
5. Perform correctly the elementary algebraic operations on complex numbers, as well as form the complex conjugate of any complex number and separate the real and imaginary parts of any complex expression.

Section 2-2. ALGEBRAIC OPERATIONS ON REAL SCALAR VARIABLES

The operations discussed in this section include the familiar arithmetic operations: addition, subtraction, multiplication, and division, and the taking of powers and roots. The numbers and variables on which we operate in this section are called *real*, because they do not include imaginary numbers such as the square root of -1. They are also called *scalars*, to distinguish them from vectors, which we discuss later in the chapter.

Real scalar numbers have *magnitude*, a specification of the size of the number, and *sign*, which can be positive or negative. You already know the rules for addition, subtraction, multiplication, and division: (a) the product of two factors of the same sign is positive, and the product of two factors of different sign is negative; (b) the difference of two numbers is the same as the sum of the first number and the negative of the second; (c) multiplication is *commutative*, which means that

$$a \times b = b \times a \qquad (2.1)$$

(d) multiplication is *associative*, which means that

$$a \times (b \times c) = (a \times b) \times c \qquad (2.2)$$

(e) multiplication and addition are *distributive*, which means that

$$a \times (b + c) = a \times b + a \times c \qquad (2.3)$$

These rules can be used to manipulate an algebraic expression into a simpler form, as in the following example.

Example 2-1

Write the following expression in a simpler form.

$$A = \frac{(2x + 5)(x + 3) - 2x(x + 5) - 14}{x^2 + 2x + 1}$$

Solution

By algebraic manipulation, we obtain

$$A = \frac{2x^2 + 11x + 15 - 2x^2 - 10x - 14}{(x + 1)(x + 1)} = \frac{x + 1}{(x + 1)^2} = \frac{1}{x + 1} \qquad \bullet$$

Problem 2-1

Write the following expression in a simpler form.

$$B = \frac{(x^2 + 2x)^2 - x^2(x - 2)^2 + 12x^4}{6x^3 + 12x^4} \qquad \bullet$$

In addition to the four simple arithmetic operations, we discuss the taking of powers and roots. If a quantity x is multiplied by itself $n - 1$ times, so that there are n factors, we represent this by the symbol x^n:

$$x^2 = x \times x, \quad x^3 = x \times x \times x, \quad \ldots, \quad x^n = x \times x \times x \times x \times \cdots \times x \qquad (2.4)$$

This equation is for $n = 1$, $n = 2$, etc. If n is not an integer, we can still define x^n, but we will talk about this in Section 2-4, when we discuss logarithms. The number n in the expression x^n is called the *exponent* of x. An exponent that is a negative number indicates the *reciprocal* of the quantity with a positive exponent:

$$x^{-1} = \frac{1}{x}, \quad x^{-3} = \frac{1}{x^3}, \quad \text{etc.} \tag{2.5}$$

Roots of real numbers are defined in an inverse way. For example, the *square root* of x is defined as the number that yields x when squared:

$$(\sqrt{x})^2 = x \tag{2.6a}$$

The *cube root* of x is denoted by $\sqrt[3]{x}$, and is defined as the number that when cubed (raised to the third power) yields x:

$$(\sqrt[3]{x})^3 = x \tag{2.6b}$$

Fourth roots, fifth roots, etc., are defined in similar ways.

There are two numbers which when squared will yield a given positive real number: for example, $2^2 = 4$ and $(-2)^2 = 4$. We can say that both 2 and -2 are square roots of 4 if we wish, but when the symbol $\sqrt{4}$ is used, generally only the positive square root, 2, is meant. To specify the negative square root of x, we should write $-\sqrt{x}$. If we confine ourselves to real numbers, there is no square root, fourth root, sixth root, etc., of a negative number. In Section 2-6, we define imaginary numbers, which can be square roots of negative quantities. Notice that both positive and negative numbers can have cube roots, fifth roots, etc., since an odd number of negative factors yields a negative product.

The square roots, cube roots, etc., of integers and other rational numbers are not always rational numbers themselves. If not, they are algebraic irrational numbers. An irrational number that does not produce a rational number when raised to any integral power is a transcendental irrational number.

The *magnitude*, or *absolute value*, of a scalar quantity is denoted by placing vertical bars before and after the symbol for the quantity. The operation means

$$|x| = \begin{cases} x & \text{if } x \geq 0 \\ -x & \text{if } x < 0 \end{cases} \tag{2.7}$$

For example,

$$|4.5| = 4.5$$
$$|-3| = 3$$

The magnitude of a number is always a positive quantity or zero.

Problem 2-2

Simplify the expression

$$\frac{3(2+4)^2 - 6(7 + |-17|)^3 + (\sqrt{37 - |-1|})^3}{(1+2^2)^4 - (|-7| + 6^3)^2 + \sqrt{12 + |-4|}}$$

Section 2-3. **TRIGONOMETRIC FUNCTIONS**

The ordinary *trigonometric functions* include the sine, the cosine, the tangent, the cotangent, the secant, and the cosecant. These are sometimes called the "circular" trigonometric functions to distinguish them from the hyperbolic trigonometric functions discussed briefly in Section 2-5.

The trigonometric functions can be defined as in Figure 2-1, which shows two angles, α_1 and α_2. Along the horizontal reference line drawn from the point E to the point D, the points C_1 and C_2 are chosen so that the triangles are right triangles. In the right triangle AB_1C_1, the radius r is called the *hypotenuse*, the vertical side of length y_1 is called the *opposite side*, and the horizontal side of length x_1 is called the *adjacent side*. We define

$$\sin(\alpha_1) = \frac{y_1}{r} \qquad \text{(opposite over hypotenuse)} \qquad \textbf{(2.8)}$$

$$\cos(\alpha_1) = \frac{x_1}{r} \qquad \text{(adjacent over hypotenuse)} \qquad \textbf{(2.9)}$$

$$\tan(\alpha_1) = \frac{y_1}{x_1} \qquad \text{(opposite over adjacent)} \qquad \textbf{(2.10)}$$

$$\cot(\alpha_1) = \frac{x_1}{y_1} \qquad \text{(adjacent over opposite)} \qquad \textbf{(2.11)}$$

$$\sec(\alpha_1) = \frac{r}{x_1} \qquad \text{(hypotenuse over adjacent)} \qquad \textbf{(2.12)}$$

$$\csc(\alpha_1) = \frac{r}{y_1} \qquad \text{(hypotenuse over opposite)} \qquad \textbf{(2.13)}$$

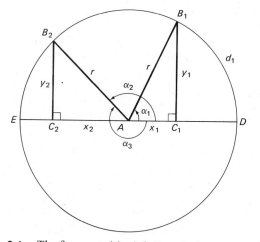

Figure 2-1. The figure used in defining trigonometric functions.

The trigonometric functions of the angle α_2 are defined in the same way, except that as drawn in Figure 2-1, the distance x_2 must be counted as negative, because the point B_2 is to the left of A. If the point B_2 were below A, then y_2 would also be counted as negative.

There are three common ways to specify the size of an angle. *Degrees* are defined so that a right angle is $90°$ (90 degrees), and a full circle contains $360°$. The *grad* is defined so that 100 grad is a right angle and a full circle contains 400 grad. In mathematics, the best way to specify the size of an angle is with *radians*. The size of an angle in radians is defined to be the length of the arc subtending the angle divided by the radius of the circle. In Figure 2-1, the arc DB_1 subtends the angle α_1, so that in radians

$$\alpha_1 = \frac{d_1}{r} \tag{2.14}$$

where d_1 is the length of the arc. The full circle contains 2π radians, and 1 radian corresponds to $57.29577\ldots$ degrees. The right angle is $\pi/2$ radians.

The trigonometric functions are examples of mathematical functions. The angle for which we evaluate the functions is called the *independent variable*, or the *argument* of the function. We choose a value for the independent variable, and the function provides a value for another variable, which we call the *dependent variable*. For example, if we write

$$f = \sin(\alpha) \tag{2.15}$$

then f is the dependent variable and α the independent variable. The trigonometric functions are *single-valued functions*. For each value of the angle α, there is one and only one value of the sine, one and only one value of the cosine, etc.

Trigonometric Identities. There are a number of relations between trigonometric functions which are true for all values of the given angles. Such relations are said to be identically true, or to be *identities*. For example, from inspection of Eqs. (2.8) through (2.13), we can write

$$\cot(\alpha) = \frac{1}{\tan(\alpha)} \tag{2.16}$$

$$\sec(\alpha) = \frac{1}{\cos(\alpha)} \tag{2.17}$$

$$\csc(\alpha) = \frac{1}{\sin(\alpha)} \tag{2.18}$$

The negative angle α_3 in Figure 2-1 has the same triangle, and therefore the same trigonometric functions as the positive angle α_2. (Negative angles are measured in a clockwise direction, whereas positive angles are measured in a counterclockwise direction.) Since α_3 is equal to $-(2\pi - \alpha_2)$ if the angles are measured in radians, we can write

$$\sin[-(2\pi - \alpha)] = \sin(\alpha) \tag{2.19}$$

with similar equations for the other trigonometric functions.

If an angle is increased by 2π radians ($360°$), the new angle corresponds to the same triangle as the old angle, and we can write

$$\sin(2\pi + \alpha) = \sin(\alpha) \tag{2.20}$$

$$\cos(2\pi + \alpha) = \cos(\alpha) \tag{2.21}$$

with similar equations for the other trigonometric functions.

Other identities can be obtained geometrically. For example,

$$\sin(\alpha) = -\sin(-\alpha) \tag{2.22}$$

$$\cos(\alpha) = \cos(-\alpha) \tag{2.23}$$

$$\tan(\alpha) = -\tan(-\alpha) \tag{2.24}$$

Equations (2.20), (2.21), and the analogous equations for the other functions express the property that the trigonometric functions are *periodic functions*, with period equal to 2π. Equations (2.22) and (2.24) express the fact that the sine and the tangent are *odd functions*, and Eq. (2.23) expresses the fact that the cosine is an *even function*.

Additional trigonometric identities are listed in Appendix 5. Equation (14) of this appendix is very useful, and can be obtained from the famous *theorem of Pythagorus*. Pythagorus drew a figure with three squares such that one side of each square formed one side of a right triangle. He then proved by geometry that the area of the square on the hypotenuse was equal to the sum of the areas of the squares on the other two sides.[1] In terms of the quantities in Figure 2-1,

$$x^2 + y^2 = r^2 \tag{2.25}$$

We divide both sides of this equation by r^2 and use Eqs. (2.8) and (2.9) to obtain[2]

$$[\sin(\alpha)]^2 + [\cos(\alpha)]^2 = \sin^2(\alpha) + \cos^2(\alpha) = 1 \tag{2.26}$$

[1] There is a joke about a polygamous chief of an American Indian tribe who had three wives, each with two sons. He gave his first wife a rug made from a hippopotamous hide, his second wife a rug made from a bear hide, and his third wife a rug made from a deer hide. In the rigid caste system of the tribe, the squaw on the hippopotamous was equal to the sons of the squaws on the other two hides.

[2] Throughout the text, equations which will frequently be used are boxed.

Notice the common notation for a power of a trigonometric function: the exponent is written just before the symbol for the independent variable.

A Useful Approximation.   Comparison of Eqs. (2.8) and (2.14) shows that for a fairly small angle, the sine of an angle and the angle itself in radians are approximately equal, since the sine differs from the angle only by having the opposite side in place of the arc length, which is approximately the same size. In fact,

$$\lim_{\alpha \to 0} \frac{\sin(\alpha)}{\alpha} = 1 \tag{2.27}$$

The symbol on the left stands for a mathematical *limit*. What it means is that we let the value of α become smaller and smaller until it becomes more and more nearly equal to zero. When we do this, the ratio of $\sin(\alpha)$ to α becomes more and more nearly equal to unity.

For fairly small angles, we write as an approximation

$$\alpha \approx \sin(\alpha) \tag{2.28}$$

The symbol $\approx$ means "approximately equal to." Since the adjacent side is nearly equal to the hypotenuse for small angles, we can also write

$$\alpha \approx \tan(\alpha) \approx \sin(\alpha) \tag{2.29}$$

If you are satisfied with an accuracy of about 1%, you can use Eq. (2.29) for angles up to about 0.2 radians (approximately 11°). However, remember that you must express the angle in radians.

Problem 2-3

For an angle that is nearly as large as $\pi/2$, find an approximate equality similar to Eq. (2.29) involving $(\pi/2) - \alpha$, $\cos(\alpha)$, and $\cot(\alpha)$.   •

General Properties of Trigonometric Functions.   To use trigonometric functions easily, you must have a clear mental picture of the way in which the sine, cosine, and tangent depend on their independent variables, or arguments. Figures 2-2, 2-3, and 2-4 show these functions. The tangent has a

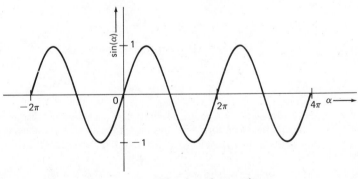

Figure 2-2.   The sine of an angle, α.

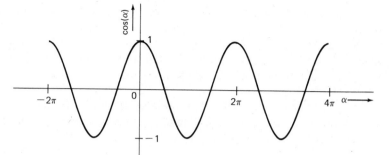

Figure 2-3. The cosine of an angle, α.

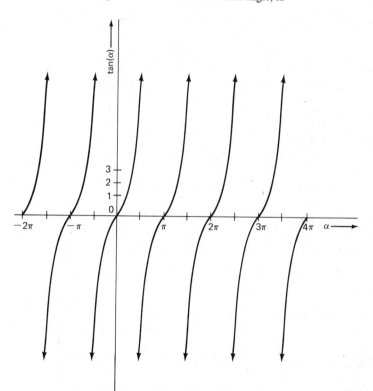

Figure 2-4. The tangent of an angle, α.

complicated behavior, becoming larger without bound as its argument approaches $\pi/2$ from the left, and becoming more negative without bound as its argument approaches the same value from the right. We can write

$$\lim_{\alpha \to \pi/2^+} [\tan(\alpha)] = -\infty$$

$$\lim_{\alpha \to \pi/2^-} [\tan(\alpha)] = \infty$$

In these equations, the superscript $+$ on the $\pi/2$ means that the value of α approaches $\pi/2$ from the right. That is, α is greater than $\pi/2$ as it becomes more and more nearly equal to $\pi/2$. The $-$ superscript means that α approaches $\pi/2$ from the left. The symbol ∞ stands for *infinity*, which is larger than any number that you or anyone else can name.

Inverse Trigonometric Functions. It is possible to think of trigonometric functions as defining a mathematical function in a reverse way from that considered thus far in this section. For example, if

$$y = \sin(x) \tag{2.30}$$

it is possible to think of Eq. (2.30) as giving a value for x as a dependent variable for each value of y, considered now to be an independent variable. We write

$$x = \arcsin(y) \tag{2.31}$$

This can be read as "x is the angle whose sine is y." The *arcsine* is also called the *inverse sine*, and another notation is also common:

$$x = \sin^{-1}(y) \tag{2.32}$$

The -1 superscript is not an exponent, even though exponents are sometimes written in the same position.

If you inspect Figure 2-2, you can see that there are lots of different angles which have the same sine. In order to make Eq. (2.31) or (2.32) into a single-valued function, we must restrict the range of x. With the arcsine function, this range is taken from $-\pi/2$ to $+\pi/2$ and is said to include the principal values of the arcsine function. The range of principal values of the *arctangent* and *acrcosecant* functions is from $-\pi/2$ to $+\pi/2$, the same as with the arcsine. The range of principal values of the *arccosine*, *arccotangent*, and *arcsecant* is taken from 0 to π.

Problem 2-4

Sketch graphs of the arcsine function, the arccosine function, and the arctangent function. Include only the principal values.   •

Section 2-4. LOGARITHMS

The operation of raising a number to an integral power is familiar to you. You know that a^2 means $a \times a$, a^3 means $a \times a \times a$, etc. In addition, you can have exponents that are not integers. If we write a^x, x can be a variable that takes on any real value. We assume a to be positive, so that a^x is positive.

The *logarithm* to the base a is defined by the relation:

If $y = a^x$, then $\boxed{x = \log_a(y)}$ (2.33)

If $a = 10$, the logarithms are called *common logarithms*: If $10^x = y$, then $x = \log_{10}(y)$, the common logarithm of y.

For integral values of x, it is easy to generate the following short table of logarithms:

y	$x = \log_{10}(y)$
1	0
10	1
100	2
1000	3
0.1	-1
0.01	-2
0.001	-3
etc.	

In order to understand logarithms that are not integers, we need to understand exponents that are not integers.

Example 2-2

Find the logarithm of $\sqrt{10}$.

Solution

The square root of 10 is the number that yields 10 when multiplied by itself:

$$(\sqrt{10})^2 = 10$$

We use the fact about exponents

$$(a^x)^z = a^{xz} \qquad \textbf{(2.34)}$$

Since 10 is the same thing as 10^1, we can see from Eq. (2.34) that

$$\sqrt{10} = 10^{1/2} \qquad \textbf{(2.35)}$$

We can thus write

$$\log_{10}(\sqrt{10}) = \log_{10}(3.162277\ldots) = \tfrac{1}{2} \qquad \bullet$$

Equation (2.34) and some other relations governing exponents can be used to generate other logarithms, as in the following problem.

Problem 2-5

Use Eq. (2.34) and the fact that $10^{-n} = 1/(10^n)$ to generate the negative logarithms in the short table of logarithms. $\qquad \bullet$

Table 2-1 lists some properties of exponents and logarithms. An additional fact is that negative numbers do not possess real logarithms.

Table 2-1. Properties of Exponents and Logarithms:
$$a^x = y, \ x = \log_a(y)$$

Exponent Fact	Logarithm Fact	Equation Number
$a^0 = 1$	$\log_a(1) = 0$	(2.36)
$a^{1/2} = \sqrt{a}$	$\log_a(\sqrt{a}) = \frac{1}{2}$	(2.37)
$a^1 = a$	$\log_a(a) = 1$	(2.38)
$a^{x_1} a^{x_2} = a^{x_1 + x_2}$	$\log_a(y_1 y_2) = \log_a(y_1) + \log_a(y_2)$	(2.39)
$a^{-x} = \dfrac{1}{a^x}$	$\log_a(1/y) = -\log_a(y)$	(2.40)
$a^{x_1}/a^{x_2} = a^{x_1 - x_2}$	$\log_a(y_1/y_2) = \log_a(y_1) - \log_a(y_2)$	(2.41)
$(a^x)^z = a^{xz}$	$\log_a(y^z) = z \log_a(y)$	(2.42)
$a^\infty = \infty$	$\log_a(\infty) = \infty$	(2.43)
$a^{-\infty} = 0$	$\log_a(0) = -\infty$	(2.44)

We will not discuss further how the logarithms of various numbers are computed, but extensive tables of logarithms are available, with up to seven or eight significant digits. Appendix 9 contains a table of common logarithms with four significant digits.

This table contains common logarithms for numbers between 1 and 10, so that the logarithms lie between 0 and 1. Each number in the table is to be read with a decimal point at the left of the number. To obtain the logarithm of a number greater than 10 or smaller than 1, we must use Eq. (2.39).

Example 2-3

Find the common logarithm of 3248.

Solution

We treat 3248 as the product of 3.248 and 10^3. The common logarithm of 10^3 is 3, and by interpolating in the table of Appendix 7, we find that the common logarithm of 3.248 is 0.5116. The logarithm of 3248 is thus 3.5116. •

Natural Logarithms. Besides 10, there is another commonly used base of logarithms. This is a transcendental irrational number called e (after Euler, a German mathematician) and equal to $2.7182818 \ldots$ If $e^x = y$,

$$x = \log_e(y) = \ln(y) \tag{2.45}$$

The notation $\ln(y)$ is more commonly than $\log_e(y)$, and logarithms to the base e are usually called *natural logarithms*. They are also occasionally called Naperian logarithms, after Napier, a French mathematician. Unfortunately, some mathematicians use the symbol $\log(y)$ without a subscript

23

§2-5 / The Exponential Function. Hyperbolic Trigonometric Functions

for natural logarithms; others use the symbol $\log(y)$ without a subscript for common logarithms.

The definition of e is

$$e = \lim_{n \to \infty} \left(1 + \frac{1}{n}\right)^n = 2.7182818\ldots \tag{2.46}$$

Natural logarithms can be used to carry out multiplications and divisions, just as common logarithms can, but they are seldom actually so used. Their principal use is in formulas, where they occur naturally as solutions of differential equations, etc.

If the common logarithm of a number is known, its natural logarithm can be computed as follows:

$$e^{\ln(y)} = 10^{\log_{10}(y)} = (e^{\ln(10)})^{\log_{10}(y)} = e^{\ln(10)\,\log_{10}(y)} \tag{2.47}$$

The natural logarithm of 10 is equal to $2.303\ldots$, so we can write

$$\ln(y) = \ln(10)\log_{10}(y) = (2.303\ldots)\log_{10}(y) \tag{2.48}$$

Problem 2-6

Using the table of common logarithms in Appendix 7, find the natural logarithms of the following numbers.

a. 2^1 b. 100   c. $10^{1/2}$ d. 3.678   e. 6.022×10^{23} •

Very few people now use logarithms for calculations, because of the availability of inexpensive electronic calculators. For use in physical chemistry calculations, a calculator should be able to calculate logarithms and exponentials and to display numbers in scientific notation.

Section 2-5. THE EXPONENTIAL FUNCTION. HYPERBOLIC TRIGONOMETRIC FUNCTIONS

The *exponential function* is one of the most commonly encountered functions of physics and chemistry. This function is given by

$$y = ae^{bx} \equiv a\exp(bx) \tag{2.49}$$

where a and b are constants and e is the base of natural logarithms. Figure 2-5 shows a graph of this function for $b > 0$.

The function shows a behavior that you should be familiar with. For $b > 0$, it doubles each time the independent variable increases by a fixed amount. If $b < 0$, the function is cut in half each time the independent variable increases by a fixed amount.

An example is in the decay of radioactive isotopes. If N_0 is the number of atoms of the isotope at time $t = 0$, the number at any other time, t, is

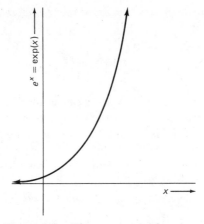

Figure 2-5. The exponential function.

given by

$$N(t) = N_0 e^{-t/\tau} \tag{2.50}$$

where τ is called the *relaxation time*. The time that is required for the number of atoms to drop to half its original value is called the half-time or *half-life*.

Example 2-4

Show that the half-life, $t_{1/2}$, is given by $t_{1/2} = \tau \ln(2)$.

Solution

If $t_{1/2}$ is the half-life, then

$$e^{-t_{1/2}/\tau} = \tfrac{1}{2}$$

Thus,

$$\frac{t_{1/2}}{\tau} = -\ln(\tfrac{1}{2}) = \ln(2) \qquad \bullet \tag{2.51}$$

Problem 2-7

A certain population is growing exponentially, so that it doubles in size each 30 years.
  a. If the population has a size of 4.00×10^6 individuals at $t = 0$, write the formula giving the population after a number of years equal to t.
  b. Find the size of the population at $t = 150$ years. $\qquad \bullet$

Hyperbolic trigonometric functions are closely related to the exponential function. The *hyperbolic sine* of x is denoted by $\sinh(x)$, and defined by

$$\sinh(x) = \tfrac{1}{2}(e^x - e^{-x}) \tag{2.52}$$

The *hyperbolic cosine* is denoted by $\cosh(x)$, and defined by

$$\cosh(x) = \tfrac{1}{2}(e^x + e^{-x}) \tag{2.53}$$

The other hyperbolic trigonometric functions are the *hyperbolic tangent*, denoted by tanh(x); the *hyperbolic cotangent*, denoted by coth(x); the *hyperbolic secant*, denoted by sech(x); and the *hyperbolic cosecant*, denoted by csch(x). These are given by the equations

$$\tanh(x) = \frac{\sinh(x)}{\cosh(x)} \tag{2.54}$$

$$\coth(x) = \frac{1}{\tanh(x)} \tag{2.55}$$

$$\mathrm{sech}(x) = \frac{1}{\cosh(x)} \tag{2.56}$$

$$\mathrm{csch}(x) = \frac{1}{\sinh(x)} \tag{2.57}$$

Problem 2-8

Find the value of each of the hyperbolic trigonometric functions for $x = 0$ and $x = \pi/2$. Compare these values with the values of the ordinary (circular) trigonometric functions for the same values of the independent variable. •

Section 2-6. VECTORS. COORDINATE SYSTEMS

Quantities that have both magnitude and direction are called *vectors*. For example, the position of an object can be represented by a vector, since it can be specified by giving the distance and the direction from a reference point. A force is also a vector, since it is not completely specified until its direction is given.

There are a number of additional vectors which are important in physical chemistry. Some examples are the dipole moments of molecules, magnetic and electric fields, angular momenta, and magnetic dipoles.

We will use a letter set in boldface type to represent a vector. For example, the force on an object is denoted by **F**. When you are writing, there is no easy way to write boldface letters, so you can use a letter with a little arrow over it (e.g., $\vec{F}$) or you can use a wavy underscore, which is the typesetter's symbol for boldface type.

Vectors in Two Dimensions. Let us begin with a restricted case, in which all vectors are required to lie in a plane. This case includes position vectors for objects that remain on a flat surface. We represent this physical surface by a mathematical plane, which is a map of the surface. To each location in the physical surface there corresponds a point of the mathematical plane. We choose some reference point as an origin. We pick some convenient line passing through the origin, and call this our *x axis*. One end of this is

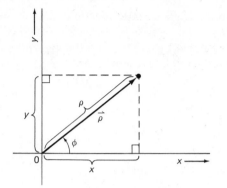

Figure 2-6. A position vector, ρ, in a plane, with plane polar coordinates and cartesian coordinates.

designated as the positive end. The line passing through the origin perpendicular to the x axis is designated as our y *axis,* and the end that is counterclockwise 90° from the positive end of the x axis is designated as its positive end. These axes are shown in Figure 2-6. In this figure, the origin is labeled as point O, and the location of some object is labeled as point P.

The directed line segment beginning at O and ending at P is the position vector for the object. We denote the position vector in two dimensions by ρ (Greek letter rho). In the figure, we draw an arrowhead on the directed line segment to make its direction clear, since the vector going from O to P is the negative of the vector going from P to O. One way to specify the location of P is to give the magnitude of ρ and the value of the angle ϕ between ρ and the positive end of the x axis, measured counterclockwise from the axis. We will denote the magnitude of any vector by the same letter in ordinary type, or by the symbol for the vector with vertical bars before it and after it, as $\rho = |\rho|$. The magnitude of a vector is a nonnegative scalar. The variables ρ and ϕ are called the *plane polar coordinates* of the point P. If we allow ρ to range from zero to ∞ and allow ϕ to range from zero to 2π radians, we can specify the location of any point in the entire plane.

There is another common way to specify the location of P. We draw two line segments from P perpendicular to the axes, as shown in Figure 2-6. The distance from the origin to the intersection on the x axis is called x, and is considered to be positive if the intersection is on the positive half of the axis, and negative if the intersection is on the negative half of the axis. The distance from the origin to the intersection on the y axis is called y, and its sign is assigned in a similar way. The variables x and y are called the *cartesian coordinates* of P, after Descartes, a French mathematician. The point P is sometimes designated by its cartesian coordinates within parentheses, as (x, y).

The values of x and y are also called the cartesian components of the position vector, and the position vector ρ is sometimes denoted by the same symbol as the point at which its head lies, as (x, y).

It is not difficult to change from polar coordinates to cartesian coordinates. This is called a *transformation of coordinates*, and can be done using the equations

$$x = \rho \cos(\phi) \tag{2.58}$$

$$y = \rho \sin(\phi) \tag{2.59}$$

Problem 2-9

Show that Eqs. (2.58) and (2.59) are correct. •

The coordinate transformation in the other direction is also possible. From the theorem of Pythagorus, Eq. (2.25),

$$\rho = \sqrt{x^2 + y^2} \tag{2.60}$$

From the definition of the tangent function, Eq. (2.10),

$$\phi = \arctan\left(\frac{y}{x}\right) \tag{2.61}$$

However, since we want ϕ to range from 0 to 2π radians, we must specify this range for the inverse tangent function, instead of using the principal value. We must decide in advance which quadrant ϕ lies in.

Problem 2-10

  a. Find x and y if $\rho = 10$ units and $\phi = \pi/6$ radians.
  b. Find ρ and ϕ if $x = -5$ units and $y = 10$ units. •

A position vector is only one example of a vector. Anything, such as a force, a velocity, or an acceleration, which has magnitude and direction is a vector. Figure 2-6 is a map of physical space, and a distance in such a diagram is measured in units of length, such as meters. Other kinds of vectors can also be represented on vector diagrams by directed line segments. However, such a diagram is not a map of physical space, and the length of a line segment representing a vector will represent the magnitude of a force, or the magnitude of a velocity, or something else.

Position vectors ordinarily remain with their tails at the origin, but since other vector diagrams do not necessarily represent a physical (geographical) space, we will consider a vector to be unchanged if it is moved from one place in a vector diagram to another, as long as its length and its direction do not change.

Vector Algebra in Two Dimensions.   We first define the sum of two vectors. Figure 2-7 is a vector diagram in which two vectors, called **A** and **B**, are shown. The sum of the two vectors is obtained as follows: (1) Move the

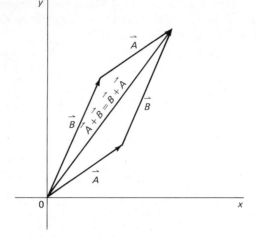

Figure 2-7. Two vectors and their sum.

second vector so that its tail coincides with the head of the first. (2) Draw the sum vector from the tail of the first vector to the head of the second. In Figure 2-7, you can see that $\mathbf{A} + \mathbf{B}$ is the same as $\mathbf{B} + \mathbf{A}$. The addition of vectors is thus commutative.

Vector addition can also be performed using the components of the vectors. The components of $\mathbf{A}$ and $\mathbf{B}$ are defined in the same way as the components of the position vector in Figure 2-6. The x components are called A_x and B_x, and the y components are called A_y and B_y. If the sum of $\mathbf{A}$ and $\mathbf{B}$ is called $\mathbf{C}$,

$$C_x = A_x + B_x \tag{2.62}$$
$$C_y = A_y + B_y \tag{2.63}$$

The difference of two vectors is the sum of the first vector and the negative of the second. The negative of $\mathbf{B}$ is denoted by $-\mathbf{B}$ and is the vector with components $-B_x$ and $-B_y$. If the vector $\mathbf{A} - \mathbf{B}$ is called $\mathbf{D}$,

$$D_x = A_x - B_y \tag{2.64}$$
$$D_y = A_x - B_y \tag{2.65}$$

The vector $\mathbf{A} - \mathbf{B}$ has its head at the head of $\mathbf{A}$ and its tail at the head of $\mathbf{B}$ if both $\mathbf{A}$ and $\mathbf{B}$ have their tails at the same place.

We next define the product of a scalar and a vector. If $\mathbf{A}$ is a vector and a is a scalar, the product $a\mathbf{A}$ has the components

$$(a\mathbf{A})_x = aA_x \tag{2.66}$$
$$(a\mathbf{A})_y = a \quad_y \tag{2.67}$$

If a is a positive scalar, the vector $a\mathbf{A}$ points in the same direction as $\mathbf{A}$, and if a is a negative scalar, the vector $a\mathbf{A}$ points in the opposite direction. The magnitude of $a\mathbf{A}$ is $|a|$ times the magnitude of $\mathbf{A}$.

The magnitude of any vector in two dimensions is obtained in the same manner as the magnitude of a position vector, as given in Eq. (2.60). Thus

$$A = |\mathbf{A}| = \sqrt{A_x^2 + A_y^2} \tag{2.68}$$

Problem 2-11

The vector A has the components $A_x = 2$, $A_y = 3$. The vector B has the components: $B_x = 3$, $B_y = 4$.
a. Find the components and the magnitude of $\mathbf{A} + \mathbf{B}$.
b. Find the components and the magnitude of $\mathbf{A} - \mathbf{B}$.
c. Find the components and the magnitude of $2\mathbf{A} - 3\mathbf{B}$. •

We next define the *scalar product* of two vectors, which is also called the *dot product*. If $\mathbf{A}$ and $\mathbf{B}$ are two vectors, and α is the angle between them, their scalar product is denoted by $\mathbf{A} \cdot \mathbf{B}$ and given by

$$\boxed{\mathbf{A} \cdot \mathbf{B} = |\mathbf{A}|\,|\mathbf{B}|\cos(\alpha)} \tag{2.69}$$

Note that the result is a scalar, as the name implies.
The following are properties of the scalar product:

1. If $\mathbf{A}$ and $\mathbf{B}$ are parallel, $\mathbf{A} \cdot \mathbf{B}$ is the product of the magnitudes of $\mathbf{A}$ and $\mathbf{B}$.
2. The scalar product of $\mathbf{A}$ with itself is the square of the magnitude of $\mathbf{A}$:

$$\mathbf{A} \cdot \mathbf{A} = |\mathbf{A}|^2 = A^2 = A_x^2 + A_y^2 \tag{2.70}$$

3. If $\mathbf{A}$ and $\mathbf{B}$ are perpendicular to each other, $\mathbf{A} \cdot \mathbf{B} = 0$. Such vectors are also said to be *orthogonal* to each other.
4. If $\mathbf{A}$ and $\mathbf{B}$ point in opposite directions (are *antiparallel*), $\mathbf{A} \cdot \mathbf{B}$ is the negative of the product of the magnitudes of $\mathbf{A}$ and $\mathbf{B}$.

There is another common product of two vectors, called the vector product. Since it is a vector that is perpendicular to the plane containing the two vectors, we defer its definition until we discuss three-dimensional vectors.

Another way to represent vectors is by using *unit vectors*. We define $\mathbf{i}$ to be a vector of unit length pointing in the direction of the positive end of the x axis, and $\mathbf{j}$ to be a vector of unit length pointing in the direction of the positive end of the y axis. These are shown in Figure 2-8. A vector $\mathbf{A}$ is represented as follows:

$$\mathbf{A} = \mathbf{i}A_x + \mathbf{j}A_y \tag{2.71}$$

A similar equation can be written for another vector, $\mathbf{B}$:

$$\mathbf{B} = \mathbf{i}B_x + \mathbf{j}B_y \tag{2.72}$$

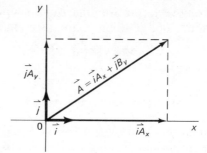

Figure 2-8. A vector in terms of the unit vectors **i** and **j**.

The scalar product $\mathbf{A} \cdot \mathbf{B}$ can be written

$$\mathbf{A} \cdot \mathbf{B} = (\mathbf{i}A_x + \mathbf{j}A_y) \cdot (\mathbf{i}B_x + \mathbf{j}B_y)$$
$$= \mathbf{i} \cdot \mathbf{i}A_xB_x + \mathbf{i} \cdot \mathbf{j}A_xB_y + \mathbf{j} \cdot \mathbf{i}A_yB_x + \mathbf{j} \cdot \mathbf{j}A_yB_y$$

From the definitions of **i** and **j**,

$$\mathbf{i} \cdot \mathbf{i} = \mathbf{j} \cdot \mathbf{j} = 1 \tag{2.73}$$
$$\mathbf{i} \cdot \mathbf{j} = \mathbf{j} \cdot \mathbf{i} = 0 \tag{2.74}$$

so that

$$\mathbf{A} \cdot \mathbf{B} = A_xB_x + A_yB_y \tag{2.75}$$

Example 2-5

Consider the following vectors: $\mathbf{A} = 2.5\mathbf{i} + 4\mathbf{j}$ and $\mathbf{B} = 3\mathbf{i} - 5\mathbf{j}$.
a. Find $\mathbf{A} \cdot \mathbf{B}$.
b. Find $|\mathbf{A}|$ and $|\mathbf{B}|$, and use them to find the angle between **A** and **B**.

Solution

a. $\mathbf{A} \cdot \mathbf{B} = (2.5)(3) + (4)(-5) = 7.5 - 20 = -12.5$
b. $|\mathbf{A}| = (6.25 + 16)^{1/2} = (22.25)^{1/2} = 4.717\ldots$
    $|\mathbf{B}| = (9 + 25)^{1/2} = (34)^{1/2} = 5.8309\ldots$

If α is the angle between **A** and **B**,

$$\cos(\alpha) = \frac{\mathbf{A} \cdot \mathbf{B}}{|\mathbf{A}||\mathbf{B}|} = \frac{-12.5}{(4.717)(5.831)} = -0.4545$$
$$\alpha = 2.043\,\mathrm{rad} = 117.0^0 = \arccos(-0.4545) \qquad \bullet$$

Problem 2-12

Consider the vectors $\mathbf{A} = 3\mathbf{i} - 4\mathbf{j}$ and $\mathbf{B} = -\mathbf{i} + 2\mathbf{j}$.
a. Find $\mathbf{A} \cdot \mathbf{B}$ and $(2\mathbf{A}) \cdot (3\mathbf{B})$.

b. Find the angle between **A** and **B**. Use the principal value of the arc-cosine, so that an angle of less than π radians (180°) results.   •

Vectors and Coordinate Systems in Three Dimensions.   Figure 2-9 shows the three-dimensional version of cartesian coordinates. The axes are viewed from the first octant, where x, y, and z are all positive. A coordinate system such as that shown is called a *right-handed coordinate system.* For such a system, the thumb, index finger, and middle finger of the right hand can be aligned with the positive ends of the x, y, and z axes, respectively. If the left hand must be used for such an alignment, the coordinate system is called a *left-handed coordinate system.*

The location of the point P is specified by x, y, and z, which are called the *cartesian coordinates of the point.* These are the distances from the origin to the points on the axes reached by moving perpendicularly from P to each axis. These coordinates can be positive or negative. The point P is sometimes denoted by its coordinates, as (x, y, z).

The directed line segment from the origin to P is the *position vector* of P, and we denote this vector by **r**. We say that x, y, and z are the cartesian components of this vector. The position vector is also represented by the symbol (x, y, z).

We can represent the position vector, or any other vector, by use of unit vectors, much as in two dimensions. In addition to the unit vectors **i** and **j**, we define **k**, a vector of unit length pointing in the direction of the positive

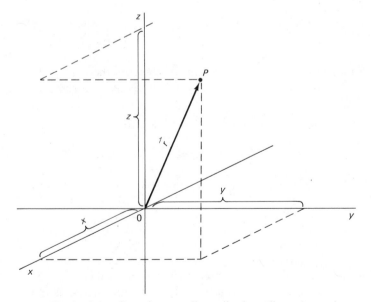

Figure 2-9.   Cartesian coordinates in three dimensions.

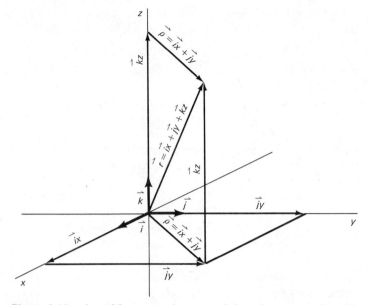

Figure 2-10. A position vector in terms of the unit vectors **i**, **j**, and **k**.

end of the z axis. Figure 2-10 shows these unit vectors and the position vector written as

$$\mathbf{r} = \mathbf{i}x + \mathbf{j}y + \mathbf{k}z \tag{2.76}$$

The magnitude of **r** can be obtained from the theorem of Pythagorus. In Figure 2-10, you can see that r is the hypotenuse of a right triangle with sides ρ and z, so that

$$r^2 = \rho^2 + z^2 = x^2 + y^2 + z^2 \tag{2.77}$$

or

$$r = |r| = \sqrt{x^2 + y^2 + z^2} \tag{2.78}$$

Figure 2-11 shows the way in which spherical polar coordinates are used to specify the location of the point P. The coordinates used are r, the angle between the positive z axis and the position vector, called θ, and the angle ϕ, defined, just as in our two-dimensional polar coordinates, as the angle between the positive x axis and $\boldsymbol{\rho}$. The angle θ is allowed to range from 0 to π

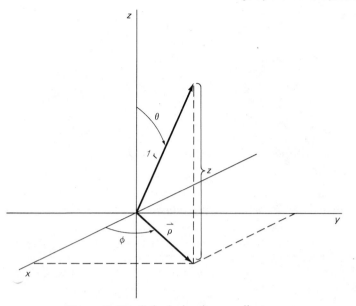

Figure 2-11. Spherical polar coordinates.

and the angle ϕ is allowed to range from 0 to 2π. The distance r is allowed to range from 0 to ∞, and these ranges allow the location of every point in the entire three-dimensional space to be given.

The following equations and Eq. (2.78) can be used to transform from cartesian coordinates to spherical polar coordinates:

$$\theta = \arccos\left(\frac{z}{r}\right) \tag{2.79}$$

and

$$\phi = \arctan\left(\frac{y}{x}\right) \tag{2.80}$$

Notice that Eq. (2.80) is the same as Eq. (2.61).

The following equations can be used to transform from spherical polar coordinates to cartesian coordinates:

$$x = r\sin(\theta)\cos(\phi) \tag{2.81}$$
$$y = r\sin(\theta)\sin(\phi) \tag{2.82}$$
$$z = r\cos(\theta) \tag{2.83}$$

Example 2-6

Find the spherical polar coordinates of the point whose cartesian coordinates are (1, 1, 1).

Solution

$$r = (1 + 1 + 1)^{1/2} = \sqrt{3} = 1.732 \ldots$$

$$\phi = \arctan\left(\frac{1}{1}\right) = \frac{\pi}{4} \text{ rad} = 45°$$

$$\theta = \arccos\left(\frac{1}{\sqrt{3}}\right) = 0.955 \ldots \text{ rad} = 54.74 \ldots °$$

Problem 2-13

Find the spherical polar coordinates of the point whose cartesian co-ordinates are $(2, -3, 4)$.

There is another three-dimensional coordinate system which is in common use. This is the *cylindrical polar coordinate system*, which uses the variables ρ, ϕ, and z, already defined and shown in Figure 2-11. The equations needed to transform from cartesian coordinates to cylindrical polar coordinates are Eqs. (2.60) and (2.61). The third coordinate, z, is the same in both cartesian and cylindrical polar coordinates. Equations (2.58) and (2.59) are used for the reverse transformation.

Example 2-7

Find the cylindrical polar coordinates of the point whose cartesian co-ordinates are $(1, -4, -2)$.

Solution

$$\rho = \sqrt{1 + 16} = \sqrt{17} = 4.123 \ldots$$

$$\phi = \arctan\left(\frac{-4}{1}\right) = 1.816 \text{ rad} = 104.0 \ldots °$$

$$z = -2 \quad \text{as before}$$

Problem 2-14

a. Find the cylindrical polar coordinates of the point whose cartesian coordinates are $(-2, -2, 3)$.
b. Find the cartesian coordinates of the point whose cylindrical polar coordinates are $\rho = 25$, $\phi = 60°$, $z = 17.5$.

Vector Algebra in Three Dimensions. The sum and scalar product of two vectors are very similar to those quantities in two dimensions. Let **A** and **B** be two vectors, represented in terms of their components and the unit vectors **i**, **j**, and **k** by

$$\mathbf{A} = \mathbf{i}A_x + \mathbf{j}A_y + \mathbf{k}A_z \qquad (2.84a)$$

$$\mathbf{B} = \mathbf{i}B_x + \mathbf{j}B_y + \mathbf{k}B_z \qquad (2.84b)$$

The sum is still obtained by placing the tail of the second vector at the head of the first and drawing the sum vector from the tail of the first to the head of the second. If $\mathbf{C} = \mathbf{A} + \mathbf{B}$, then

$$C_x = A_x + B_x \tag{2.85}$$
$$C_y = A_y + B_y$$
$$C_z = A_z + B_z$$

The product of a vector $\mathbf{A}$ and a scalar a is

$$C = aA = \mathbf{i}aA_x + \mathbf{j}aA_y + \mathbf{k}aA_z \tag{2.86}$$

The magnitude of a vector is

$$|\mathbf{A}| = A = (A_x^2 + A_y^2 + A_z^2)^{1/2} \tag{2.87}$$

The scalar product of two vectors is still given by

$$\mathbf{A} \cdot \mathbf{B} = |\mathbf{A}| |\mathbf{B}| \cos(\alpha) \tag{2.88}$$

where α is the angle between the vectors.

Analogous to Eq. (2.75), we have

$$\mathbf{A} \cdot \mathbf{B} = A_x B_x + A_y B_y + A_z B_z \tag{2.89}$$

Example 2-8

Let $\mathbf{A} = 2\mathbf{i} + 3\mathbf{j} + 7\mathbf{k}$ and $\mathbf{B} = 7\mathbf{i} + 2\mathbf{j} + 3\mathbf{k}$. Find $\mathbf{A} \cdot \mathbf{B}$ and the angle between $\mathbf{A}$ and $\mathbf{B}$. Find $(3\mathbf{A}) \cdot \mathbf{B}$.

Solution

$$\mathbf{A} \cdot \mathbf{B} = 14 + 6 + 21 = 41$$

Let α be the angle between $\mathbf{A}$ and $\mathbf{B}$.

$$\alpha = \arccos\left(\frac{\mathbf{A} \cdot \mathbf{B}}{|A| |B|}\right)$$

$$= \arccos\left(\frac{41}{\sqrt{(62)}\sqrt{(62)}}\right) = \arccos(0.6613) = 0.848 \text{ rad} = 48.6°$$

$$(3\mathbf{A}) \cdot \mathbf{B} = 3 \times 14 + 3 \times 6 + 3 \times 21 = 123$$

Notice that $(3\mathbf{A}) \cdot \mathbf{B} = 3(\mathbf{A} \cdot \mathbf{B})$. ●

Problem 2-15

a. Find the position vector of the point whose spherical polar coordinates are $r = 2$, $\theta = 90°$, $\phi = 0°$. Call this vector $\mathbf{A}$.
b. Find the scalar product of the vector $\mathbf{A}$ from part a and the vector $\mathbf{B}$ whose components are $(1, 2, 3)$.
c. Find the angle between these two vectors. ●

We now introduce another kind of a product between two vectors, called the *vector product*, or *cross product*, and denoted by $\mathbf{A} \times \mathbf{B}$.

If $\mathbf{C} = \mathbf{A} \times \mathbf{B}$, then $\mathbf{C}$ is defined to be perpendicular to the plane containing $\mathbf{A}$ and $\mathbf{B}$. Its magnitude is given by

$$C = |\mathbf{C}| = |\mathbf{A}||\mathbf{B}|\sin(\alpha) \qquad (2.90)$$

where α is the angle between $\mathbf{A}$ and $\mathbf{B}$, measured so that it is less than or equal to $180°$.

The direction of the cross product $\mathbf{A} \times \mathbf{B}$ is defined as follows. If the first vector listed, $\mathbf{A}$ in this case, is rotated through the angle α so that its direction coincides with that of $\mathbf{B}$, then $\mathbf{C}$ points in the direction that an ordinary (right-handed) screw thread would move with this rotation. Another rule to obtain the direction is a "right-hand rule." If the thumb of the right hand points in the direction of $\mathbf{A}$ and the index finger points in the direction of $\mathbf{B}$, the middle finger points in the direction of the cross product.

Problem 2-16

From the geometrical definition just given, show that

$$\mathbf{A} \times \mathbf{B} = -\mathbf{B} \times \mathbf{A} \qquad \bullet \quad (2.91)$$

In Problem 2-16, you have shown that the vector product of two vectors is not commutative. From Eq. (2.90),

$$\mathbf{A} \times \mathbf{A} = 0 \qquad (2.92)$$

where 0 is the null vector, which has zero magnitude and no particular direction.

To express the cross product in terms of components, we can write

$$\mathbf{i} \times \mathbf{i} = \mathbf{j} \times \mathbf{j} = \mathbf{k} \times \mathbf{k} = 0 \qquad (2.93)$$

and

$$\mathbf{i} \times \mathbf{j} = \mathbf{k} \qquad (2.94a)$$
$$\mathbf{j} \times \mathbf{i} = -\mathbf{k} \qquad (2.94b)$$
$$\mathbf{i} \times \mathbf{k} = -\mathbf{j} \qquad (2.94c)$$
$$\mathbf{k} \times \mathbf{i} = \mathbf{j} \qquad (2.94d)$$
$$\mathbf{j} \times \mathbf{k} = \mathbf{i} \qquad (2.94e)$$
$$\mathbf{k} \times \mathbf{j} = -\mathbf{i} \qquad (2.94f)$$

By use of these relations, we obtain

$$\mathbf{C} = \mathbf{A} \times \mathbf{B} = \mathbf{i}(A_y B_z - A_z B_y) + \mathbf{j}(A_z B_x - A_x B_z) + \mathbf{k}(A_x B_y - A_y B_x)$$

$$(2.95)$$

Problem 2-17

Show that Eq. (2.95) follows from Eq. (2.94). •

Example 2-9

Find the cross product $\mathbf{A} \times \mathbf{B}$, where $\mathbf{A} = (1, 2, 3)$ and $\mathbf{B} = (1, 1, 1)$.

Solution

Let $\mathbf{C} = \mathbf{A} \times \mathbf{B}$.

$$\mathbf{C} = \mathbf{i}(2 - 3) + \mathbf{j}(3 - 1) + \mathbf{k}(1 - 2)$$
$$= -\mathbf{i} + 2\mathbf{j} - \mathbf{k}$$ •

Example 2-10

Show that the vector $\mathbf{C}$ obtained in Example 2-11 is perpendicular to $\mathbf{A}$.

Solution

We do this by showing that $\mathbf{A} \cdot \mathbf{C} = 0$.

$$\mathbf{A} \cdot \mathbf{C} = A_x C_x + A_y C_y + A_z C_z$$
$$= -1 + 4 - 3 = 0$$ •

Problem 2-18

Show that the vector $\mathbf{C}$ obtained in Example 2-9 is perpendicular to $\mathbf{B}$, and that Eq. (2.90) is satisfied. Do this by finding the angle between $\mathbf{A}$ and $\mathbf{B}$ through calculation of $\mathbf{A} \cdot \mathbf{B}$. •

An example of a vector product is the force on a moving charged particle due to a magnetic field. If q is the charge on the particle in coulombs, $\mathbf{v}$ is the velocity of the particle in meters per second, and $\mathbf{B}$ is the magnetic induction in tesla

$$\mathbf{F} = q\mathbf{v} \times \mathbf{B} \tag{2.96}$$

Since this force is perpendicular to the velocity, it causes the trajectory of the particle to curve, rather than speeding it up or slowing it down.

Example 2-11

Find the force on an electron in a magnetic field if

$$\mathbf{v} = \mathbf{i}(1.000 \times 10^5 \text{ m s}^{-1})$$

and

$$\mathbf{B} = \mathbf{j}(1.000 \times 10^{-4} \text{ T})$$

Solution

The value of q is -1.602×10^{-19} C (note the negative sign).

$$\mathbf{F} = -(\mathbf{i} \times \mathbf{j})(1.602 \times 10^{-19} \text{ C})(1.000 \times 10^5 \text{ m s}^{-1})(1.000 \times 10^{-4} \text{ T})$$
$$= -\mathbf{k}(1.602 \times 10^{-18} \text{ A s m s}^{-1} \text{ kg s}^{-2} \text{ A}^{-1})$$
$$= \mathbf{k}(-1.602 \times 10^{-18} \text{ kg m s}^{-2}) = \mathbf{k}(-1.602 \times 10^{-18} \text{ N})$$ •

The force on a charged particle due to an electric field is

$$\mathbf{F} = q\mathscr{E} \tag{2.97}$$

where $\mathscr{E}$ is the electric field. If the charge is measured in coulombs and the field in volts per meter, the force is in newtons.

Problem 2-19

Find the direction and the magnitude of the electric field necessary to provide a force on the electron in Example 2-11 which is equal in magnitude to the force due to the magnetic field but opposite in direction. If both these forces act on the particle, what will be their effect? •

Section 2-7. COMPLEX NUMBERS

The imaginary unit is called i (not to be confused with the unit vector $\mathbf{i}$) and is defined to be the square root of -1:

$$i = \sqrt{-1} \tag{2.98}$$

If b is a real number, the quantity ib is said to be pure imaginary, and if a is also real, the quantity

$$c = a + ib \tag{2.99}$$

is said to be complex. The real number a is called the *real part* of c and is denoted by

$$a = R(c) \tag{2.100}$$

The real number b is called the *imaginary part* of c and is denoted by

$$b = I(c) \tag{2.101}$$

All of the rules of ordinary arithmetic apply with complex numbers. Some of the rules follow.

Addition and multiplication are both associative. If A, B, and C are complex numbers

$$A + (B + C) = (A + B) + C \tag{2.102}$$

$$A(BC) = (AB)C \tag{2.103}$$

Addition and multiplication are distributive.

$$A(B + C) = AB + AC \tag{2.104}$$

Addition and multiplication are both commutative.

$$A + B = B + A \tag{2.105}$$

$$AB = BA \tag{2.106}$$

Subtraction is simply the addition of a number whose real and imaginary parts are the negatives of the number to be subtracted, and division is the multiplication by the reciprocal of a number. If

$$z = x + iy \tag{2.107}$$

then the reciprocal of z, called z^{-1}, is given by

$$z^{-1} = \frac{x}{x^2 + y^2} - i\frac{y}{x^2 + y^2} \tag{2.108}$$

Example 2-12

Show that $z(z^{-1}) = 1$.

Solution

$$z(z^{-1}) = (x + iy)\left(\frac{x}{x^2 + y^2} - i\frac{y}{x^2 + y^2}\right)$$

$$= \frac{1}{x^2 + y^2}(x^2 + ixy - ixy - i^2y^2) = \frac{1}{x^2 + y^2}(x^2 - i^2y^2)$$

$$= 1 \qquad \bullet$$

Problem 2-20

Show that

$$(a + ib)(c + id) = ac - bd + i(bc + ad) \qquad \bullet \quad (2.109)$$

Problem 2-21

Show that

$$\frac{a + ib}{c + id} = \frac{(ac + bd - iad + ibc)}{c^2 + d^2} \qquad \bullet \quad (2.110)$$

Problem 2-22

Find the value of

$$(4 + 6i)(3 + 2i) + 4i - \frac{1 + i}{3 - 2i} \qquad \bullet$$

Since the real and imaginary parts of a complex number are independent of each other, specifying a complex number is equivalent to specifying two real numbers, such as the coordinates of a point in a plane or the two components of a two-dimensional vector. We can therefore represent a complex number by the location of a point in a plane, as shown in Figure 2-12. This kind of a figure is called an *Argand diagram*, and the plane of the figure is called the *Argand plane* or the *complex plane*. The horizontal coordinate represents the real part of the number and the vertical coordinate represents the imaginary part. The horizontal axis, labeled R, is called the *real axis*, and the vertical axis, labeled I, is called the *imaginary axis*.

The location of the point in the Argand plane can be given by polar coordinates, just as in Section 2-6. We now use the symbol r for the distance from the origin to the point, and the symbol ϕ for the angle in radians between the positive real axis and the line segment joining the origin and

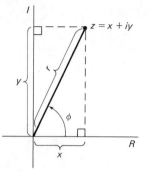

Figure 2-12. Representation of the complex number $z = x + iy$ by the Argand diagram.

the point. The quantity r is the magnitude, or absolute value, or modulus, and ϕ is the argument, or phase. From Eqs. (2.58) and (2.59),

$$x + iy = r\cos(\phi) + ir\sin(\phi) \tag{2.111}$$

There is a remarkable theorem, known as *Euler's formula*, which is

$$re^{i\phi} = r\cos(\phi) + ir\sin(\phi) = x + iy \tag{2.112}$$

where e is the base of natural logarithms, $e = 2.7182818\ldots$.

In the polar representation, the product of two complex numbers, say $z_1 = r_1 e^{i\phi_1}$ and $z_2 = r_2 e^{i\phi_2}$, is given by

$$z_1 z_2 = r_1 r_2 e^{i(\phi_1 + \phi_2)} \tag{2.113}$$

The quotient z_1/z_2 is given by

$$\frac{z_1}{z_2} = \left(\frac{r_1}{r_2}\right) e^{i(\phi_1 - \phi_2)} \tag{2.114}$$

In all of these formulas, ϕ must be measured in radians.

DeMoivre's formula gives the result of raising a complex number to a given power:

$$(re^{i\phi})^n = r^n e^{in\phi} = r^n[\cos(n\phi) + i\sin(n\phi)] \tag{2.115}$$

Example 2-13

Evaluate the following.

a. $(4e^{i\pi})(3e^{2i\pi})$

b. $(8e^{2i\pi})/(2e^{i\pi/2})$

c. $(8e^{4i})^2$

Solution

a. $(4e^{i\pi})(3e^{2i\pi}) = 12e^{3i\pi} = 12e^{i\pi}$

b. $(8e^{2i\pi})/(2e^{i\pi/2}) = 4e^{3i\pi/2}$

c. $(8e^{4i})^2 = 64e^{8i}$ •

In part a of Example 2-13, we have used the fact that an angle of 2π radians gives the same point in the complex plane as an angle of 0, so that

$$e^{2\pi i} = 1 \qquad (2.116)$$

and

$$e^{3\pi i} = e^{\pi i} \qquad (2.117)$$

If a number is given in the form $z = x + iy$, we can find the magnitude and the phase as follows:

$$r = \sqrt{x^2 + y^2} \qquad (2.118)$$

$$\phi = \arctan\left(\frac{y}{x}\right) \qquad (2.119)$$

Just in transforming to ordinary polar coordinates, we cannot simply use the principal value of the arctangent function, but must obtain an angle in the proper quadrant, with ϕ ranging from 0 to 2π.

Problem 2-23

Express the following complex numbers in the form $re^{i\phi}$.

a. $4 + 4i$ b. -1 c. 1 d. $1 - i$

Express the following complex numbers in the form $x + iy$.

a. $e^{i\pi}$ b. $3e^{\pi i/2}$ c. $e^{3\pi i/2}$ •

The *complex conjugate* of a number is defined as the number that has the same real part and an imaginary part which is the negative of that of the original number. The complex conjugate is denoted by an asterisk or by a bar over the letter for the number: If $z = x + iy$,

$$\boxed{\bar{z} = z^* = (x + iy)^* = x - iy} \qquad (2.120)$$

Figure 2-13 shows the location of a complex number and of its complex conjugate in the Argand plane. The phase of the complex conjugate is $-\phi$ if the phase of the original number is ϕ. The magnitude is the same, so

$$(re^{i\phi})^* = re^{-i\phi} \qquad (2.121)$$

The following fact is useful in obtaining the complex conjugate of a complex quantity:

Fact: *The complex conjugate of any expression is obtained by changing the sign in front of every i that occurs.*

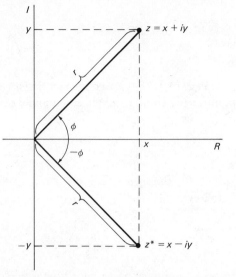

Figure 2-13. A complex number, $z = x + iy$, and its complex conjugate, $z^* = x - iy$, in the Argand plane.

Example 2-14

Find the complex conjugates of the following.
a. $A = (1 + 2i)^{3/2} + \exp(3 + 4i)$
b. $B = a(b + ic)^2 + 4(c - id)^{-1}$

Solution

a. $A^* = (1 - 2i)^{3/2} + \exp(3 - 4i)$
b. $B^* = a(b - ic)^2 + 4(c + id)^{-1}$ •

Problem 2-24

Find the complex conjugates of the following
a. $A = (x + iy)^2 - 4e^{ixy}$
b. $B = (3 + 7i)^3 - (7i)^2$ •

Once we have an expression for the complex conjugate of a quantity, we can use it to express the real and imaginary parts separately:

$$R(z) = \frac{z + z^*}{2} \tag{2.122}$$

$$I(z) = \frac{z - z^*}{2i} \tag{2.123}$$

Problem 2-25

a. Use Eq. (2.120) to show that Eqs. (2.122) and (2.123) are correct.
b. Obtain the famous formulas

$$\cos(\phi) = \frac{e^{i\phi} + e^{-i\phi}}{2} = R(e^{i\phi}) \qquad (2.124)$$

$$\sin(\phi) = \frac{e^{i\phi} - e^{-i\phi}}{2i} = I(e^{i\phi}) \qquad \bullet \ (2.125)$$

The magnitude of an expression can also be obtained by using the complex conjugate. We find that

$$\boxed{zz^* = (re^{i\phi})(re^{-i\phi}) = r^2} \qquad (2.126)$$

so that

$$r = \sqrt{zz^*} \qquad (2.127)$$

where the positive square root is to be taken. Note that zz^* is always real and nonnegative.

Example 2-15

If $z = 4e^{3i} + 6i$, find $R(z)$, $I(z)$, r, and ϕ.

Solution

$$R(z) = \frac{z + z^*}{2} = \frac{4e^{3i} + 6i + 4e^{-3i} - 6i}{2}$$

$$= 2(e^{3i} + e^{-3i}) = 4\cos(3) = -3.960$$

$$I(z) = \frac{z - z^*}{2i} = \frac{4e^{3i} + 6i - 4e^{-3i} + 6i}{2i}$$

$$= \frac{4(e^{3i} - e^{-3i})}{2i} + 6 = 4\sin(3) + 6 = 4\sin(171.89°) + 6$$

$$= 6.5645$$

$$r = (zz^*)^{1/2} = (x^2 + y^2)^{1/2} = [(-3.960)^2 + (6.5645)^2]^{1/2}$$

$$= 7.666$$

$$\phi = \arctan\left(\frac{I}{R}\right) = \arctan\left(\frac{-6.5645}{3.960}\right) = \arctan(-1.6577)$$

The principal value of this arctangent is $-58.90°$. However, since $R(z)$ is negative and $I(z)$ is positive, we require an angle in the second quadrant.

$$\phi = 180° - 58.90° = 121.10° = 2.114 \text{ rad} \qquad \bullet$$

Problem 2-26

If $z = \left(\dfrac{3 + 2i}{4 + 5i}\right)^2$, find $R(z)$, $I(z)$, r, and ϕ.

The *square root of a complex number* is a number which when multiplied by itself will yield the first number. Just as with real numbers, there are two square roots of a complex number. If $z = re^{i\phi}$, one of the square roots is given by

$$\sqrt{re^{i\phi}} = \sqrt{r}\, e^{i\phi/2} \tag{2.128}$$

The other square root is obtained by realizing that if ϕ is increased by 2π, the same point in the Argand plane is represented. Therefore, the square root of $re^{i(2\pi + \phi)}$ is the same as the other square root of $re^{i\phi}$:

$$\sqrt{re^{i\phi}} = \sqrt{r}\, e^{i(\pi + \phi/2)} \tag{2.129}$$

Example 2-16

Find the square roots of $3e^{i\pi/2}$.

Solution

One square root is, from Eq. (2.128),

$$\sqrt{3e^{i\pi/2}} = \sqrt{3}\, e^{i\pi/4}$$

The other square root is, from Eq. (2.129),

$$\sqrt{3e^{i(\pi + \pi/4)}} = \sqrt{3}\, e^{i5\pi/4} \qquad \bullet$$

Problem 2-27

Find the square roots of $4 + 4i$. Sketch an Argand diagram and locate the roots on it. $\qquad \bullet$

There are three *cube roots of a complex number*. These can be found by looking for the numbers which when cubed yield $re^{i\phi}$, $re^{i(2\pi + \phi)}$, and $re^{i(4\pi + \phi)}$. These numbers are

$$\sqrt[3]{re^{i\phi}} = \sqrt[3]{r}e^{i\phi/3}, \quad \sqrt[3]{r}e^{i(2\pi + \phi)/3}, \quad \sqrt[3]{r}e^{i(4\pi + \phi)/3}$$

Higher roots are obtained similarly.

ADDITIONAL PROBLEMS

2.28

A Boy Scout finds a tall tree while hiking and wants to estimate its height. He walks away from the tree and finds that when he is 100 m from the tree, he must look upward at an angle of $35°$ to look at the top of the tree. His eye is 1.50 m from the ground, which is perfectly level. How tall is the tree?

2.29

Evaluate

$$\frac{(0.213)(1.76 \times 10^{-46})}{(7.321)(8.67 \times 10^{76})}$$

using a table of logarithms.

2.30

The equation $x^2 + y^2 + z^2 = c^2$, where c is a constant, represents a surface in three dimensions. Express the equation in spherical polar coordinates. What is the shape of the surface?

2.31

Express the equation $y = b$, where b is a constant, in plane polar coordinates.

2.32

Find the values of the plane polar coordinates that coorespond to $x = -2$, $y = 4$.

2.33

Find the values of the cartesian coordinates that correspond to $r = 10$, $\theta = 45°$, $\phi = 135°$.

2.34

Find $\mathbf{A} - \mathbf{B}$ if $\mathbf{A} = 2\mathbf{i} + 3\mathbf{j}$ and $\mathbf{B} = \mathbf{i} + 3\mathbf{j} - \mathbf{k}$.

2.35

Find $\mathbf{A} \cdot \mathbf{B}$ if $\mathbf{A} = (0, 2)$ and $\mathbf{B} = (2, 0)$.

2.36

Find $|\mathbf{A}|$ if $\mathbf{A} = 3\mathbf{i} + 4\mathbf{j} - \mathbf{k}$.

2.37

Find $\mathbf{A} \times \mathbf{B}$ if $\mathbf{A} = (0, 1, 2)$ and $\mathbf{B} = (2, 1, 0)$.

2.38

Find the angle between $\mathbf{A}$ and $\mathbf{B}$ if $\mathbf{A} = \mathbf{i} + 2\mathbf{j} + \mathbf{k}$ and $\mathbf{B} = \mathbf{i} + \mathbf{j} + \mathbf{k}$.

2.39

A spherical object falling in a fluid has three forces acting upon it:

1. The gravitional force, whose magnitude is $F_g = mg$, where m is the mass of the object.
2. The buoyant force, whose magnitude is $F_b = m_f g$, where m_f is the mass of the displaced fluid, and whose direction is upward.
3. The frictional force, given by $\mathbf{F}_f = -6\pi\eta$ r$\mathbf{v}$, where r is the radius of the object, $\mathbf{v}$ its velocity, and η the coefficient of viscosity of the fluid.

If the object is falling at a constant speed of 0.500 m s^{-1}, has a radius of 0.15 m and a mass of 0.0600 kg, and displaces a mass of fluid equal to 0.01257 kg, find the value of η and its units.

2.40

The solutions to the Schrödinger equation for the electron in a hydrogen atom have three quantum numbers associated with them, called n, l, and m, and these solutions are often denoted by ψ_{nlm}. One of the solutions is

$$\psi_{211} = \frac{1}{4\sqrt{4\pi}} \left(\frac{1}{a_0}\right)^{3/2} \frac{r}{a_0} e^{-r/2a_0} \sin(\theta)e^{i\phi}$$

where a_0 is a distance equal to 0.529×10^{-10} m, called the Bohr radius.

a. Write this function in terms of cartesian coordinates.
b. Write an expression for the magnitude of this complex function.
c. This function is sometimes called ψ_{2p_1}. Write expressions for the real and imaginary parts of the function, which are proportional to the related functions called ψ_{2px} and ψ_{2py}.

2.41

Find the sum of $4e^{3i}$ and $5e^{2i}$.

2.42

Find the three cube roots of $3 - 2i$.

2.43

Find the real and imaginary parts of

$$\sqrt{3 + 2i} + (7 + 5i)^2$$

Obtain a separate answer for each of the two square roots.

ADDITIONAL READING

Sherman K. Stein, *Calculus and Analytic Geometry*, 2nd ed., McGraw-Hill Book Company, New York, 1977.

  This is a calculus textbook that uses quite a few examples from physics in its discussions. You can read about coordinate systems, vectors, and complex numbers in almost any calculus textbook, including this one.

Herbert B. Dwight, *Tables of Integrals and Other Mathematical Data*, 4th ed., Macmillan Publishing Co., New York, 1962.

  This book is a very useful small compilation of formulas, including trigonometric identities, derivatives, definite and indefinite integrals, and infinite series.

D. D. Fitts, *Nonequilibrium Thermodynamics*, McGraw-Hill Book Company, New York, 1962.

  After discussing equilibrium thermodynamics in your physical chemistry course, you may be interested in glancing through this book, which presents the nonequilibrium version of the subject. There is also a brief discussion of dyadics, a kind of product of vectors which we did not discuss in this chapter.

CHAPTER THREE

Mathematical Functions and Differential Calculus

Section 3-1. INTRODUCTION

The purpose of this chapter is to survey those parts of differential calculus which are most useful in the study of physical chemistry. No attempt at mathematical rigor is made. The emphasis is on facts that may be useful.

The concept of a mathematical function is presented, and its relationship to physical variables in actual systems is discussed. The derivative of a function of one independent variable is introduced, and derivatives of various specific functions are presented. The use of derivatives in finding minimum and maximum values of functions is discussed.

After studying this chapter, you should be able to do the following:

1. **Obtain the derivative of any fairly simple function without consulting a table.**
2. **Draw a rough graph of any fairly simple function and locate important features on the graph.**
3. **Find maximum and minimum values of a function.**

Section 3-2. MATHEMATICAL FUNCTIONS

The idea of a *mathematical function* is one of the most important mathematical concepts. The simplest definition of a function is that it is a set of ordered pairs of numbers. This means that if you have a table with two columns of numbers in it, each number in the first column is associated with the number on the same line in the other column, and you must keep track of which number is in the first column.

For example, a physical chemistry student might measure the vapor pressure of liquid ethanol at several temperatures and present the results in such a table. In the first column, he puts the values of the temperature, and on each line in the second column he puts the observed vapor pressure for the value of the temperature on that line.

One variable, say the temperature, is chosen to be the *independent variable*. The vapor pressure is then the *dependent variable*. This means that if we choose a value of the temperature, the function provides the corresponding value of the vapor pressure. This is the important property of a function. It is as though the function says: "You give me a value for the independent variable, and I'll give you a value of the dependent variable."

Mathematical functions are useful in physics and chemistry because physical systems do the same thing that mathematical functions do. The physical chemistry student chooses a value of the temperature by setting the control on a thermoregulator, and the system of liquid and vapor assumes a value of the vapor pressure which is determined by the temperature, and which will always be the same for the same value of the temperature.

If we have a list of 10 temperatures and the corresponding 10 values of the vapor pressure, this qualifies as a mathematical function. However, such a function is not very useful, because it can be used for only these temperatures. The temperature of the liquid–vapor system can take on any value from the triple point to the critical point, and there will be a value of the vapor pressure for each value of the temperature.

We make the following assumptions about the actual behavior of physical systems:

1. Of the macroscopic variables such as temperature, pressure, volume, density, entropy, energy, etc., only a certain number (depending on circumstances) can be independent variables; the others are dependent variables, governed by mathematical functions.
2. These mathematical functions are *single-valued*, except possibly at isolated points.
3. These mathematical functions are *continuous* and *differentiable*, except possibly at isolated points.

Our first goal is to provide a way to generate values for the dependent variable corresponding to new values of the independent variable. If we have a precise way to do this, the rule used to generate new values can be considered to be equivalent to the function itself. Approximate ways to generate new values include interpolation between values in a table, use of a graph on which a curve is drawn passing through or near data points, and use of a mathematical formula chosen to represent the data points. In all these cases, we hope to make our rule for generating new values give nearly the same values as the "correct" function.

One of the properties that we assume our "correct" function possesses is that of being single-valued. If a function is single-valued, it will deliver one and only one value of the dependent variable for any given value of the independent variable. In discussing real functions of real variables, mathematicians will usually not call something a function unless it is single-valued.

The other important property of our functions is continuity. Roughly speaking, if a function is continuous, the dependent variable does not change abruptly if the independent variable changes gradually. If you are drawing the graph of a continuous function, you will not have to lift your pencil from the paper or draw a vertical step in your curve. This can be expressed mathematically by saying that the function

$$y = f(x) \tag{3.1}$$

is continuous at $x = a$ if

$$\lim_{x \to a} f(x) = f(a) \tag{3.2}$$

In words, as x draws close to a, from either direction, $f(x)$ draws close to $f(a)$, the value that y has at $x = a$.

The functions that represent physical variables are piecewise continuous in some cases, and in other cases are continuous over their entire range of values. Figure 3-1 shows schematically the density of a pure substance as a function of temperature at fixed pressure. The vertical axis is compressed in one place, and you can see that there is a large discontinuity at the boiling temperature, T_b, and a smaller discontinuity at the freezing temperature, T_f. The function is not continuous at these values of T, the independent variable. We say that the density is piecewise continuous, which means that it is continuous except at isolated points.

With the density, and some other physical variables, the situation is more complicated than with most simple mathematical functions. The density is double-valued at $T = T_f$ and at $T = T_b$. At these temperatures, two phases

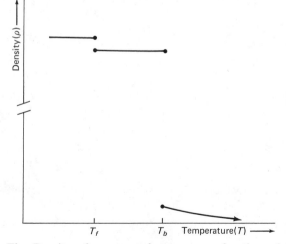

Figure 3-1. The Density of a pure substance as a function of temperature (schematic).

can coexist, each having a different value of the density. Again, we assume that this happens only at isolated points.

There is a comment on the notation used in Eq. (3.1) which is worth making. In this equation, we have used the notation of a mathematician. In this case, the mathematician thinks of the letter y as representing the dependent variable and of the letter f as representing the rule that provides us with values of y. These are, of course, conceptually different things. A physicist or chemist will usually not use two letters, but will write for the density, for example,

$$\rho = \rho(T) \tag{3.3}$$

where the letter ρ has to stand both for the density and for the function that provides values of the density. The reason for this is that physicists and chemists have a lot of different variables to discuss, and only a limited number of letters in the alphabet.

Graphical Representation of Functions. The best way to communicate quickly the general behavior of a function is with a graph. A graph containing data points and a curve drawn through or near the points can reveal the presence of an inacurrate data point, as well as providing values of the dependent variable.

Problem 3-1

The following is a set of data for the vapor pressure of ethanol. Plot these points on a graph, with the temperature on the horizontal axis (the abscissa) and the vapor pressure on the vertical axis (the ordinate). Decide if there are any bad data points. Draw a smooth curve nearly through the points, disregarding any bad points.

Celsius Temperature	Vapor Pressure (torr)
25.00	55.9
30.00	70.0
35.00	97.0
40.00	117.5
45.00	154.1
50.00	190.7
55.00	241.9

There are a number of families of functions which occur frequently in physical chemistry, and you should be familiar with them. The following is one such family. It is called a *linear function* or *first-order polynomial*:

$$y = mx + b \tag{3.4}$$

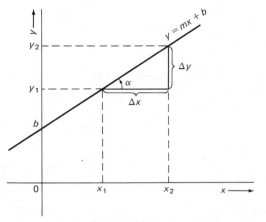

Figure 3-2. The graph of the linear function $y = mx + b$.

A family of functions is a set of related functions. In this family, we have a different function for each value of m and each value of b. The graph of such a function is a straight line. The constant b is called the *intercept*, because it is the value of the function for $x = 0$. The constant m is called the *slope*, because it gives the steepness of the line. Figure 3-2 shows two particular values of x, called x_1 and x_2, and their corresponding values of y, y_1 and y_2. If $m > 0$, then $y_2 > y_1$ and the line slopes upward to the right. If $m < 0$, then $y_2 < y_1$, and the line slopes downward to the right.

Example 3-1

Show that the slope is given by

$$m = \frac{y_2 - y_1}{x_2 - x_1} = \frac{\Delta y}{\Delta x} \tag{3.5}$$

Solution

$$y_2 - y_1 = mx_2 + b - (mx_1 + b) = m(x_2 - x_1)$$

or

$$m = \frac{y_2 - y_1}{x_2 - x_1}$$

This result, that the slope is equal to an increment in y divided by the corresponding increment in x, is important. Another important fact is that

$$m = \tan(\alpha) \tag{3.6}$$

where α is the angle between the horizontal axis and the straight line of the function. The angle α is taken to be between $-90°$ and $90°$, and the slope can range from $-\infty$ to ∞.

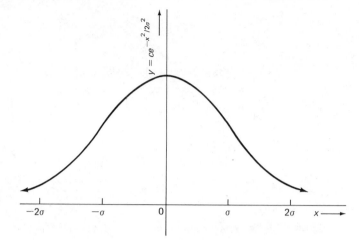

Figure 3-3. The graph of the gaussian function.

Other families of functions whose graphs should be familiar to you include the *quadratic function* or *second-order polynomial*:

$$y = ax^2 + bx + c \tag{3.7}$$

This graph of this function is a parabola. You should also be familiar with the graphs of the trigonometric functions and the exponential function, which are shown in Figures 2-2 through 2-5.

Another important function is the *gaussian function*:

$$y = ce^{-x^2/2\sigma^2} \tag{3.8}$$

This is the function whose graph is the *bell-shaped curve* shown in Figure 3-3. This function is proportional to the probability that a value of x will occur in a number of statistical applications and is discussed in Chapter 10. The constant σ is called the *standard deviation* and is a measure of the width of the "hump" in the curve.

The functions just listed form a repertoire of functions which you can use to generate approximate graphs of some other functions.

Example 3-2

Sketch a rough graph of the function

$$y = x\cos(\pi x)$$

Solution

The given function is a product of the function x, which is shown in Figure 3-4a, and $\cos(\pi x)$, which is shown in Figure 3-4b.

Where one factor vanishes, the product vanishes. Since the cosine oscillates between -1 and $+1$, the product oscillates between $-x$ and $+x$. A rough graph of the product is shown in Figure 3-4c. •

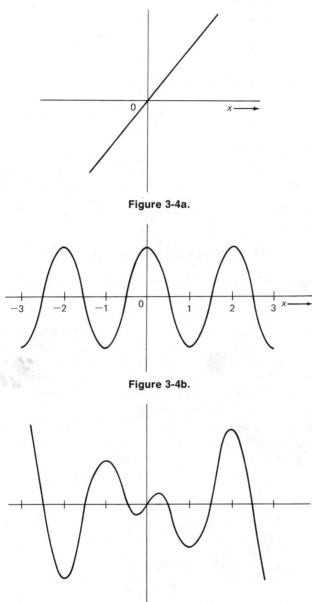

Figure 3-4a.

Figure 3-4b.

Figure 3-4c.

Problem 3-2

Draw rough sketches of the graphs representing the following functions.

a. $e^{-x}\sin(x)$

b. $\sin^2(x) = \sin(x)^2$

c. $x^2 e^{-x^2/2}$

   d. $1/x^2$

   e. $(1 - x)e^{-x}$ ●

A graph of a function is often useful in solving equations. If an equation to be solved can be written

$$f(x) = 0 \tag{3.9}$$

then a graph of $f(x)$ will cross the x axis at the real values of x which are solutions to the equation.

Problem 3-3

Draw a graph of the function and use it to find an approximate solution to the equation

$$e^x - 3x = 0 \qquad ●$$

Section 3-3. THE DERIVATIVE OF A FUNCTION

The Line Tangent to a Curve.   A function other than a linear function has a graph with a curve other than a straight line. Such a curve has a different direction, or different steepness, at different points on the curve. In order to describe the steepness of a curve, we introduce the idea of a line tangent to a curve. At most points on the curve, the tangent line is the line that has the point in common with the curve but does not cross it at that point. This is shown in Figure 3-5. We define the direction of the curve at $x = x_1$ as the direction of the line which is tangent to the curve at that point.

In the figure, we have labeled another point at x_2, a finite distance away from x_1. At $x = x_2$, the vertical distance from the horizontal line to the tangent line is given by $m(x_2 - x_1)$. This is not necessarily equal to the

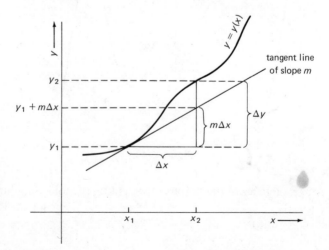

Figure 3-5.   The curve representing the function $y = y(x)$ and its tangent line.

distance from the horizontal line to the curve, which is given by $y(x_2) - y(x_1) = y_2 - y_1 = \Delta y$.

Problem 3-4

Using graph paper, plot the curve representing $y = \sin(x)$ for values of x lying between 0 and $\pi/2$ radians. Using a ruler (a transparent one if possible) draw the tangent line at $x = \pi/4$. By drawing a right triangle on your graph and measuring its sides, find the slope of the tangent line. •

There is a case in which the definition of the tangent line is more complicated than in the case shown in Figure 3-5. In this case, the point x_1 lies between a region in which the curve is concave downward and a region in which the curve is concave upward. Such a point is called an *inflection point*. For such a point, we must consider tangent lines at points which are taken closer and closer to x_1. As we approach closer and closer to x_1, the tangent line will approach more and more closely to a line which is the tangent line at x_1. This line is the same whether we approach from the left or the right, and it does cross the curve at the point which it shares with the curve.

If Δx is not too large, we can write as an approximation

$$\Delta y \approx m \, \Delta x \tag{3.10}$$

We divide both sides of Eq. (3.10) by Δx and write

$$m \approx \frac{\Delta y}{\Delta x} = \frac{y_2 - y_1}{x_2 - x_1} = \frac{y(x_2) - y(x_1)}{x_2 - x_1} \tag{3.11}$$

If the curve is smooth, this equation becomes a better and better approximation as Δx becomes smaller, and if we take the mathematical limit as $\Delta x \to 0$, it becomes exact.

$$m = \lim_{\Delta x \to 0} \frac{\Delta y}{\Delta x} = \left(\frac{dy}{dx}\right)_{x_1} \tag{3.12}$$

The limit in Eq. (3.12) is called the *derivative*, if the limit exists, and is given the symbol shown in the final equality in this equation. The subscript is written to show the value of x for which the derivative is evaluated and is not a necessary part of the derivative symbol.

If the function giving y as a function of x is discontinuous at $x = x_1$, the limit will not exist, and we say that a function is not differentiable at a discontinuity. There are also cases in which the limit exists but has a different value if $x_2 > x_1$ than it does if $x_2 < x_1$. In this case, we also say that the function is not differentiable at $x = x_1$. Figure 3-6 shows the graph of a function that is not differentiable at $x = b$, because the function is discontinuous at $x = b$ and is also not differentiable at $x = a$, even though the function is continuous at $x = a$, because the limit has different values for $x_2 > x_1$ and for $x_2 < x_1$.

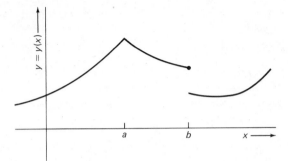

Figure 3-6. A function which is not differentiable at $x = a$ and at $x = b$.

Example 3-3

Decide where the following functions are differentiable.

a. $y = |x|$

b. $y = \sqrt{x}$

Solution

a. Differentiable everywhere except at $x = 0$, where the limit has different values when $x = 0$ is approached from the two different directions.

b. This function has real values only for $x \geq 0$. It is differentiable for all positive values of x, but at $x = 0$ the limit does not exist, so it is not differentiable at $x = 0$. ●

Problem 3-5

Decide where the following functions are differentiable.

a. $y = \ln(x)$

b. $y = \tan(x)$

c. $y = \sec(x)$ ●

Now that we have defined the derivative, let us see how it is applied to a particular function.

Example 3-4

Find the derivative of the function

$$y = y(x) = ax^2$$

Solution

We find the quotient $\Delta y / \Delta x$:

$$\Delta y = y_2 - y_1 = ax_2^2 - ax_1^2 = a(x_1 + \Delta x)^2 - ax_1^2$$
$$= a[x_1^2 + 2x_1 \Delta x + (\Delta x)^2] - ax_1^2 = 2ax_1 \Delta x + (\Delta x)^2$$

$$\frac{\Delta y}{\Delta x} = 2ax_1 + \Delta x$$

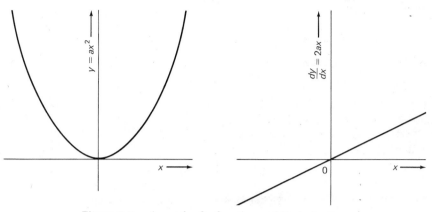

Figure 3-7. A graph of a function and its derivative.

We now take the limit as $x_2 \to x_1$ or $\Delta x \to 0$. The first term, $2ax_1$, is not affected. The second term, Δx, vanishes. Thus, if we use the symbol x instead of x_1,

$$\frac{dy}{dx} = \frac{d(ax^2)}{dx} = \lim_{\Delta x \to 0} \frac{\Delta y}{\Delta x} = 2ax \qquad \bullet \quad (3.13)$$

In Example 3-4, the function is differentiable for all values of the independent variable.

Figure 3-7 shows a graph of the function $y = ax^2$ and a graph of its derivative, $dy/dx = 2ax$. Notice the following facts. (1) When the function has a horizontal tangent line, at $x = 0$, the derivative is equal to zero. (2) When the function increases as x increases, the derivative is positive and is larger when the tangent line is steeper. (3) When the function decreases as x increases, as it does in this example for negative values of x, the derivative is negative and has a larger magnitude when the tangent line is steeper. These are general properties of the derivative of any function, and you should remember them.

Table 3-1 gives the derivatives of other simple functions, derived in much the same way as Eq. (3.13). You should memorize these formulas. Additional derivatives are given in Appendix 2.

Table 3-1 Some Simple Functions and Their Derivatives*

Function, $y = y(x)$	Derivative, dy/dx
ax^n	nax^{n-1}
ae^{bx}	abe^{bx}
a	0
$a \sin(bx)$	$ab \cos(bx)$
$a \cos(bx)$	$-ab \sin(bx)$
$a \ln(bx)$	a/x

* In these formulas, a, b, and n are constants, not necessarily integers.

Problem 3-6

Make rough graphs of several functions from Table 3-1. Below each graph, on the same sheet of paper, make a rough graph of the derivative of the same function. •

It is important that you have a good intuitive grasp of the meaning of the derivative of a function. It is the rate of change of the dependent variable with respect to the independent variable. That is, because it is the slope of the tangent line, it has a large magnitude when the curve is steep and a small magnitude when the curve is nearly horizontal.

Section 3-4. DIFFERENTIALS

In Section 3-3, we talked about a change in a dependent variable produced by a change in an independent variable. If y is a function of x, we wrote in Eq. (3.10)

$$\Delta y \approx m \, \Delta x \tag{3.14}$$

where $\Delta y = y(x_2) - y(x_1)$, $\Delta x = x_2 = x_1$, and m was the slope of the tangent line at $x = x_1$. Using Eq. (3.12), this becomes

$$\Delta y \approx \left(\frac{dy}{dx}\right)\Delta x \tag{3.15}$$

This approximate equality will generally be more nearly correct when Δx is made smaller, but may be quite badly wrong if Δx is fairly large.

Example 3-5

Using Eq. (3.15), estimate the change in the pressure of 1.000 mol of an ideal gas at $0\,°C$ when its volume is changed from 22.414 liters to 21.414 liters.

Solution

An ideal gas obeys

$$P = \frac{nRT}{V} \tag{3.16}$$

so that if n and T are kept fixed, P depends only on V and

$$\frac{dP}{dV} = \frac{-nRT}{V^2} \tag{3.17}$$

$$\frac{dP}{dV} = \frac{-(1.000 \text{ mol})(0.082057 \text{ liter atm K}^{-1} \text{ mol}^{-1})(273.15 \text{ K})}{(22.414 \text{ liters})^2}$$

$$= -0.0446 \text{ atm liter}^{-1}$$

Thus,

$$\Delta P \approx \left(\frac{dP}{dV}\right)\Delta V = (-0.0446 \text{ atm liter}^{-1})(-1.0000 \text{ liter})$$

$$\approx 0.0446 \text{ atm}$$

●

Example 3-6

Determine accuracy of the result of Example 3-5.

Solution

$$P = P(21.414 \text{ liters}) - P(22.414 \text{ liters})$$

$$= 1.0468 \text{ atm} - 1.0000 \text{ atm} = 0.0468 \text{ atm}$$

Our estimate in Example 3-5 was wrong by about 5%. If the change in volume had been 0.1 liter, the error would have been about 0.5%. ●

Since Eq. (3.15) becomes more nearly exact as Δx is made smaller, we make it into an exact equation by making Δx become smaller than any finite quantity that you or anyone else can name. We do not make Δx strictly vanish, but we say that we make it become *infinitesimal*. In this limit, we say that Δx becomes the *differential*, dx, and write

$$dy = \left(\frac{dy}{dx}\right)dx \qquad \qquad \textbf{(3.18)}$$

The infinitesimal quantity dy is the result of the infinitesimal increment dx. Notice that it is proportional to dx and to the slope of the tangent line, which is equal to dy/dx. Remember that x is an independent variable, so that dx is arbitrary, or subject to our control, and that y is a dependent variable, so that its differential, dy, is determined by dx.

Equation (3.18) has the appearance of an equation in which the dx in the denominator is canceled by the dx in the numerator. This is not the case. The symbol dy/dx is *not* a fraction or a ratio. It is the limit which a ratio approaches, and that is not the same thing.

In exact numerical calculations, differentials are not of direct use, since they are smaller than any finite quantities that you can name. Their use lies in the construction of formulas, especially through the process of integration, in which infinitely many infinitesimal quantities are added up to produce something finite. We discuss this in Chapter 4.

Problem 3-7

The number of atoms of a radioactive substance at time t is given by

$$N(t) = N_0 e^{-t/\tau}$$

where N_0 is the initial number of atoms and τ is the relaxation time. For ^{14}C, $\tau = 8320$ years. Estimate $N(t = 1\text{ year})/N_0$. •

Section 3-5 SOME USEFUL FACTS ABOUT DERIVATIVES

In this section we present some useful theorems about derivatives, which, together with the formulas for the derivatives of simple functions presented in Table 3-1, will enable you to obtain the derivative of almost any function that you will encounter in physical chemistry.

The Derivative of a Product of Two Functions. If y and z are both functions of x, then

$$\frac{d(yz)}{dx} = y\frac{dz}{dx} + z\frac{dy}{dx} \tag{3.19}$$

The Derivative of the Sum of Two Functions. If y and z are both functions of x, then

$$\frac{d(y+z)}{dx} = \frac{dy}{dx} + \frac{dz}{dx} \tag{3.20}$$

The Derivative of the Difference of Two Functions. If y and z are both functions of x,

$$\frac{d(y-z)}{dx} = \frac{dy}{dx} - \frac{dz}{dx} \tag{3.21}$$

The Derivative of the Quotient of Two Functions. If y and z are both functions of x, and z is not zero,

$$\frac{d(y/z)}{dx} = \frac{x\left(\dfrac{dy}{dx}\right) - y\left(\dfrac{dz}{dx}\right)}{z^2} \tag{3.22}$$

An equivalent result can be obtained by considering y/z to be a product of y and $1/z$, and using Eq. (3.19).

The Derivative of a Constant.   If c is a constant,

$$\frac{dc}{dx} = 0 \qquad (3.23)$$

This can be seen from the fact that a constant is a linear function of any variable, with zero slope. From this follows the important but simple fact

$$\frac{d(y + c)}{dx} = \frac{dy}{dx} \qquad (3.24)$$

If we add any constant to a function, we do not change the derivative at all.

The Derivative of a Function Times a Constant.   If y is a function of x and c is a constant,

$$\frac{d(cy)}{dx} = c\frac{dy}{dx} \qquad (3.25)$$

This can be deduced by substituting into the definition of the derivative, or by using Eqs. (3.19) and (3.23).

The Derivative of a Function of a Function (The Chain Rule).   If u is a differentiable function of x, and f is a differentiable function of u,

$$\frac{df}{dx} = \frac{df}{du}\frac{du}{dx} \qquad (3.26)$$

The function f is sometimes referred to as a *composite function*. It is a function of x, because if a value of x produces a value of u, this value of u produces a value of f. This is written

$$f(x) = f[u(x)] \qquad (3.27)$$

Here we have used the same letter for the quantity f whether it is expressed as a function of u or of x. In the two cases, it would represent two different mathematical formulas which would give the same numerical value.

We now illustrate how these facts about derivatives can be used to obtain formulas for the derivatives of various functions.

Example 3-7

Find the derivative of $\tan(ax)$.

Solution

$$\frac{d}{dx}\tan(ax) = \frac{d}{dx}\left[\frac{\sin(ax)}{\cos(ax)}\right] = \frac{\cos(ax)\,a\cos(ax) + \sin(ax)\,a\sin(ax)}{\cos^2(ax)}$$

$$= a\left[\frac{\cos^2(ax) + \sin^2(ax)}{\cos^2(ax)}\right] = \frac{a}{\cos^2(ax)}$$

$$= a\sec^2(ax) \qquad \bullet$$

Note the use of several trigonometric identities in the solution.

Example 3-8

Find dP/dT if $P(T) = ke^{-Q/T}$.

Solution

Let $u = -Q/T$.

$$\frac{dP}{dT} = \frac{dP}{du}\frac{du}{dT}$$

$$= ke^u\frac{Q}{T^2} = ke^{-Q/T}\left(\frac{Q}{T^2}\right) \qquad \bullet$$

Problem 3-8

Find the following derivatives.

a. dy/dx, where $y = (ax^2 + bx + c)^{-3/2}$

b. $\dfrac{d\ln(P)}{dT}$, where $P = ke^{-Q/T}$

c. dy/dx, where $y = a\cos(bx^3)$

d. $d(yz)/dx$, where $y = ax^2$, $z = \sin(bx)$ $\qquad \bullet$

Section 3-6. HIGHER-ORDER DERIVATIVES

Since the derivative of a function is itself a function, the derivative can usually be differentiated. The derivative of a derivative is called a *second derivative*, and the derivative of a second derivative is called a *third derivative*, etc. We use the notation

$$\frac{d^2y}{dx^2} = \frac{d}{dx}\left(\frac{dy}{dx}\right) \qquad (3.28)$$

and

$$\frac{d^3y}{dx^3} = \frac{d}{dx}\left(\frac{d^2y}{dx^2}\right) \qquad (3.29)$$

etc. The nth-order derivative is

$$\frac{d^n y}{dx^n} = \frac{d}{dx}\left(\frac{d^{n-1}y}{dx^{n-1}}\right)$$ (3.30)

Example 3-9

Find $d^2 y/dx^2$ if $y = a\sin(bx)$.

Solution

$$\frac{d^2 y}{dx^2} = \frac{d}{dx}[ab\,\cos(bx)] = -ab^2\sin(bx)$$ •

Remember the result of Example 3-9. The sine is proportional to the negative of its second derivative. The cosine has the same behavior. The exponential function is proportional to all of its derivatives.

Problem 3-9

Find the second and third derivatives of the following functions.
a. $y = ax^n$
b. $y = ae^{bx}$
c. $y = x^{1/2}$ •

The Curvature of a Function. Figure 3-8 shows a rough graph of a function, a rough graph of the first derivative of the function, and a rough graph of the second derivative of the function in the interval $a < x < g$ or (a, g). Where the function is concave downward, the second derivative is negative, and where the function is concave upward, the second derivative is positive. Where the graph of the function is more sharply curved, the magnitude of the second derivative is larger. This suggests that the second derivative can provide a measure of the curvature of the function curve, and this is the case.

For any function that possesses a second derivative, the curvature is given by the formula

$$K = \frac{d^2 y/dx^2}{[1 + (dy/dx)^2]^{3/2}}$$ (3.31)

The magnitude of the curvature is equal to the reciprocal of the radius of the circle which fits the curve at that point, in the sense of having the same curvature as the curve of the function.

Since the denominator in Eq. (3.31) is always positive, the curvature has the same sign as the second derivative, and is therefore positive when the curve is concave upward and negative when it is concave downward.

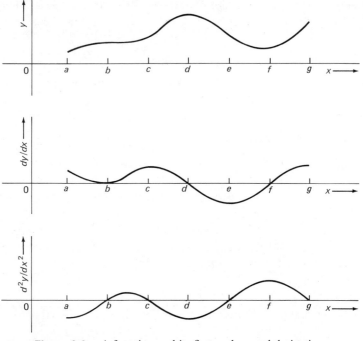

Figure 3-8. A function and its first and second derivatives.

Example 3-10

Find the curvature of the function $y = x^2$ at $x = 0$ and at $x = 2$.

Solution

$$\frac{d^2y}{dx^2} = \frac{d}{dx}(2x) = 2$$

$$K = \frac{2}{(1 + 4x^2)^{3/2}} = \begin{cases} 2 & \text{at } x = 0 \\ 0.0285 & \text{at } x = 2 \end{cases}$$

●

Problem 3-10

Find the curvature of the function $y = \sin(x)$ at $x = 0$ and at $x = \pi/2$. ●

Section 3-7. MAXIMUM–MINIMUM PROBLEMS

Sometimes it is necessary to find the largest or smallest value that a function attains or approaches in a certain interval, or to find the value of the independent variable at which this happens. In Figure 3-8, the maximum value of the function in the interval shown is at $x = d$ and the minimum value of the function in the interval is at the left end of the interval.

We now state the fact that enables us to find maximum and minimum values of a function:

Fact: *The maximum or minimum value of a differentiable function in an interval will either occur at an end of the interval or at a place where the first derivative of the function vanishes.*

The minimum value of a function means the most negative value, not necessarily the smallest magnitude. The maximum value means the most positive value, not necessarily the largest magnitude. If a function happens to be negative in an entire interval, the maximum will correspond to the smallest magnitude and the minimum to the largest magnitude.

In Figure 3-8, there are two points of horizontal tangent (where the first derivative vanishes) in addition to the point $x = d$ at which the maximum occurs. At $x = f$ we have a "relative minimum" or "local minimum." At such a point the function has a smaller value than at any other point in the immediate vicinity. At a local maximum, the function has a larger value than at any other point in the immediate vicinity. At $x = b$, there is an inflection point with a horizontal tangent line, at which the first derivative also vanishes.

Relative minima, relative maxima, and such inflection points can be distinguished from each other by finding the sign of the curvature from the second derivative. At a relative minimum, the second derivative is positive. At a relative maximum, the second derivative is negative. At an inflection point with horizontal tangent line, the second derivative vanishes.

These facts suggest the following procedure for finding the maximum and minimum value of a function in an interval:

1. Find all the points in the interval at which the first derivative vanishes.
2. Evaluate the function at these points and at the ends of the interval. The largest value in the list is the maximum value and the smallest value is the minimum.
3. If you want to find whether a point at which the first derivative vanishes is a relative maximum, a relative minimum, or an inflection point, evaluate the second derivative and use the fact that the second derivative is positive at a minimum, negative at a maximum, and zero at an inflection point with horizontal tangent line.

Example 3-11

Find the maximum and minimum values of the function

$$y = x^2 - 4x + 6$$

in the interval $0 < x < 5$.

Solution

The derivative is

$$\frac{dy}{dx} = 2x - 4$$

Let the value of x which satisfies the equation $dy/dx = 0$ be called x_m.

$$2x_m - 4 = 0 \quad \text{or} \quad x_m = 2$$

We evaluate the function at the ends of the interval and at $x = x_m$:

$$y(0) = 6$$
$$y(2) = 2$$
$$y(5) = 11$$

The maximum value of the function is at $x = 5$, the end of the interval. The minimum value is at $x = 2$. $\qquad\bullet$

Problem 3-11

a. For the interval $-10 < x < 10$, find the maximum and minimum values of

$$y = -x^3 + 3x^2 - 3x + 8$$

b. The probability that a molecule in a gas will have a speed v is proportional to the function

$$f_v(v) = 4\pi \left(\frac{m}{2\pi kT}\right)^{3/2} v^2 \exp\left(\frac{-v^2}{2mkT}\right)$$

where m is the mass of the molecule, k is Boltzmann's constant, and T is the temperature on the Kelvin scale. The most probable speed is the speed for which this function is at a maximum. Find the expression for the most probable speed, and find its value for N_2 molecules at $T = 298$ K. $\qquad\bullet$

Section 3–8. LIMITING VALUES OF FUNCTIONS. L'HÔPITAL'S RULE

We have already mentioned mathematical limits. The limit of y as x approaches a is denoted by

$$\lim_{x \to a} [y(x)] \qquad (3.32)$$

and is defined as the number that y approaches ever more closely as x approaches ever more closely to a, if such a number exists. The number a is not required to be finite, and a limit such as

$$\lim_{x \to \infty} [y(x)] \qquad (3.33)$$

exists if $y(x)$ approaches more closely to some number as x is made larger and larger without bound.

An example of a limit that does not exist is

$$\lim_{x \to a} \left(\frac{1}{x - a}\right) \qquad (3.34)$$

67

§3-8 / Limiting Values of Functions. L'Hôpital's Rule

As x approaches closer to a, $1/(x - a)$ becomes larger without bound if x approaches a from the right (from values larger than a) and $1/(x - a)$ becomes more negative without bound if x approaches a from the left.

Another limit that does not exist is

$$\lim_{x \to \infty} [\sin(x)] \tag{3.35}$$

which continues to oscillate between -1 and 1 as x continues to become larger and larger.

However, a limit that does exist as x approaches infinity is

$$\lim_{x \to \infty} (1 - e^{-x}) = 1 \tag{3.36}$$

Problem 3-12

Decide which of the following limits exist, and find the values of the ones that do.

a. $\lim_{x \to 0} (1 - e^x)$

b. $\lim_{x \to \infty} (e^{-x^2})$

c. $\lim_{x \to \pi/2} [x \tan(x)]$

d. $\lim_{x \to 0} [\ln(x)]$ •

Sometimes a limit exists but cannot be evaluated in a straightforward way by substituting into the expression the limiting value of the independent variable. For example, if we try to determine the limit

$$\lim_{x \to 0} \left[\frac{\sin(x)}{x} \right] \tag{3.37}$$

we find that for $x = 0$, both the numerator and denominator of the expression vanish. If an expression appears to approach $0/0$, it might approach 0, it might approach a finite constant of either sign, or it might diverge in either direction (approach $-\infty$ or $+\infty$).

The *rule of l'Hôpital* provides a way to determine the limit in such cases, and also for cases where the limit appears to approach ∞/∞. This rule can be stated: *If the numerator and denominator of a quotient both approach zero or both approach infinity in some limit, the limit of the quotient is equal to the limit of the quotient of the derivatives of the numerator and denominator, if this limit exists.*

That is, if the limits exist, then

$$\lim_{x \to a} \left[\frac{f(x)}{g(x)} \right] = \lim_{x \to a} \left[\frac{df/dx}{dg/dx} \right] \tag{3.38}$$

You should note that the limit must appear to approach 0/0 or ∞/∞ for the rule to apply. One author put it: "As a rule of thumb, l'Hôpital's rule applies when you need it, and not when you do not need it."[1]

Example 3-12

Find the value of the limit in Eq. (3.37) by use of l'Hôpital's rule.

Solution

$$\lim_{x \to 0} \left[\frac{\sin(x)}{x} \right] = \lim_{x \to 0} \left[\frac{d\sin(x)/dx}{dx/dx} \right] = \lim_{x \to 0} \left[\frac{\cos(x)}{1} \right] = 1 \quad \bullet \quad \textbf{(3.39)}$$

Sometimes the rule must be applied more than once in order to find the value of the limit, as in the following example.

Example 3-13

Find the limit

$$\lim_{x \to \infty} (x^3 e^{-x})$$

Solution

$$\lim_{x \to \infty} \left(\frac{x^3}{e^x} \right) = \lim_{x \to \infty} \left(\frac{3x^2}{e^x} \right) = \lim_{x \to \infty} \left(\frac{6x}{e^x} \right) = \lim_{x \to \infty} \left(\frac{6}{e^x} \right) = 0 \qquad \textbf{(3.40)}$$

In this example, the expression was not in the form of a quotient which appeared to approach either 0/0 or ∞/∞, but we were able to change it into the required form by using the identity

$$e^{-x} = \frac{1}{e^x} \qquad \bullet \quad \textbf{(3.41)}$$

By applying l'Hôpital's rule n times, we can show the following:

$$\lim_{x \to \infty} (x^n e^{-x}) = 0 \qquad \textbf{(3.42)}$$

for any finite value of n. This is an important result.

Another interesting limit is that of the next example.

Example 3-14

Find the limit

$$\lim_{x \to \infty} \left[\frac{\ln(x)}{x} \right]$$

Solution

$$\lim_{x \to \infty} \left[\frac{\ln(x)}{x} \right] = \lim_{x \to \infty} \left(\frac{1/x}{1} \right) = 0 \qquad \bullet \quad \textbf{(3.43)}$$

[1] Sherman K. Stein, *Calculus and Analytic Geometry*, 2nd ed., McGraw-Hill Book Company, New York, 1977, p. 239.

Problem 3-13

Find the limit

$$\lim_{x \to \infty} \left[\frac{\ln(x)}{\sqrt{x}} \right].$$

•

Problem 3-14

A mole of harmonic oscillators at thermal equilibrium at absolute temperature T is shown by statistical mechanics to have the energy

$$E = \frac{Nh\nu}{e^{h\nu/kT} - 1} \qquad (3.44)$$

where N is Avogadro's number, k is Boltzmann's constant, h is Planck's constant, and ν is the vibrational frequency.

a. Find the limit of E as $\nu \to 0$.
b. Find the limit of E as $T \to 0$.

•

There are a number of applications of limits in physical chemistry, and l'Hôpital's rule is useful in some of them. There is an article by Missen which lists some of these applications.[2]

Problem 3-15

Draw a rough graph of the function

$$y = \frac{\sin(x)}{x}$$

in the interval $-\pi < x < \pi$. Do not attempt to locate the relative extrema exactly (this would be difficult), but use l'Hôpital's rule to evaluate the function at $x = 0$.

•

ADDITIONAL PROBLEMS

3.16

Using the definition of the derivative and Eq. (20) of Appendix 6, show that

$$\frac{d \ln(x)}{dx} = \frac{1}{x}.$$

3.17

Find the first and second derivatives of the following functions.

a. $P = P(V) = RT \left(\dfrac{1}{V} + \dfrac{B}{V^2} + \dfrac{C}{V^3} \right)$, where R, T, B, and C are constants

b. $G = G(x) = G^\circ + RTx \ln(x) + RT(1 - x) \ln(1 - x)$, where G^0, R, and T are constants

[2] Ronald W. Missen, "Applications of the l'Hôpital–Bernoulli Rule in Chemical Systems," *J. Chem. Educ.* **54**, 448 (1977).

c. $y = a \ln(x^{1/3})$, where a is a constant

d. $y = 3x^3 \ln(x)$

e. $y = \dfrac{1}{(c-x)^2}$, where c is a constant

f. $y = \ln[\tan(2x)]$

g. $y = \dfrac{1}{x}\dfrac{1}{1+x}$

h. $f = f(v) = ce^{-mv^2/(2kT)}$, where m, c, k, and T are constants

i. $y = 3\sin^2(2x) = 3\sin(2x)^2$

j. $y = a_0 + a_1 x + a_2 x^2 + a_3 x^3 + a_4 x^4 + a_5 x^5$, where a_0, a_1, etc., are constants

3.18

Find the following derivatives and evaluate them at the points indicated.

a. $(dy/dx)_{x=0}$ if $y = \sin(bx)$, where b is a constant

b. $(df/dt)_{t=0}$ if $f = Ae^{-kt}$, where A and k are constants

c. $(dy/dx)_{x=1}$ if $y = (ax + bx^2)^{-1/2}$, where a and b are constants

d. $(d^2y/dx^2)_{x=0}$ if $y = ae^{-bx}$, where a and b are constants

3.19

The volume of a cube is given by

$$V = V(a) = a^3$$

where a is the length of a side. Estimate the error in the volume if a 1% error is made in measuring the length, using the formula

$$\Delta V \approx \left(\frac{dV}{da}\right)\Delta a$$

Check the accuracy of this estimate by computing $V(a_0)$ and $V(a_0 + 0.01a_0)$.

3.20

a. Draw a rough graph of the function

$$y = y(x) = e^{-|x|}$$

b. Is the function differentiable at $x = 0$?

c. Draw a rough graph of the derivative of the function.

3.21

Show that the function

$$\psi = \psi(x) = A\sin(kx)$$

satisfies the equation

$$\frac{d^2\psi}{dx^2} = -k^2\psi$$

if A and k are constants. This is an example of a differential equation.

3.22

Draw rough graphs of the third and fourth derivatives of the function whose graph is given in Figure 3-8.

3.23

The Gibbs free energy of a mixture of two enantiomorphs (optical isomers of the same substance) is given by

$$G = G(x) = G^0 + RTx\ln(x) + RT(x_0 - x)\ln(x_0 - x)$$

where x_0 is the sum of the concentrations of the enantiomorphs and x is the concentration of one of them. G^0 is a constant, R is the gas constant, and T is the temperature. If the temperature is maintained constant, what is the concentration of each enantiomorph when G has its minimum value? What is the maximum value of G in the interval $0 \leq x \leq x_0$?

3.24

a. A rancher wants to enclose a rectangular part of a large pasture so that 1 km^2 is enclosed with the minimum amount of fence. Find the shape and size of the rectangle that he should choose. The area is

$$A = xy$$

but A is fixed at 1 km^2, so that

$$y = \frac{A}{x}$$

b. The rancher now decides that the fenced area must lie along a road and finds that the fence costs $5 per meter along the road and $3 per meter along the other edges. Find the shape and size of the rectangle that would minimize the cost of the fence.

3.25

Using

$$\Delta y \approx \left(\frac{dy}{dx}\right)\Delta x$$

show that

$$e^{\Delta x} - 1 \approx \Delta x \qquad \text{if } \Delta x \ll 1$$

3.26

The sum of two nonnegative numbers is 100. Find their values if their product plus twice the square of the first is to be a maximum.

3.27

A cylindrical tank in a chemical factory is to contain a corrosive liquid. Because of the cost of the material, it is necessary to minimize the area of the tank, which must contain 1.000 m^3. Find the optimum radius and height, and find the resulting area.

3.28

Find the following limits.

a. $\lim\limits_{x \to \infty} \left[\dfrac{\ln(x)}{x^2} \right]$

b. $\lim\limits_{x \to 3} \left(\dfrac{x^3 - 27}{x^2 - 9} \right)$

c. $\lim\limits_{x \to 0^+} \left[\sin(x) \ln(x) \right]$

d. $\lim\limits_{x \to \infty} \left(\dfrac{e^{-x^2}}{e^{-x}} \right)$

e. $\lim\limits_{x \to 0} \left[\dfrac{x^2}{(1 - \cos(2x))} \right]$

f. $\lim\limits_{x \to \pi} \left[\dfrac{\sin(x)}{\sin(3x/2)} \right]$

3.29

If a hydrogen atom is in a $2s$ state, the probability of finding the electron at a distance r from the nucleus is proportional to $4\pi r^2 \psi_{2s}^2$, where

$$\psi_{2s} = \frac{1}{4\sqrt{2\pi}} \left(\frac{1}{a_0} \right)^{3/2} \left(2 - \frac{r}{a_0} \right) e^{-r/a_0}$$

where a_0 is a constant, the Bohr radius, equal to 0.529×10^{-10} m.

a. Draw a rough graph of ψ_{2s}, locating maxima and minima correctly.

b. Draw a rough graph of $4\pi r^2 \psi_{2s}^2$, locating maxima and minima correctly.

3.30

The energy of a mole of harmonic oscillators (approximate representations of molecular vibrations) is given in Problem 3-14.

a. Draw a rough sketch of the energy as a function of T, the temperature.

b. The heat capacity at constant volume, C_v, is given by

$$C_v = \frac{dE}{dT}$$

Show that the heat capacity is given by

$$C_v = Nk \left(\frac{hv}{kT} \right)^2 \frac{e^{hv/kT}}{(e^{hv/kT} - 1)^2}$$

c. Find the limit of the heat capacity as $T \to 0$ and as $T \to \infty$. Note that the limit as $T \to \infty$ is the same as the limit $v \to 0$.

d. Draw a rough graph of C_v as a function of T.

ADDITIONAL READING

James Newton Butler and Daniel Gureasko Bobrow, *The Calculus of Chemistry*, W. A. Benjamin, New York, 1965.

This is an elementary introduction to calculus aimed particularly at students of chemistry.

Daniel A. Greenberg, *Mathematics for Introductory Science Courses*, W. A. Benjamin, New York, 1965.

This book is also an elementary introduction to calculus.

Clifford E. Swartz, *Used Math for the First Two Years of College Science*, Prentice-Hall, Englewood Cliffs, N.J., 1973.

This book is a survey of various mathematical topics at the beginning college level.

Sherman K. Stein, *Calculus and Analytic Geometry*, 2nd. ed., McGraw-Hill Book Company, New York, 1977.

This is one of a number of standard introductory calculus textbooks. You can read about mathematical functions and differential calculus in this or whatever introductory calculus textbook you may have access to.

CHAPTER FOUR

Integral Calculus

Section 4-1. INTRODUCTION

In this chapter we survey the parts of integral calculus that are most useful in physical chemistry. The material is presented from the point of view of application rather than from a fundamental mathematical point of view.

We begin with the simple concept of finding a function from knowledge of its derivative and proceed through the definition of integration as a process to applications of integration to specific kinds of problems.

After studying this chapter, you should be able to do all of the following:

1. **Obtain the antiderivative or indefinite integral of a function using a table.**
2. **Calculate a definite integral using the antiderivative.**
3. **Calculate mean values of quantities using a probability distribution with one random variable.**
4. **Obtain an approximate value for a definite integral using numerical analysis.**
5. **Manipulate integrals into tractable forms by use of partial integration, the method of substitution, and the method of partial fractions.**

Section 4-2. THE ANTIDERIVATIVE OF A FUNCTION

In Chapter 3, we discussed the derivative of a function. Let us now review a little of that discussion, using a particular example.

If a particle moves vertically, we can express its position as a function of time by

$$z = z(t) \tag{4.1}$$

where we use the coordinate z to measure the height above the origin. The coordinate z is really the component of a position vector, but if the other two components remain fixed, we do not need to mention them.

The velocity is the derivative of the position vector, so that

$$v_z = v_z(t) = \frac{dz}{dt} \tag{4.2}$$

The acceleration is the derivative of the velocity, or the second derivative of the position vector

$$a_z = \frac{dv_z}{dt} = \frac{d^2z}{dt^2} \tag{4.3}$$

Example 4-1

A particle falling in a vacuum near the surface of the earth has a position given by

$$z = z(t) = z(0) + v_z(0)t - \frac{gt^2}{2} \tag{4.4}$$

where $z(0)$ is the position at $t = 0$, $v_z(0)$ is the velocity at $t = 0$, and g is a constant called the *acceleration due to gravity*, equal to 9.80 m s^{-2}. Find the velocity and the acceleration.

Solution

$$v_z(t) = \frac{dz}{dt} = v_z(0) - gt \tag{4.5}$$

$$a_z(t) = \frac{d^2z}{dt^2} = -g \qquad \bullet \tag{4.6}$$

Now consider the reverse problem from that of Example 4-1. If we are given the acceleration as a function of time, how do we find the velocity? If we are given the velocity as a function of time, how do we find the position?

Example 4-2

Given only that the acceleration of a particle is

$$a_z = a_z(t) = -g$$

and that the particle moves only in the z direction, find the velocity and the acceleration.

Solution

We know that a function which possesses $-g$ as its derivative is $-gt$, so we might have

$$v_z(t) = -gt$$

However, the derivative of any constant is zero, so that we can write

$$v_z(t) = v_z(0) - gt \tag{4.7}$$

where $v_z(0)$ is a constant. Equation (4.7) represents a family of functions, one for each value of this constant.

To find the position, we seek a function that has $v_z(0) - gt$ as its derivative. From Table 3-1,

$$\frac{d}{dx}(ax^2) = 2ax \quad \text{and} \quad \frac{d}{dx}(ax) = a \tag{4.8}$$

so that

$$z = z(t) = z(0) + v_z(0)t - \frac{gt^2}{2} \tag{4.9}$$

where $z(0)$ is again a constant. Equation (4.9) includes all of the functions which give the position z as a function of time such that $-g$ is the acceleration. The constants $z(0)$ and $v_z(0)$ would be assigned values to correspond to any particular case.

In Example 4-2, we found two *antiderivative functions*. An antiderivative function is a function that possesses a particular derivative.

Example 4-3

Find the antiderivative of

$$f(x) = a\sin(bx)$$

where a anb b are constants.

Solution

The antiderivative function is, from Table 3-1,

$$F(x) = -\frac{a}{b}\cos(bx) + c \tag{4.10}$$

where c is another constant. You can check by differentiating that

$$dF/dx = f(x) \qquad \bullet \tag{4.11}$$

Problem 4-1

Find the family of functions whose derivative is ae^{bx}.

Example 4-4

Find the expression for the velocity of a particle falling near the surface of the earth in a vacuum, given that the velocity at $t = 1.000$ s is 10.00 m s^{-1} and that $g = 9.80$ m s^{-2}.

Solution

The necessary family of functions is given by Eq. (4.7). We find that $v_z(0) = 19.80$ m s^{-1}. Thus

$$v_z(t) = (19.80 - 9.80t) \text{ m s}^{-1} \qquad \bullet \tag{4.12}$$

Problem 4-2

Find the function whose derivative is $-10e^{-5x}$ and whose value at $x = 0$ is 20. •

Section 4-3. INTEGRATION

Consider the general version of the problem discussed in Section 4-2. Say that we have a function $f = f(x)$, and we want to find its antiderivative function, which we call $F = F(x)$. That is,

$$\frac{dF}{dx} = f(x) \tag{4.13}$$

Assume that we know the value of F at some value of x, say at $x = x_0$.

In Section 3-4, we discussed the approximate calculation of an increment in a function, using the slope of the tangent line, which is equal to the derivative of the function. Equation (3.15) is

$$\Delta F = F(x_1) - F(x_0) \approx \left(\frac{dF}{dx}\right)_{x=x_0} \Delta x \tag{4.14}$$

This approximation becomes more and more nearly exact as Δx is made smaller.

If we want the value of F at some point that is not close to $x = x_0$, we can get a better approximation by repeating the use of Eq. (4.14) several times. Say that we want the value of F at $x = x'$. We divide the interval (x_0, x') into n equal subintervals. This is shown in Figure 4-1, with n equal to 3. We now write

$$F(x') - F(x_0) \approx \left(\frac{dF}{dx}\right)_{x=x_0} \Delta x + \left(\frac{dF}{dx}\right)_{x=x_1} \Delta x$$
$$+ \left(\frac{dF}{dx}\right)_{x=x_2} \Delta x + \cdots + \left(\frac{dF}{dx}\right)_{x=x_{n-1}} \Delta x \tag{4.15}$$

where Δx is the length of each subinterval:

$$\Delta x = x_1 - x_0 = x_2 - x_1 = x_3 - x_2 = \cdots = x' - x_{n-1} \tag{4.16}$$

This approximation to $F(x') - F(x_0)$ is generally a better approximation than the one obtained by multiplying the slope at the beginning of the interval by the length of the whole interval, as you can see in Figure 4-1. In fact, if we make n fairly large, we can make the approximation nearly exact.

Let us now rewrite Eq. (4.15), using the symbol $f = f(x)$ instead of dF/dx.

$$F(x') - F(x_0) \approx f(x_0)\,\Delta x + f(x_1)\,\Delta x + f(x_2)\,\Delta x + \cdots + f(x_n)\,\Delta x \tag{4.17}$$
$$\approx \sum_{k=0}^{n-1} f(x_k)\,\Delta x \tag{4.18}$$

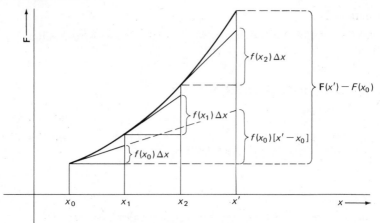

Figure 4-1. Figure to illustrate Eq. (4-15).

We use the standard notation for a sum. The letter k is called the *summation index*. Its initial value is given under the capital Greek letter Σ (sigma) and its final value is given above it.

We can use Eq. (4.16) to write

$$x_k = x_0 + k\,\Delta x \tag{4.19}$$

and use this in Eq. (4.18):

$$F(x') - F(x_0) \approx \sum_{k=0}^{n-1} f(x_0 + k\,\Delta x)\,\Delta x \tag{4.20}$$

We now make Eq. (4.20) become exactly correct by taking the limit as n becomes larger and larger without bound, meanwhile making Δx smaller and smaller so that $n\,\Delta x$ remains fixed and equal to $x' - x_0$.

$$F(x') - F(x_0) = \lim_{\substack{\Delta x \to 0 \\ n \to \infty \\ (n\,\Delta x = x' - x_0)}} \left[\sum_{k=0}^{n-1} f(x_0 + k\,\Delta x)\,\Delta x \right] \tag{4.21}$$

The limit on the right-hand side of Eq. (4.21) is called an *integral*. The function $f(x)$ is called the *integrand function*. The notation in this equation is cumbersome, so another symbol is used:

$$\lim_{\substack{\Delta x \to 0 \\ n \to \infty \\ (n\,\Delta x = x' - x_0)}} \left[\sum_{k=0}^{n-1} f(x_0 + k\,\Delta x)\,\Delta x \right] = \int_{x_0}^{x'} f(x)\,dx \tag{4.22}$$

The integral sign on the right-hand side of Eq. (4.22) is a stretched-out letter S, standing for a sum. However, an integral is not an ordinary sum. It is the limit that a sum approaches as the number of terms in the sum becomes infinite in a particular way. The number x_0 at the bottom of the integral sign is called the *lower limit of integration*, and the number x' at the top is called the *upper limit of integration*. Equation (4.21) is now

$$F(x') - F(x_0) = \int_{x_0}^{x'} f(x)\, dx \qquad (4.23)$$

The finite increment $F(x) - F(x_0)$ is equal to the sum of infinitely many infinitesimal increments, each given by the differential $dF = (dF/dx)\, dx = f(x)\, dx$ at the appropriate value of x. The integral on the right-hand side of Eq. (4.23) is called a *definite integral*, because the limits of integration are specific (definite) numbers. Equation (4.23) is a very important equation. Some authors call it one of the fundamental theorems of calculus.

Example 4-5

Find the value of the definite integral

$$\int_0^\pi \sin(x)\, dx = I$$

Solution

The antiderivative function is

$$-\cos(x) + C = F(x)$$

where C is an arbitrary constant.

$$I = F(\pi) - F(0) = -\cos(\pi) + C - [-\cos(0) + C]$$
$$= -\cos(\pi) + \cos(0) = -(-1) + 1 = 2 \qquad \bullet$$

Notice an obvious but important fact: The definite integral is not a function of x. It depends only on what numbers are chosen for the limits and on what function occurs under the integral sign.

Problem 4-3

Find the value of the definite integral

$$I = \int_0^1 e^x\, dx \qquad \bullet$$

The Definite Integral As an Area. We will now show that a definite integral is equal in value to an area between the curve and the x axis in a graph of the integrand function. We return to Eq. (4.18), which gives an approximation to a definite integral. Figure 4-2 shows the situation graphically, with $n = 3$.

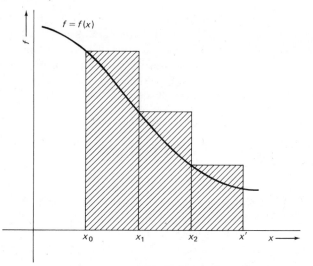

Figure 4-2. The area under a bar graph and the area under a curve.

Each term in the sum is equal to the area of a rectangle with height $f(x_k)$ and width Δx, so that the sum is equal to the shaded area in the figure, which is the area under a bar graph. Note that the curve in the figure is the curve representing the integrand function $f(x)$, not the antiderivative function $F(x)$.

As the limit of Eq. (4.21) is taken, the number of bars between $x = x_0$ and $x = x'$ becomes larger and larger, while the width Δx becomes smaller and smaller. The roughly triangular areas between the bar graph and the curve become smaller and smaller, and although there are more and more of them, their total area shrinks to zero as the limit is taken. The integral thus becomes equal to the area bounded by the x axis, the curve of the integrand function, and the vertical lines at $x = x_0$ and $x = x'$.

Figure 4-2 shows the simplest case, in which all values of the integrand function are positive. If $x_0 < x'$, the increment Δx is always positive, so in those regions where the integrand function is negative, we must count the area as negative. If the integrand function is negative in some regions, we must take the area above the x axis minus the area below the x axis as equal to the value of the integral.

Example 4-6

Find the area bounded by the x axis and the curve representing

$$f(x) = \sin(x)$$

a. between $x = 0$ and $x = \pi$
b. between $x = 0$ and $x = 2\pi$

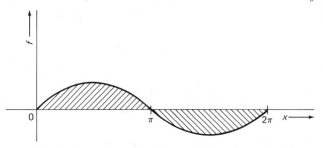

Figure 4-3. A graph of $f = \sin(x)$ for Example 4-6.

Solution

The graph of the function is shown in Figure 4-3.

a. $$\text{area} = \int_0^\pi \sin(x)\,dx = -\cos(\pi) - [-\cos(0)]$$
$$= 2$$

This integral was obtained in Example 4-5.

b. $$\text{area} = \int_0^{2\pi} \sin(x)\,dx = -\cos(2\pi) - [-\cos(0)]$$
$$= -(1) - [-(1)] = 0 \qquad \bullet$$

Notice that when we found the value of the integral by finding the anti-derivative function in this example, we omitted the constant term that generally must be present. This constant canceled out in Example 4-5, and would have canceled out in Example 4-6, so we left it out from the beginning.

Problem 4-4

Find the following areas by computing the values of definite integrals:

a. The area bounded by the curve representing $y = x^3$, the x axis, and the line $x = 3$.

b. The area bounded by the straight line $y = 2x + 3$, the x axis, the line $x = 1$, and the line $x = 4$.

c. The area bounded by the parabola $y = 4 - x^2$ and the x axis. You will have to find the limits of integration. $\qquad \bullet$

The relation between a definite integral and an area in the graph of the integrand function is useful for two reasons. First, since we can compute the integral from knowledge of the antiderivative function, we can use this to determine an area. Also, there are integrand functions for which no anti-derivative function can be written in a usable form. If you must obtain the value of such a definite integral, you can sometimes do this by making a graph of the integrand function and actually measuring the appropriate area in the graph. There are three practical ways to do this. One is by counting squares on your graph paper. Another is by using a uniform graph paper, cutting out the area to be determined, and weighing this piece of graph paper and another piece of known area from the same sheet. A third is by using a mechanical device called a planimeter, which registers an area on a dial after a stylus is moved around the boundary of the area.

Problem 4-5

Find the approximate value of the integral

$$\int_0^1 e^{-x^2} \, dx$$

by making a graph of the integrand and measuring an area. •

Facts About Integrals. The following facts can be understood by considering the relation between integrals and areas in graphs of integrand functions:

1. A definite integral over the interval (a, c) is the sum of the definite integrals over the intervals (a, b) and (b, c), where $a < b < c$:

$$\int_a^c f(x) \, dx = \int_a^b f(x) \, dx + \int_b^c f(x) \, dx \qquad (4.24)$$

This is illustrated in Figure 4-4. The integral on the left-hand side of Eq. (4.24) is equal to the entire area shown, and each of the two terms on the right-hand side is equal to one of the two differently shaded areas which combine to make the entire area.

2. The presence of a finite step discontinuity in an integrand function does not affect the process of integration. In this regard, integration differs from differentiation. Figure 4-5 illustrates the situation. If the discontinuity is at $x = b$, we simply apply Eq. (4.24) and find that the integral is given by the integral up to $x = b$ plus the integral from $x = b$ to the end of the interval.

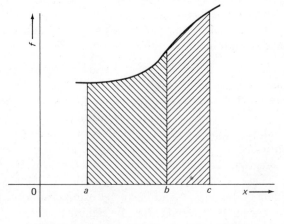

Figure 4-4. Figure to illustrate Eq. (4-24).

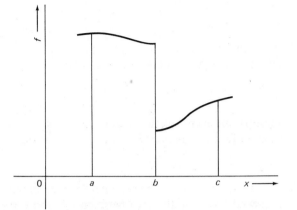

Figure 4-5. An integrand function which is discontinuous at $x = b$.

3. If the two limits of integration are interchanged, the resulting integral is the negative of the original integral. In Eq. (4.23), we assumed that $x_0 < x'$, so that Δx would be positive. If the lower limit is larger than the upper limit of integration, Δx must be taken as negative, reversing the sign of the area in the graph. Therefore,

$$\int_a^b f(x)\,dx = -\int_b^a f(x)\,dx \tag{4.25}$$

Use of this fact makes Eq. (4.24) usable for any real values of a, b, and c.

4. If an integrand consists of a constant times some function, the constant can be factorized out of the integral:

$$\int_a^b cf(x)\,dx = c\int_a^b f(x)\,dx \tag{4.26}$$

5. The integral of an odd function from $-c$ to c vanishes.
An odd function is one that obeys the relation

$$f(-x) = -f(x) \tag{4.27}$$

The sine function and the tangent function are examples of odd functions, as stated in Eqs. (2.22) and (2.24). In this case, the area above the axis exactly cancels the area below the axis, so that

$$\int_{-c}^c f(x)\,dx = 0 \qquad (f(x)\ \text{odd}) \tag{4.28}$$

where c is any real constant.

Problem 4-6

Draw a rough graph of $f(x) = xe^{-x^2}$ and satisfy yourself that this is an odd function. Identify the area in this graph that is equal to the following integral, and satisfy yourself that the integral vanishes:

$$\int_{-2}^{2} xe^{-x^2}\,dx = 0 \qquad \bullet$$

6. The integral of an even function from $-c$ to c is twice the integral from 0 to c. An even function is one that obeys the relation

$$f(-x) = f(x) \tag{4.29}$$

The cosine function is an example of an even function, as stated in Eq. (2.23). For such an integrand function, the area between $-c$ and 0 is equal to the area between 0 and c, so that

$$\int_{-c}^{c} f(x)\,dx = 2 \int_{0}^{c} f(x)\,dx \qquad (f(x)\ \text{even}) \tag{4.30}$$

where c is any real constant.

Problem 4-7

Draw a rough graph of $f(x) = e^{-x^2}$. Satisfy yourself that this is an even function. Identify the area in the graph that is equal to the definite integral

$$I_1 = \int_{-1}^{1} e^{-x^2}\,dx$$

and satisfy yourself that this integral is equal to twice the integral

$$I_2 = \int_{0}^{1} e^{-x^2}\,dx \qquad \bullet$$

If you have an integrand that is a product of several factors, you can use the following facts:

1. The product of two even functions is an even function.
2. The product of two odd functions is an even function.
3. The product of an odd function and an even function is an odd function.

Equations (4.28) and (4.30) are useful results. They represent two cases in which consideration of the symmetry of integrand functions simplifies the evaluation of integrals. See Chapter 9 for a discussion of symmetry.

7. If the limits of a definite integral are considered to be variables, we can write

$$\frac{d}{db}\left[\int_{a}^{b} f(x)\,dx\right] = f(b) \tag{4.31a}$$

where a is considered to be a constant, and

$$\frac{d}{da}\left[\int_a^b f(x)\,dx\right] = -f(a) \tag{4.31b}$$

where b is considered to be a constant. If a and b are both functions of some variable c (not the variable of integration, x), then

$$\frac{d}{dc}\left[\int_a^b f(x)\,dx\right] = f(b)\frac{db}{dc} - f(a)\frac{da}{dc} \tag{4.31c}$$

These equations follow from Eq. (4.23), and Eq. (4.31c) also comes from the chain rule, Eq. (3.26).

Section 4-4. INDEFINITE INTEGRALS. TABLES OF INTEGRALS

Let us consider the upper limit in Eq. (4.23) to be variable and the lower limit to be fixed and equal to a:

$$\int_a^{x'} f(x)\,dx = F(x') - C$$

The quantity C is equal to $f(a)$ and is arbitrary if a is arbitrary. We omit mention of a and write

$$\int^{x'} f(x)\,dx = F(x') - C \tag{4.32}$$

This is called an *indefinite integral*, since the lower limit is unspecified and the upper limit is variable.

Large tables of indefinite integrals have been compiled. Appendix 3 is a brief version of such a table. In most such tables, the notation of Eq. (4.32) is not maintained. The entries are written in the form

$$\int f(x)\,dx = F(x) \tag{4.33}$$

You should understand that this is an abbreviation for Eq. (4.32). The upper limit and the integration variable are not the same thing, even though the same symbol is used for both.

The function $F(x)$ is of course the antiderivative function of $f(x)$. It would seem that the same information contained in a table of indefinite integrals is contained in a table of derivatives, and this is true. However, we can get by with a fairly short table of derivatives, since we have the chain rule and other facts listed in Section 3-5. Antiderivatives are harder to find, so it is good to have a separate table of antiderivatives (indefinite integerals), arranged so that similar integrand functions occur together.

Example 4-7

Using a table, find the indefinite integrals.

a. $\displaystyle\int \frac{dx}{a^2 + x^2}$

b. $\displaystyle\int x \sin^2(x)\, dx$

c. $\displaystyle\int x e^{ax}\, dx$

Solution

From Appendix 3, or any other table of indefinite integrals,

a. $\displaystyle\int^{x'} \frac{dx}{a^2 + x^2} = \frac{1}{a}\arctan\left(\frac{x'}{a}\right) + C$

b. $\displaystyle\int^{x'} x \sin^2(x)\, dx = \frac{x'^2}{4} - \frac{x' \sin(2x')}{4} - \frac{\cos(2x')}{8} + C$

c. $\displaystyle\int^{x'} x e^{ax}\, dx = \frac{e^{ax'}}{a^2}(ax' - 1) + C$

Notice that we have written $+C$ instead of $-C$ as in Eqs. (4.31) and (4.32). Since C is an arbitrary constant, this makes no difference.   •

Problem 4-8

Show by differentiation that the functions on the right-hand sides of the equations in Example 4-7 yield the integrand functions when differentiated. •

Since the indefinite integral is just the antiderivative function, it is used to find a definite integral in the same way as in Section 4-3. If x_1 and x_2 are the limits of the definite integral,

$$\int_{x_1}^{x_2} f(x)\, dx = \int_a^{x_2} f(x)\, dx - \int_a^{x_1} f(x)\, dx$$

$$= F(x_2) - C - [F(x_1) - C] = F(x_2) - F(x_1) \qquad \textbf{(4.34)}$$

Notice that Eq. (4.34) is the same as Eq. (4.23).

Example 4-8

Using a table of indefinite integrals, find the definite integral

$$\int_0^{\pi/2} \sin(x)\cos(x)\, dx$$

Solution

From Appendix 3, Eq. (36), we find that the indefinite integral is $\sin^2(x)/2$.

$$\int_0^{\pi/2} \sin(x)\cos(x)\, dx = \left.\frac{\sin^2(x)}{2}\right|_0^{\pi/2} = \frac{1}{2}\left[\sin^2\left(\frac{\pi}{2}\right) - \sin^2(0)\right]$$

$$= \tfrac{1}{2}(1 - 0) = \tfrac{1}{2}$$

   •

We have used the common notation

$$F(x)\Big|_a^b = F(b) - F(a) \tag{4.35}$$

Problem 4-9

Using a table of indefinite integrals, find the definite integrals.

a. $\displaystyle\int_0^3 \cosh(2x)\,dx$

b. $\displaystyle\int_1^2 \frac{\ln(3x)}{x}\,dx$

c. $\displaystyle\int_0^5 4^x\,dx$ •

In addition to tables of indefinite integrals, there exist tables of definite integrals. Some are listed at the end of the chapter, and Appendix 4 is a short version of such a table. Some of the entries in these tables are integrals that could be worked out by using a table of indefinite integrals, but others are integrals that cannot be obtained as indefinite integrals, but by some particular method can be worked out for one set of limits. An example of such an integral is worked out in Appendix 7.

Tables of definite integrals usually include only sets of limits such as $(0, 1)$, $(0, \pi)$, $(0, \pi/2)$, and $(0, \infty)$. The last set of limits corresponds to an improper integral, which is discussed in the next section.

Section 4-5. IMPROPER INTEGRALS

Our discussion of integrals has thus far been based on the following assumptions: (1) that both limits of integration are finite numbers, and (2) that the integrand function does not become infinite inside the interval of integration. If either of these conditions is not met, an integral is said to be an *improper integral*. For example,

$$I = \int_0^\infty f(x)\,dx \tag{4.36}$$

is an improper integral because its upper limit is infinite.

We must decide what is meant by the infinite upper limit in Eq. (4.36), because Eq. (4.21) cannot be modified by simply inserting ∞ into the equation instead of x'. We define

$$\boxed{\int_0^\infty f(x)\,dx = \lim_{b \to \infty} \int_0^b f(x)\,dx} \tag{4.37}$$

That is, we let the upper limit of integration become larger and larger without bound. If the integral approaches more and more closely to some finite number as this is done, we say that the limit exists and that the improper integral is equal to this limit. In this case, the improper integral converges. However, in some cases, the magnitude of the integral will become larger and larger without bound as the limit of integration is made larger and larger. In other cases, the integral will continue to oscillate in value as the limit of integration is made larger and larger. We say in both of these cases that the integral diverges.

In some improper integrals, the lower limit of integration is made to approach $-\infty$, and in some improper integrals, the lower limit is made to approach $-\infty$ while the upper limit is made to approach $+\infty$. Just as in the case of Eq. (4-37), if the integral approaches a definite number more and more closely as the limit or limits approach infinite magnitude, the improper integral is said to converge to that number.

Another kind of improper integral has an integrand function that becomes infinite somewhere in the interval of integration. For example,

$$I = \int_0^1 \frac{1}{x} dx \tag{4.38}$$

is an improper integral because the integrand function becomes infinite at $x = 0$. In this case, the improper integral is defined by

$$\int_0^1 \frac{1}{x} dx = \lim_{a \to 0^+} \int_a^1 \frac{1}{x} dx \tag{4.39}$$

Just as in the other cases, if the integral grows larger and larger in magnitude as the limit is taken, or if it oscillates in value without approaching closer and closer to some number, we say that it diverges. If the limit exists, we say that the improper integral converges to that limit. The situation is similar if the point at which the integrand becomes infinite is at the upper limit of integration or within the interval of integration. (In this case, break the interval into two subintervals.)

The two principal questions that we need to ask about an improper integral are:

1. Does it converge?
2. If so, what is its value?

Example 4-9

Determine whether the following improper integral converges, and if so, find its value:

$$\int_0^\infty e^{-x} dx$$

Solution

$$\int_0^\infty e^{-x} dx = \lim_{b \to \infty} \int_0^b e^{-x} dx = \lim_{b \to \infty} \left[-e^{-x} \right]_0^b$$
$$= \lim_{b \to \infty} - (e^{-b} - 1) = 0 + 1 = 1 \qquad \bullet$$

The integral converges to the value 1.

Problem 4-10

Determine whether the following improper integrals converge, and if so, determine their values:

a. $\displaystyle\int_0^\infty \sin(x)\,dx$

b. $\displaystyle\int_0^\infty \frac{1}{1+x}\,dx$

c. $\displaystyle\int_0^\infty \frac{1}{x^3}\,dx$

d. $\displaystyle\int_{-\infty}^0 e^x\,dx$ •

Section 4-6. PROBABILITY DISTRIBUTIONS AND MEAN VALUES

In this section, we discuss how the process of integration is used in calculating mean values when a probability distribution is known.

The *mean* of N numbers is defined as

$$\bar{x} = \frac{1}{N}(x_1 + x_2 + x_3 + x_4 + \cdots + x_N) \tag{4.40}$$

where we use the symbols x_1, x_2, x_3, etc., for the numbers. The mean is one kind of average.

Problem 4-11

Calculate the mean of the integers beginning with 10 and ending with 20. •

There is another way to write the mean of a set of numbers, which is convenient if several of the members of the list are equal. Let us arrange the members of our list so that the first M members of the list have different values (are distinct from each other), and each of the other $N - M$ members is equal to some member of the first subset. Let N_i be the total number of members of the entire list that are equal to x_i, where x_i is one of the distinct numbers in the first subset. Equation (4.40) can be rewritten

$$\bar{x} = \frac{1}{N}(N_1 x_1 + N_2 x_2 + N_3 x_3 + \cdots + N_M x_M) \tag{4.41}$$

$$\bar{x} = \frac{1}{N}\sum_{i=1}^M N_i x_i = \sum_{i=1}^M p_i x_i \tag{4.42}$$

We again use the standard notation for a sum, introduced in Eq. (4.18). The quantity p_i is equal to N_i/N and is the fraction of the members of the entire list that are equal to x_i. If we were to sample the entire list by choosing a member at random, the probability that this member would equal x_i is given by p_i.

Example 4-10

A quiz was given in a class with 100 members. The scores were as follows:

Score	Number of Students
100	8
90	11
80	35
70	23
60	14
50	9

Find the mean score.

Solution

$$\bar{s} = (0.08)(100) + (0.11)(90) + (0.35)(80) + (0.23)(70)$$
$$+ (0.14)(60) + (0.09)(50) = 74.9 \qquad \bullet$$

As you can see, it is somewhat easier to take six products and to add them up than to add up 100 numbers, so if you have a set of numbers with considerable duplication, Eq. (4.42) is easier to use than Eq. (4.40).

It is also possible to take the mean of a function of the numbers in our set. For example, to form the mean of the squares of the numbers, we have

$$\overline{x^2} = \frac{1}{N}(x_1^2 + x_2^2 + x_3^2 + \cdots + x_N^2) \qquad (4.43)$$

$$= \frac{1}{N}\sum_{i=1}^{M} N_i x_i^2 = \sum_{i=1}^{M} p_i x_i^2 \qquad (4.44)$$

Similarly, if $g = g(x)$ is any function defined for all values of x that occur in our list, the mean value of this function is given by

$$\overline{g(x)} = \sum_{i=1}^{M} p_i g(x_i) \qquad (4.45)$$

Problem 4-12

Find the mean of the squares of the scores given in Example 4-10. Find also the square root of this mean, which is called the *root-mean-square* score. $\bullet$

In gas kinetic theory, the mean speed of molecules in a gas is discussed. The calculation of such a mean is a little more complicated than Eq. (4.44).

The reasons for this are: (1) the speeds of the molecules can take on any real nonnegative value, and (2) there are so many molecules in almost any sample of a gas that we can proceed as though the number of molecules were infinite. Because of these facts, we cannot simply sum over a finite number of possible speeds as was done in Example 4-10.

Let us develop formulas to handle cases like this. Consider a variable x, which can take on any real values between $x = a$ and $x = b$. Let us divide the interval (a, b) into n subintervals. Look at one of the subintervals, say (x_i, x_{i+1}), which can also be written $(x_i, x_i + \Delta_x)$, where

$$\Delta x = x_{i+1} - x_i \tag{4.46}$$

Let the fraction of all members of our sample that have values of x lying between x_i and x_{i+1} be called p_i. If Δx is quite small, p_i will be approximately proportional to Δx. We write

$$p_i = f_i \Delta x \tag{4.47}$$

The quantity f_i will not depend strongly on Δx. The mean value of x is given by Eq. (4.42):

$$\bar{x} \approx \sum_{i=0}^{n-1} p_i x_i = \sum_{i=0}^{n-1} x_i f_i \Delta x \tag{4.48}$$

This equation is only approximately true, because we have multiplied the probability that x is in the subinterval $(x_i, x_i + \Delta x)$ by x_i, which is only one of the values of x in the subinterval. However, Eq. (4.48) can be made more and more nearly exact by making n larger and larger, and Δx smaller and smaller so that $n \Delta x$ is constant. In this limit, f_i becomes independent of Δx. We replace the symbol f_i by $f(x_i)$, and assume that this is an integrable function of x_i. It must be at least piecewise continuous. Our formula is now very similar to Eq. (4.22):

$$\bar{x} = \lim_{\substack{n \to \infty \\ \Delta x \to 0 \\ n \Delta x = b - a}} \sum_{i=0}^{n-1} x_i f(x_i) \Delta x = \int_a^b x f(x) \, dx \tag{4.49}$$

The function $f(x)$ is called the *probability density*, or *probability distribution*, or sometimes the *distribution function*.

If we desire the mean value of a function of x, say $g(x)$, the formula is analogous to Eq. (4.45).

$$\overline{g(x)} = \int_a^b g(x) f(x) \, dx \tag{4.50}$$

For example, the mean of x^2 is given by

$$\overline{x^2} = \int_a^b x^2 f(x) \, dx \tag{4.51}$$

As defined above, the probability density must obey the condition

$$\int_a^b f(x)\,dx = 1 \tag{4.52}$$

A probability density that obeys this condition is said to be *normalized*.

It is possible to use a probability density that is not normalized, but if you do this, you must modify Eqs. (4.49) and (4.50). Equation (4.50) becomes

$$\overline{g(x)} = \frac{\int_a^b g(x)f(x)\,dx}{\int_a^b f(x)\,dx} \tag{4.53}$$

The probability that x lies in the infinitesimal interval $(x, x + dx)$ is $f(x)\,dx$, which is the probability per unit length times the length of the infinitesimal interval. This is the reason for using the name "probability density" for $f(x)$.

Example 4-11

If all values of x between a and b are equally probable, find the mean value of x and the root-mean-square value of x.

Solution

The probability density is a constant, and in order to be normalized, it is

$$f(x) = \frac{1}{b - a}$$

so that

$$\bar{x} = \int_a^b x\frac{1}{b - a}\,dx = \tfrac{1}{2}(b + a)$$

Also,

$$\overline{x^2} = \int_a^b x^2\frac{1}{b - a}\,dx = \tfrac{1}{3}(b^2 + ab + a^2)$$

The root-mean-square value of x is

$$(\overline{x^2})^{1/2} = [\tfrac{1}{3}(b^2 + ab + a^2)]^{1/2} \qquad \bullet \ \ (4.54)$$

Problem 4-13

From the results of Example 4-11, find the numerical values of $\bar{x}$ and $(\overline{x^2})^{1/2}$ for $a = 0$ and $b = 10$. Comment on the difference between the values. $\bullet$

Problem 4-14

If x ranges from 0 to 10 and $f(x) = cx^2$, find the value of c so that $f(x)$ is normalized. Find the mean value of x and the root-mean-square value of x. $\bullet$

In gas kinetic theory, the speed of a molecule is denoted by v, and the probability density for the speed is given by

$$f_v(v) = 4\pi\left(\frac{m}{2\pi kT}\right)^{3/2} v^2 \exp\left(-\frac{mv^2}{2kT}\right) \tag{4.55}$$

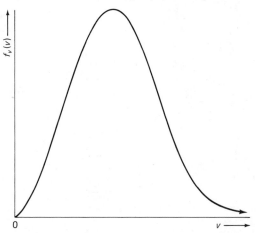

Figure 4-6. A graph of the probability density for speeds of molecules in a gas.

where m is the mass of the molecule, T the temperature on the Kelvin scale, and k is Boltzmann's constant. A rough graph of this function is shown in Figure 4-6. The speed is never negative, so that the graph does not extend to the left of the origin. In ordinary gas kinetic theory, the requirements of special relativity are ignored, and speeds approaching infinity are included. The error is not large, because of the low probability ascribed to high speeds.

The mean speed is given by

$$v = \int_0^\infty v f_v(v)\, dv \tag{4.56}$$

Example 4-12

Obtain a formula for the mean speed of molecules in a gas.

Solution

$$\bar{v} = 4\pi \left(\frac{m}{2\pi kT}\right)^{3/2} \int_0^\infty v^3 \exp\left(-\frac{mv^2}{2kT}\right) dv \tag{4.57}$$

This integral can be obtained from Eq. (7) of Appendix 7.

$$\bar{v} = 4\pi \left(\frac{m}{2\pi kT}\right)^{3/2} \frac{(2kT)^2}{2m^2} = \left(\frac{8kT}{\pi m}\right)^{1/2} \qquad \bullet \tag{4.58}$$

Problem 4-15

Find the value of $\bar{v}$ for N_2 gas at 298 K. $\qquad\qquad\bullet$

Example 4-13

Find a formula for the mean of the square of the speed of molecules in a gas, and for the root-mean-square speed.

Solution

$$\overline{v^2} = 4\pi \left(\frac{m}{2\pi kT}\right)^{3/2} \int_0^\infty v^4 \exp\left(-\frac{mv^2}{2kT}\right) dv$$

$$= \frac{3kT}{m}$$

$$(\overline{v^2})^{1/2} = \left(\frac{3kT}{m}\right)^{1/2}$$

Problem 4-16

A measure of the "spread" of a probability distribution is the standard deviation, given by

$$\sigma_x = [\overline{x^2} - (\overline{x})^2]^{1/2} \tag{4.60}$$

For molecules in a gas, find the formula for σ_v and find its value for N_2 gas at 298 K.
•

Quite frequently, time averages of time-dependent quantities are required. If $g = g(t)$, we define the time average of g as

$$\overline{g(t)} = \int_{t_1}^{t_2} g(t)f(t)\,dt \tag{4.61}$$

where we call $f(t)$ the weighting function, which plays the same role as a probability density. Most time averages are unweighted, which means that $f(t)$ is a constant.

$$\overline{g(t)} = \frac{1}{t_1 - t_2} \int_{t_1}^{t_2} g(t)\,dt \tag{4.62}$$

Equation (4.62) is a version of the *mean value theorem* of integral calculus, which states that the mean value of a function is equal to the integral of the function divided by the length of the interval over which the mean is taken.

Example 4-14

A particle falls in a vacuum near the surface of the earth. Find the average speed (magnitude of the velocity) during the first 10.00 s of fall if the initial speed is zero.

Solution

From Eq. (4.7),

$$v = -gt$$

where g is the acceleration due to gravity, 9.80 m s^{-2}.

$$\overline{v} = -\tfrac{1}{10} \int_0^{10} gt\,dt = -\frac{g}{10}\frac{t^2}{2}\bigg|^{10} = -5g = -49.0 \text{ m s}^{-1}$$
•

Problem 4-17

Find the time-average value of the z coordinate of the particle in Example 4-14 for the first 10.00 s of fall if the initial position is $z = 0.00$ m.   ●

Section 4-7. ANALYTICAL METHODS OF INTEGRATION

In this section, we discuss three methods that can be used to find an integral: the method of substitution, the method of partial integration (integration by parts), and the method of partial fractions.

The Method of Substitution. In this method, a change of variables is performed. The integrand function is expressed in terms of a different independent variable, which becomes the variable of integration.

Example 4-15

Find the integral

$$\int_0^\infty x e^{-x^2}\, dx$$

without using a table of integrals.

Solution

We have x^2 in the exponent, which suggests using $y = x^2$ as a new variable. If $y = x^2$, then $dy = 2x\, dx$, or $x\, dx = \frac{1}{2} dy$.

$$\int_0^\infty x e^{-x^2}\, dx = \frac{1}{2}\int_{x=0}^{x=\infty} e^{-y}\, dy = \frac{1}{2}\int_0^\infty e^{-y}\, dy$$

$$= -\frac{1}{2}e^{-y}\Big|_0^\infty = -\frac{1}{2}(0-1) = \frac{1}{2} \qquad ●$$

This example illustrates the method of substitution. The first thing to do is to reexpress the integrand in terms of the new variable, choosing a variable that looks as though it will give a simpler integrand function. The differential of the integration variable must also be reexpressed. The second thing to do is to reexpress the limits of integration, making the limits equal to the values of the new variable that correspond to the values of the old variable at the old limits. The final thing is to compute the new integral, which is equal to the old integral.

Example 4-16

Find the integral

$$\int_0^{1/2} \frac{dx}{2 - 2x}$$

without using a table of integrals.

Solution

We let $y = 2 - 2x$, in order to get a simple denominator. With this, $dy = -2dx$, or $dx = -dy/2$. When $x = 0$, $y = 2$, and when $x = \frac{1}{2}$, $y = 1$, so

$$\int_0^{1/2} \frac{1}{2 - 2x} dx = -\frac{1}{2} \int_2^1 \frac{1}{y} dy = \frac{1}{2} \int_1^2 \frac{1}{y} dy = \frac{1}{2} \ln(y) \Big|_1^2$$

$$= \frac{1}{2} [\ln(2) - \ln(1)] = \frac{1}{2} \ln(2) \qquad \bullet$$

Problem 4-18

Find the integral

$$\int_0^\pi e^{\cos(\theta)} \sin(\theta) \, d\theta$$

without using a table of integrals. $\qquad \bullet$

Integration by Parts. This method, which is also called *partial integration*, consists of application of the formula

$$\int u \frac{dv}{dx} dx = uv - \int v \frac{du}{dx} + C \qquad (4.63)$$

or the corresponding formula for definite integrals

$$\int_a^b u \frac{dv}{dx} dx = u(x)v(x) \Big|_a^b - \int_a^b v \frac{du}{dx} dx \qquad (4.64)$$

In both these formulas, u and v must be functions of x which are differentiable everywhere in the interval of integration.

We can derive Eq. (4.63) by use of Eq. (3.19), which gives the derivative of the product of two functions:

$$\frac{d}{dx}(uv) = u \frac{dv}{dx} + v \frac{du}{dx}$$

The antiderivative of either side of this equation is just $uv + C$, where C is an arbitrary constant, so we can write the indefinite integral

$$\int \frac{d(uv)}{dx} dx = \int u \frac{dv}{dx} dx + \int v \frac{du}{dx} dx = u(x)v(x) + C$$

This is the same as Eq. (4.63).

Example 4-17

Find the indefinite integral

$$\int x \sin(x)\,dx$$

without using a table.

Solution

Let $u(x) = x$ and $\sin(x) = dv/dx$. We make this choice because the anti-derivative of x is $x^2/2$, which will lead to a worse integral than the one containing x, making it better to differentiate x and integrate $\sin(x)$. With this choice

$$\frac{du}{dx} = 1 \quad \text{and} \quad v = -\cos(x)$$

where we omit the arbitrary constant of integration. Our result is

$$\int x \sin(x)\,dx = -x\cos(x) + \int \cos(x)\,dx$$
$$= -x\cos(x) + \sin(x) + C \qquad \bullet$$

Problem 4-19

Find the integral

$$\int_0^\pi x^2 \sin(x)\,dx$$

without using a table. You will have to apply partial integration twice. $\quad \bullet$

The fundamental equation of partial integration, Eq. (4.63), is sometimes written with differentials:

$$\int u\,dv = uv - \int v\,du \qquad \textbf{(4.65)}$$

The Method of Partial Fractions. This method consists of an algebraic procedure for turning a difficult integrand into a sum of two or more easier functions. It works with an integral of the type

$$I = \int \frac{P(x)}{Q(x)}\,dx \qquad \textbf{(4.66)}$$

where $P(x)$ and $Q(x)$ are polynomials in x. The highest power of x in P must be lower than the highest power of x in Q. However, if this is not the case, you can proceed by performing a long division, obtaining a polynomial plus a remainder, which will be a quotient of polynomials that does obey the condition. The polynomial can be integrated easily, and the remainder quotient can be handled with the method of partial fractions.

The first step in the procedure is to factorize the denominator, $Q(x)$, into a product of polynomials of order 1 and 2. A polynomial of order 1 is just $ax + b$, and a polynomial of order 2 is $ax^2 + bx + c$. Let us first assume that all of the factors are of order 1, so that

$$Q(x) = (a_1x + b_1)(a_2x + b_2)(a_3x + b_3) \cdots (a_nx + b_n) \qquad \textbf{(4.67)}$$

where all the a's and b's are constants.

The fundamental formula of the method of partial fractions is a theorem of algebra which says that if $Q(x)$ is given by Eq. (4.67) and $P(x)$ is of lower order than $Q(x)$, then

$$\frac{P(x)}{Q(x)} = \frac{A_1}{a_1x + b_1} + \frac{A_2}{a_2x + b_2} + \cdots + \frac{A_n}{a_nx + b_n} \qquad \textbf{(4.68)}$$

where $A_1, A_2, \ldots, A_n$ are all constants.

Equation (4.68) is applicable only if all the factors in $Q(x)$ are distinct from each other. If the same factor occurs more than once, Eq. (4.68) must be modified. If the factor $a_1x + b_1$ occurs m times, we must write

$$\frac{P(x)}{Q(x)} = \frac{A_1}{a_1x + b_1} + \frac{A_2}{(a_1x + b_1)^2} + \cdots + \frac{A_m}{(a_1x + b_1)^m}$$

$$+ \text{ terms for other factors as in Eq. (4.68)} \qquad \textbf{(4.69)}$$

Sometimes one or more factors of order 2 occur which themselves cannot be factorized. If Q contains a factor $a_1x^2 + b_1x + c_1$, we must write

$$\frac{P(x)}{Q(x)} = \frac{A_1x + B_1}{a_1x^2 + b_1x + c_1} + \text{ terms for other factors as in Eqs. (4.68) and (4.69)}$$

$$\textbf{(4.70)}$$

Example 4-18

Apply Eq. (4.68) to

$$\int \frac{6x - 30}{x^2 + 3x + 2} dx$$

Solution

$$\frac{6x - 30}{x^2 + 3x + 2} = \frac{A_1}{x + 2} + \frac{A_2}{x + 1}$$

We need to solve for A_1 and A_2 so that this equation will be satisfied for all values of x. We multiply both sides of the equation by $(x + 2)(x + 1)$:

$$6x - 30 = A_1(x + 1) + A_2(x + 2)$$

Since this equation must be valid for all values of x, we can get a different equation for each value of x. If we let $x = 0$, we get

$$-30 = A_1 + 2A_2 \qquad \textbf{(4.71)}$$

If we let x become very large, so that the constant terms can be neglected, we obtain

$$6x = A_1x + A_2x$$

or
$$6 = A_1 + A_2 \qquad\qquad \textbf{(4.72)}$$

Equations (4.71) and (4.72) can be solved simultaneously to obtain
$$A_1 = 42, \qquad A_2 = -36 \qquad\qquad \textbf{(4.73)}$$

Our result is
$$\int \frac{6x - 30}{x^2 + 3x + 2}\,dx = \int \frac{42}{x + 2}\,dx - \int \frac{36}{x + 1}\,dx \qquad \bullet \;\; \textbf{(4.74)}$$

Problem 4-20

Solve Eqs. (4.71) and (4.72) simultaneously to obtain Eq. (4.73). •

Problem 4-21

Find the indefinite integrals on the right-hand side of Eq. (4.74). •

The methods presented thus far in this chapter provide an adequate set of tools for the calculation of most integrals that will be found in a physical chemistry course. In applying these methods, it is probably best to proceed as follows:

1. If the limits are 0 and ∞ or 0 and π, or something else quite simple, look first in a table of definite integrals.
2. If this does not work, or if the limits were not suitable, look in a table of indefinite integrals.
3. If you do not find the integral in a table, see if the method of partial fractions is applicable, and use it if you can.
4. If you still have not gotten the integral, try the method of substitution.
5. If this did not work, manipulate the integrand into a product of two factors and try the method of partial integration.
6. If all these things have failed, or if they could not be attempted, do a numerical approximation to the integral. This is discussed in the next section.

Section 4-8. NUMERICAL METHODS OF INTEGRATION

In Section 4-3, we mentioned graphical methods for finding a numerical approximation to a definite integral. In this section, we discuss purely numerical methods, which have become very widely used since the advent of rapid computers.

There are two cases for which a numerical approximation to an integral must be sought. In one case the integrand function does not possess a usable antiderivative function. One such integrand function is e^{-x^2}. (see Appendix 7.) In the other case the integrand function is not known exactly but is represented approximately by a set of data points.

We begin with Eq. (4.22):

$$\int_a^b f(x)\,dx = \lim_{\substack{\Delta x \to 0 \\ n \to \infty \\ n\,\Delta x = b - a}} \sum_{j=0}^{n-1} f(a + j\,\Delta x)\,\Delta x \qquad (4.75)$$

If we do not take the limit but allow n to be some convenient finite number, this equation will still be approximately correct.

Let us consider an example, the integral of e^{-x^2} from 1 to 2. We apply an approximate version of Eq. (4.75) with $n = 10$. The result is

$$\int_1^2 e^{-x^2}\,dx \approx \sum_{j=0}^{9} \exp[-(1 + 0.1j)^2](0.1) = 0.15329 \qquad (4.76)$$

This result is represented by the area under a bar graph such as in Figure 4-2, and we call this approximation the bar-graph approximation. It is in error by the area of the roughly triangular areas between the bar graph and the curve of the integrand function. The error in Eq. (4.76) is about 15%, since the correct value of this integral is 0.13525726 to eight significant digits.

The bar-graph approximation can be made more nearly accurate by increasing the number of terms, but the rate of improvement is quite slow. For example, if we take $n = 20$ and $\Delta x = 0.05$, we get 0.14413 for the integral in Eq. (4.76), which is still in error by about 6%. If we take $n = 100$ and $\Delta x = 0.01$, we get 0.13701, which is still wrong by about 1%.

The Trapezoidal Approximation. One way to improve on the bar-graph approximation is to take the height of the rectangles as equal to the value of the integrand function somewhere near the middle of the bar. For a single rectangle, this will give two smaller triangular areas, as shown in Figure 4-7. One of these areas is included, but should not be, and the other

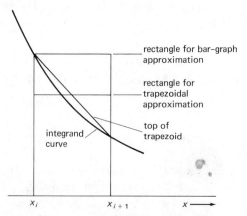

Figure 4-7. Figure to illustrate the trapezoidal approximation (enlarged view of one panel shown).

is excluded, but should not be. If these two areas are equal, their contributions to the error will cancel.

In the trapezoidal approximation, the height of the bar is taken as the average of the values of the function at the two sides of the rectangle. This gives

$$\int_a^b f(x)\,dx \approx \frac{f(a)\Delta x}{2} + \sum_{k=1}^{n-1} f(a + k\Delta x)\Delta x + \frac{f(b)\Delta x}{2} \qquad (4.77)$$

As expected, the trapezoidal approximation gives more nearly correct values than does the bar-graph approximation, for the same number of panels. (The bars in the bar graph or the trapezoids are called *panels*, so n is the number of panels.) For $n = 10$, the trapezoidal approximation gives a result of 0.135810 for the integral in Eq. (4.76). For $n = 100$, the trapezoidal approximation is correct to five significant digits.

Problem 4-22

Using Eq. (4.77), find the approximate value of the integral

$$\int_{10}^{20} x^2\,dx$$

Use $n = 5$, and calculate the exact value of the integral for comparison. ●

Simpson's Rule. In the bar-graph approximation, we used only one value of the integrand for each panel. In the trapezoidal approximation, we used two values for each panel, to determine a line fitting the integrand curve at both ends of the panel. If three points in a plane are given, there is one and only one parabola that can be drawn through all three. In *Simpson's rule*, we taken the panels two at a time, construct a parabola through the three points, find the area under the parabola, and use this to approximate the integral.

A parabolic curve is likely to fall closer to the integrand curve than a straight line, so we expect this to give a better approximation than the trapezoidal approximation, and it usually does. However, we do have the requirement that the number of panels be an even number.

We let $f_0 = f(a)$, $f_1 = f(a + \Delta x)$, $f_2 = f(a + 2\Delta x)$, etc., and use the formula for the area under a parabola to obtain as our final result

$$\boxed{\int_a^b f(x)\,dx \approx \frac{(f_0 + 4f_1 + 2f_2 + 4f_3 + \cdots + 4f_{n-1} + f_n)\Delta x}{3}} \qquad (4.78)$$

Notice the pattern, with alternating coefficients of 2 and 4, except for the first and last values of the integrand.

The version of Simpson's rule that we are discussing is sometimes called Simpson's *one-third rule* because of the 3 in the denominator. There is another version, called Simpson's *five-eighths rule*, which corresponds to fitting third-order polynomials to four points at a time.

Problem 4-23

Apply Simpson's rule to the integral of Problem 4-22, using $n = 2$. Since the integral curve is a parabola, your result should be exactly correct. •

There is another widely used way to obtain a numerical approximation to a definite integral, known as *Gauss quadrature*. In this method, the integrand function must be evaluated at particular unequally spaced points on the interval of integration. We will not discuss this method, but you can read about it in almost any book on numerical analysis.

So far, we have assumed that the integrand function was known, so that it could be evaluated at the required points. Most of the applications of numerical integration in physical chemistry are to integrals where the integrand function is not known exactly, but is known only approximately from experimental measurements at a few points on the interval of integration. If there are an odd number of data points (an even number of panels) which are equally spaced, we can apply Simpson's rule.

Problem 4-24

In thermodynamics, it is shown that the entropy change of a system which is heated from temperature T_1 to temperature T_2 at constant pressure without going through a phase change is given by

$$\Delta S = S(T_2) - S(T_1) = \int_{T_1}^{T_2} \frac{C_p}{T} dT \tag{4.79}$$

where C_p is the constant-pressure heat capacity. The temperature must be measured on the Kelvin scale. Calculate ΔS for the heating of 1 mol of solid zinc from 20 K to 100 K, using the following data:

Temperature (K)	C_p(cal K^{-1} mol^{-1})
20	0.406
30	1.187
40	1.953
50	2.671
60	3.250
70	3.687
80	4.031
90	4.328
100	4.578

•

Section 4-9. A NUMERICAL INTEGRATION COMPUTER PROGRAM

This entire section is an example of the construction of a computer program in the BASIC language to carry out a numerical calculation. If you are not yet familiar with computer programming, you should skip this section until you have studied Chapter 11.

Assume that you are working on a time-sharing system, using a remote terminal. You can also adapt the following discussion to apply to other kinds of systems.

You first turn on the terminal and use whatever little ritual is required to "log in." For the DEC 11/70 RSTS/E system which the author is accustomed to, type the word HELLO and press the carriage return. The computer then prints a message, after which you type an account number and press the carriage return. The computer prints another message. You type a "password" and press the carriage return. The computer prints another message, and you are "logged in."

You next type in SWITCH BASIC to call in the compiler, and type

NEW NUMIT

to notify the computer that you are writing a new program that you want to name NUMIT.

If someone else is going to use the program, or if you will use it again after an extended period of time, it is good to put in a few PRINT statements to have the computer print out an explanation of what the program is. You type in the statements

```
100 PRINT "THIS PROGRAM CALCULATES AN INTEGRAL BY
          SIMPSON'S ONE-THIRD RULE."
110 PRINT "IT PERMITS UP TO 101 DATA POINTS, WHICH
          MUST BE EQUALLY SPACED."
120 PRINT "THE NUMBER OF DATA POINTS MUST BE ODD."
```

Next, you need a DIMENSION statement to reserve storage space for the arrays for the independent variable and the integrand function. You type

```
130     DIM X(101), Y(101)
```

You are now ready to write the part of the program that will read in the necessary numbers. For a program like this one, INPUT statements are more convenient than READ and DATA statements. You type

```
140     INPUT "TYPE IN THE NUMBER OF DATA POINTS. "; N
150     INPUT "TYPE IN THE LOWER LIMIT. "; X0
160     INPUT "TYPE IN THE UPPER LIMIT. "; X9
```

These statements will allow values for these quantities to be read in when the program is run. You now need a loop to read in the data points. Since the values of the independent variable must be equally spaced, it is not necessary to read them in. You type in the statements for the loop:

```
170 FOR I = 1 TO N STEP 1
180 PRINT "TYPE IN THE VALUE OF THE INTEGRAND AT
          POINT NUMBER"; I
190 INPUT Y(I)
200 NEXT I
```

Use the symbol W for the spacing between the values of the independent variable, and type in the statement that will evaluate this:

$$210 \ W = (X9 - X0)/(N - 1)$$

Notice that the number of panels is $N - 1$, not N. You can now have the program cause the values of the integrand and the independent variable to be printed out for each point, so that the user of the program can see if he entered his data properly. You type in the statements:

```
220 INPUT "DO YOU WANT TO SEE A TABLE OF THE DATA
          POINTS (YES OR NO)"; B$
230 IF B$ = "YES" THEN 240 ELSE 300
240 FOR I = 1 TO N STEP 1
250 J = I - 1
260 X(I) = X0 + J*W
270 PRINT "X("; I; ") = "; X(I), "Y("; I; ") = "; Y(I)
280 NEXT I
```

You are now ready to write the part of the program to compute the integral. In Eq. (4.78), the points were numbered from 0 to n. However, most computers will not allow 0 to be used as an index for an array, so we are numbering the points from 1 to N in the program. The value of I in the program is greater by unity than the subscripts in Eq. (4.78). You can now type in statements to form the sum in Eq. (4.78):

```
300 S = Y(1)
310 FOR I = 2 TO N - 1 STEP 2
320 S = S + 4*Y(I) + 2*Y(I + 1)
330 NEXT I
340 S = (S - Y(N))*W/3
```

You type in a statement to print out the answer, and an END statement, and the program is complete:

```
350 PRINT "THE VALUE OF THE INTEGRAL IS"; S
999 END
```

Notice statement number 340. After the loop, the last point will have been added in with a coefficient of 2. The formula requires that the last point be put in with a coefficient of 1. Statement 340 subtracts off Y(N) to restore it to the proper contribution.

The program is now ready to run. In order to keep a copy of the program on file on the disk of the computer, you type

SAVE

and then if you are ready to run the program for the first time, you type

RUN

and the execution begins.

Example 4-19

Modify the program NUMIT so that it will perform the integration in Problem 4-22, including the division of C_p by T.

Solution

The loop beginning with statement 240 can be used to carry out the divisions. We proceed by calling up the program by typing

OLD NUMIT

We need to delete the statements that make the loop optional, so we type

220
230

This replaces statements 220 and 230 by blanks. Unless we type in a REPLACE statement, this replacement will be temporary, affecting only the working version of the program, not the version stored in the disk file. Let us leave statement 270 in the program, so that we will see our data to make sure that it is entered correctly. We insert a statement to carry out the divisions by typing

275   Y(I) = Y(I)/X(I)

This will be inserted into the program between statement 270 and statement 280. The program is now modified. If we want to make the modifications permanent, we type

REPLACE   ●

Problem 4-25

Modify the program NUMIT so that it will evaluate the integral

$$\int_1^2 e^{-x^2}\, dx$$

This will require replacement of the part of the program that reads in data points by statements that will evaluate the integrand function for each value of the independent variable. ●

A slightly more sophisticated numerical integration program, called NUMINT, is listed in Appendix 8. One of the options in this program provides for a modified Simpson's rule which allows for unequally spaced values of the independent variable and does not require the number of data points to be odd.

ADDITIONAL PROBLEMS

4.26

Find the indefinite integrals, without using a table.

a. $\int x \ln(x) \, dx$

b. $\int x \sin^2(x) \, dx$

c. $\int \frac{1}{x(x-a)} \, dx$

d. $\int x^3 \ln(x^2) \, dx$

4.27

Find the definite integrals, without using a table.

a. $\int_0^{2\pi} \sin(x) \, dx$

b. $\int_1^2 x \ln(x^2) \, dx$

c. $\int_0^{\pi/2} \sin^2(x) \cos(x) \, dx$

d. $\int_0^{\pi^2} x \sin(x^2) \cos(x^2) \, dx$

4.28

Evaluate the improper integrals.

a. $\int_1^{\infty} \frac{1}{x^2} \, dx$

b. $\int_0^{\pi/2} \tan(x) \, dx$

4.29

At 298 K, what fraction of nitrogen molecules have speeds lying between 0 and the mean speed? Do a numerical integration.

4.30

Approximate the integral

$$\int_0^\infty e^{-x^2}\, dx$$

using Simpson's rule. You will have to take a finite upper limit, choosing a value large enough so that the error caused by using the wrong limit is negligible. The correct answer is $\sqrt{\pi}/2 = 0.886226926\ldots.$

4.31

Find the integrals

a. $\int \sin[x(x+1)](2x+1)\, dx$

b. $\int_0^\pi \sin[\cos(x)]\sin(x)\, dx$

4.32

When a gas expands reversibly, the work that it does against its surroundings is given by the integral

$$w = \int_{V_1}^{V_2} P\, dV$$

where V_1 is the initial volume, V_2 the final volume, and P the pressure.

Certain nonideal gases are described quite well by the van der Waals equation of state,

$$\left(P + \frac{a}{V^2}\right)(V - b) = RT$$

where V is the volume of 1 mol of the gas, T is the temperature on the Kelvin scale, and a, b, and R are constants. R is usually taken to be the ideal gas constant, $8.3144\ \text{J K}^{-1}\ \text{mol}^{-1}$.

Obtain a formula for the work done if 1 mol of such a gas expands reversibly at constant temperature from volume V_1 to volume V_2. Find the value of this for 1 mol of CO_2, which has

$$a = 3.59\ \text{liters}^2\ \text{atm mol}^{-2}$$
$$b = 42.7\ \text{cm}^3\ \text{mol}^{-1} = 0.0427\ \text{liter mol}^{-1}$$
$$R = 8.3144\ \text{J K}^{-1}\ \text{mol}^{-1} = 0.082057\ \text{liter atm K}^{-1}\ \text{mol}^{-1}$$

if $T = 298$ K, $V_1 = 1.00$ liter $= 1.000 \times 10^{-3}$ m^3, and $V_2 = 100$ liters $= 0.1$ m^3

4.33

The entropy change to bring a sample from 0 K (absolute zero) to some state is called the *absolute entropy* of the sample in that state. Using Eq. (4.79) calculate the absolute entropy of 1.000 mol of solid silver at 270 K. For the region 0 to 30 K, use the approximate relation

$$C_p = aT^3$$

where a is a constant that you can evaluate from the value of C_p at 30 K. For the region 30 to 270 K, use the following data:[1]

T (K)	C_p(J K^{-1} mol^{-1})
30	4.77
50	11.65
70	16.33
90	19.13
110	20.96
130	22.13
150	22.97
170	23.61
190	24.09
210	24.42
230	24.73
250	25.03
270	25.31

ADDITIONAL READING

Sherman K. Stein, *Calculus and Analytic Geometry*, 2nd ed., McGraw-Hill Book Company, New York, 1977.

> *This is one of a number of standard calculus textbooks. All of them discuss integration in much more detail than the present book.*

Herbert B. Dwight, *Tables of Integrals and Other Mathematical Data*, 4th ed., Macmillan Publishing Co., New York, 1962.

> *This is a very good small compilation of definite and indefinite integrals, and also contains a lot of other useful formulas and tables.*

I. S. Gradshteyn and I. M. Ryzhik, *Tables of Integrals, Series and Products*, 4th ed., prepared by Yu. V. Geronimus and M. Yu. Tseytlin, translated by Alan Jeffreys, Academic Press, New York, 1965.

> *This is a large book with lots of definite and indefinite integrals in it.*

D. Bierens de Haan, *Nouvelles tables d'intégrales définies* (New Tables of Definite Integrals), Hafner Publishing Co., New York, 1956.

> *This book is a corrected reprint of an edition of 1867 and is an excellent large collection of definite integrals. However, some entries use an unfamiliar notation which is unfortunately not explained in the book.*

Milton Abramowitz and Irene A. Stegun, eds., *Handbook of Mathematical Functions with Formulas, Graphs and Mathematical Tables*, National Bureau of Standards Applied Mathematics Series No. 55, U.S. Government Printing Office, Washington, D.C., 1964 (also reprinted by Dover Publications, New York).

> *This large book contains a variety of different things, including integrals and useful formulas. It is very inexpensive.*

---

[1] Meads, Forsythe, and Giaque, *J. Am. Chem. Soc.* **63**, 1902 (1941).

Robert W. Hornbeck, *Numerical Methods*, Quantum Publishers, New York, 1975.
This book discusses many topics of numerical analysis, including numerical integration, and appears as an inexpensive paperback. There are a number of quite similar books, entitled "Numerical Analysis" or something similar, and you can read about numerical integration in any of them.

CHAPTER FIVE

Mathematical Series

Section 5-1. INTRODUCTION

A *series* is a sum of terms. A *finite series* has a finite number of terms, and an *infinite series* has a infinite number of terms. A series is not the same as a *sequence*, which is a set of quantities with a rule for generating one member of the set from the previous one. If the members of a sequence are added together, the result is a series.

If a series has terms that are constants, it is a *constant series*. Such a series is

$$s = a_0 + a_1 + a_2 + a_3 + a_4 + \cdots + a_n + \cdots \tag{5.1}$$

For an infinite series, we define the nth *partial sum* as the sum of the first n terms:

$$S_n = a_0 + a_1 + a_2 + a_3 + \cdots + a_{n-1} \tag{5.2}$$

The entire infinite series is

$$s = \lim_{n \to \infty} S_n \tag{5.3}$$

If this limit exists and is finite, we say that the series converges. If the magnitude of S_n becomes larger and larger without bound as n becomes large, or if S_n continues to oscillate as n becomes larger, we say that the series diverges.

A series might have terms that are functions, in which case the series can be used to represent a function. Such a series is

$$f(x) = s(x) = a_0 g_0(x) + a_1 g_1(x) + a_2 g_2(x) + a_3 g_3(x) + \cdots \tag{5.4}$$

We call the set of quantities a_0, a_1, a_2, etc., the *coefficients* of the series and the set of functions $g_0, g_1, g_2, g_3, \ldots$ the *basis functions*. Just as with constant series, a functional series such as that of Eq. (5.4) might converge, or it might diverge. Furthermore, it might converge for some values of x and diverge for others. If there is an interval of values of x such that the series converges for all values of x in that interval, we say that the series is uniformly convergent in that interval.

A common example of a functional series is the *power series*

$$f(x) = a_0 + a_1(x - h) + a_2(x - h)^2 + a_3(x - h)^3 + \cdots \qquad \textbf{(5.5)}$$

Such a series is called a *Taylor series*, and if $h = 0$ it is called a *Maclaurin series*.

Another common functional series is the *Fourier series*, in which the basis functions are sine and cosine functions. Such a series is

$$
\begin{aligned}
f(x) = a_0 + a_1 \cos\left(\frac{\pi x}{L}\right) + a_2 \cos\left(\frac{2\pi x}{L}\right) + \cdots \\
+ b_1 \sin\left(\frac{\pi x}{L}\right) + b_2 \sin\left(\frac{2\pi x}{L}\right) + \cdots
\end{aligned}
\qquad \textbf{(5.6)}
$$

A finite power series is a polynomial, and polynomial functions do occur in physical chemistry, as for example in the solution of the Schrödinger equation for the hydrogen atom and the harmonic oscillator.

After studying this chapter, you should be able to do all of the following:

1. **Determine whether a constant series converges.**
2. **Determine how large a partial sum must be taken to approximate a series by a partial sum.**
3. **Compute the coefficients for a power series to represent a function.**
4. **Determine the region of uniform convergence of a power series.**
5. **Determine the coefficients for a Fourier series to represent a given function.**

Section 5-2.   SERIES WITH CONSTANT TERMS

The two questions that we generally ask about an infinite constant series are: (1) Does the series converge? and (2) What is the value of the series if it does converge? Sometimes it is impossible or inconvenient to find the value of the series, and we then might ask how well we can approximate the series with a partial sum.

Let us consider a well-known series, which is known to be convergent:

$$s = 1 + \frac{1}{2} + \frac{1}{4} + \frac{1}{8} + \cdots + \frac{1}{2^n} + \cdots \qquad (5.7a)$$

$$= \sum_{n=0}^{\infty} \frac{1}{2^n} \qquad (5.7b)$$

There is no general method that is capable of finding the value of every series,[1] but the value of this series can be obtained as in the following example.

Example 5-1

Find the value of the series in Eq. (5.7b).

Solution

We write the sum as the first term plus the other terms, with a factor of $\frac{1}{2}$ factorized out of all the other terms:

$$s = 1 + \tfrac{1}{2}(1 + \tfrac{1}{2} + \tfrac{1}{4} + \tfrac{1}{8} + \cdots)$$

The series in the parentheses is just the same as the original series. There is no problem due to the apparent difference that the series in the parentheses seems to have one less term than the original series, because both series have an infinite number of terms. We now write

$$s = 1 + \tfrac{1}{2}s$$

which can be solved to give

$$s = 2 \qquad \bullet \quad (5.8)$$

Problem 5-1

Show that in the series of Eq. (5.7) any term of the series is equal to the sum of all the terms following it. (*Hint:* Factorize a factor out of all of the following terms so that they will equal this factor times the original series, whose value is now known.) $\qquad \bullet$

The result of Problem 5-1 is of interest in seeing how a series can be approximated by a partial sum. For the series of Eq. (5.7), we can write

$$s = S_n + a_{n-1} \qquad (5.9)$$

where the symbols of Eqs. (5.2) and (5.3) are used. This equation is of course applicable only to this one series. In some cases it is necessary to approximate a series with a partial sum, and in such cases Eq. (5.9) may be applied to other series as an approximation.

Example 5-2

Determine which partial sums approximate the series of Eq. (5.7) to (a) 1% and (b) 0.001%

---

[1] See A. D. Wheelon, "On the Summation of Infinite Series," *J. Appl. Phys.* **25**, 113 (1954), for a method that can be applied to a large number of series.

Solution

a. 1% of 2 is 0.02, so we find the first term of the series that is equal to or smaller than 0.02, and take the partial sum which ends with that term. We have

$$\frac{1}{2^6} = \frac{1}{64} = 0.015625$$

so the partial sum required is the one ending with $\frac{1}{64}$, or S_7. Its value is

$$S_7 = 1.984375$$

b. 0.001% of 2 is 2×10^{-5}, and $1/2^n$ has the value 1.5259×10^{-5} when $n = 16$, so we need the partial sum S_{17}, which has the value

$$S_{17} = 1.999984741$$ •

Problem 5-2

Consider the series

$$s = 1 + \frac{1}{2^2} + \frac{1}{3^2} + \frac{1}{4^2} + \cdots + \frac{1}{n^2} + \cdots$$

which is known to be convergent. Using Eq. (5.9) as an approximation, determine which partial sum approximates the series to (a) 1% and (b) 0.001%.
•

The series of Eq. (5.7) is an example of a *geometric series*, which is

$$s = a + ar + ar^2 + ar^3 + ar^4 + \cdots + ar^n + \cdots \quad \textbf{(5.10a)}$$
$$= a(1 + r + r^2 + r^3 + \cdots + r^n + \cdots) \quad \textbf{(5.10b)}$$

The value of this series can be obtained in the same way as the value of the series in Eq. (5.7) was obtained. We write

$$s = a + r(a + ar + ar^2 + ar^3 + \cdots)$$
$$= a + rs$$

or

$$s = \frac{a}{1 - r} \quad \textbf{(5.11)}$$

The partial sums of the geometric series are given by

$$S_n = a + ar + ar^2 + \cdots + ar^{n-1} = a\frac{1 - r^n}{1 - r} \quad \textbf{(5.12)}$$

In order for the series of Eq. (5.10) to converge, the magnitude of r must be less than unity, However, r can be positive or negative. Equation (5.12) is correct for any value of r.

Example 5-3

The partition function is defined in the statistical mechanics of non-interacting molecules as the sum over all the states of one molecule

$$q = \sum_{i=0}^{\infty} \exp\left(\frac{-E_i}{kT}\right) \tag{5.13}$$

where i is an index specifying the state, E_i is the energy which the molecule has when in that state, k is Boltzmann's constant, and T is the absolute temperature. If we consider only the vibration of a diatomic molecule, to a good approximation

$$E_i = E_v = hv(v + \tfrac{1}{2}) \tag{5.14}$$

where v is the vibrational frequency; v is the vibrational quantum number, which can take on the integral values 0, 1, 2, 3, etc.; and h is Planck's constant.

Use Eq. (5.11) to find the value of the partition function for vibration.

Solution

$$q_{\text{vib}} = \sum_{v=0}^{\infty} \exp\left[\frac{-hv(v + \tfrac{1}{2})}{kT}\right]$$

$$= \exp\left(\frac{-hv}{2kT}\right) \sum_{v=0}^{\infty} \left[\exp\left(\frac{hv}{kT}\right)\right]^v$$

$$= e^{-x/2} \sum_{v=0}^{\infty} (e^{-x})^v$$

where $hv/kT = x$.

The sum is a geometric series, so

$$q_{\text{vib}} = \frac{e^{-x/2}}{1 - e^{-x}} \tag{5.15}$$

The series is convergent, because e^{-x} is smaller than unity for all positive values of x and is never negative. •

Let us now look at a divergent series called the *harmonic series*.

$$s = 1 + \frac{1}{2} + \frac{1}{3} + \frac{1}{4} + \cdots + \frac{1}{n} + \cdots$$

Here are a few partial sums of this series:

$$S_1 = 1$$
$$S_2 = 1.5$$
$$S_{200} = 6.87803$$
$$S_{1000} = 8.48547$$
$$S_{100,000} = 13.0902$$

However,

$$s = \lim_{n \to \infty} S_n = \infty$$

Many people are surprised when they first learn that this series diverges, because the terms keep on getting smaller as you go further into the series. This is a necessary condition for a series to converge, but it is obviously not sufficient.

Problem 5-3

Write a computer program in the BASIC language to evaluate partial sums of the harmonic series, and use it to verify the foregoing values. (Example 11-10 has the body of a similar program.) •

There are a number of constant series listed in Appendix 6, and additional series can be found in the references listed at the end of the chapter.

Tests for Convergence. We now discuss several tests that will usually tell us whether an infinite series converges or not.

1. *The Comparison Test.* If a series has terms each of which is smaller in magnitude than the corresponding term of a series known to converge, it is convergent. If a series has terms each of which is larger in magnitude than the corresponding term of a series known to diverge, it is divergent.

2. *The Alternating Series Test.* If a series has terms that alternate in sign, it is convergent if the terms approach zero as you go further into the series and if each term is smaller in magnitude than the previous term.

3. *The nth-Term Test.* If the terms of a series approach some limit other than zero or do not approach any limit as you go further into the series, the series diverges.

4. *The Integral Test.* If a formula can be written to deliver the terms of a series

$$a_n = f(n) \tag{5.16}$$

   then the series will converge if the improper integral

$$\int_1^\infty f(x)\,dx \tag{5.17}$$

   converges, and will diverge if the improper integral diverges.

5. *The Ratio Test.* For a series of positive terms, consider the limit

$$r = \lim_{n \to \infty} \frac{a_{n+1}}{a_n} \tag{5.18}$$

   If $r < 1$, the series converges. If $r > 1$, the series diverges. If $r = 1$, the test fails, and the series might either converge or diverge. If the ratio does not approach any limit but does not increase without bound, the test also fails.

Example 5-4

Apply the ratio test and the integral test to the harmonic series, Eq. (5.14).

Solution

$$r = \lim_{n \to \infty} \left[\frac{1/n}{1/(n-1)} \right] = \lim_{n \to \infty} \frac{n-1}{n} = 1$$

The ratio test fails.

$$\int_1^\infty \frac{1}{x} dx = \ln(x) \Big|_1^\infty = \lim_{b \to \infty} \left[\ln(b) - \ln(1) \right] = \infty$$

The series diverges, by the integral test. •

Example 5-5

Determine whether the series

$$s = 1 - \frac{1}{2} + \frac{1}{3} - \frac{1}{4} + \frac{1}{5} - \cdots = \sum_{n=1}^\infty \frac{(-1)^{n-1}}{n} \qquad (5.19)$$

converges.

Solution

This is an alternating series, so the alternating series test applies. Since every term approaches more closely to zero than the previous term, the series is convergent. •

Example 5-6

Determine whether the series converges:

$$s = 1 - \tfrac{1}{2} + \tfrac{2}{2} - \tfrac{1}{3} + \tfrac{2}{3} - \tfrac{1}{4} + \tfrac{2}{4} - \tfrac{1}{5} + \tfrac{2}{5} - \cdots$$

Solution

This is a tricky series, because it is an alternating series, and the nth term approaches zero as n becomes large. However, the series diverges. The alternating series test does not apply, because it requires that each term be closer to zero than the previous term. Half the time as you go from one term to the next in this series, the magnitude increases instead of decreasing. Let us manipulate the series by subtracting each negative term from the following positive term to obtain

$$s = 1 + \tfrac{1}{2} + \tfrac{1}{3} + \tfrac{1}{4} + \tfrac{1}{5} + \cdots$$

This is the harmonic series, which we already found to diverge. •

Problem 5-4

Show that the series in Eq. (5.7) converges. •

Problem 5-5

Show that the geometric series converges if $r^2 < 1$. •

Problem 5-6

Test the following series for convergence.

a. $\displaystyle\sum_{n=0}^{\infty} \frac{1}{n^2}$

c. $\displaystyle\sum_{n=1}^{\infty} \frac{(-1)^n(n-1)}{n^2}$

b. $\displaystyle\sum_{n=0}^{\infty} \frac{1}{n!}$

d. $\displaystyle\sum_{n=1}^{\infty} \frac{(-1)^n n}{n!}$

Note for part b: $n!$ (n factorial) is defined to be $n(n-1)(n-2)\cdots(2)(1)$ for $n > 0$, but $0!$ is defined to equal 1. •

Section 5-3. POWER SERIES

A power series in x (see Eq. (5.5)) can represent a function of x if it is convergent. We will consider two problems in finding a series to represent a function. The first is finding the values of the coefficients so that the function will be represented, and the second is finding the interval in which the series is uniformly convergent (the region in which it can represent the function).

Let us consider first a specific example, that of finding a Maclaurin series to represent $\sin(x)$:

$$\sin(x) = a_0 + a_1 x + a_2 x^2 + \cdots = s(x) \tag{5.20}$$

In order for the function and the series to be equal at all values of x, they must be equal at $x = 0$. Since $\sin(x) = 0$, a_0 must equal 0.

In order for the series and the function to be equal at all values of x, they must have the same value of the first derivative at all points. At $x = 0$, we must have

$$\left(\frac{ds}{dx}\right)_0 = a_1 = \left[\frac{d}{dx}\sin(x)\right]_0 = \cos(0) = 1 \tag{5.21}$$

For sufficiently small values of x, we can approximate $\sin(x)$ by the second partial sum, S_2, giving the same approximation as Eq. (2.28). Figure 5-1 shows the function $\sin(x)$ and the approximation, $S_2 = x$.

We continue in this way, requiring all derivatives of the function and the series to be equal at $x = 0$. This means that only a function which has derivatives of all orders at $x = 0$ can be represented by a Maclaurin series. Such a function is called an *analytic function* at $x = 0$. The nth derivative of the series is, at $x = 0$,

$$\left(\frac{d^n s}{dx^n}\right)_0 = n!\, a_n \tag{5.22}$$

where $n!$ (n factorial) is, for $n \geq 1$

$$n! = n(n-1)(n-2)\cdots(2)(1) \tag{5.23}$$

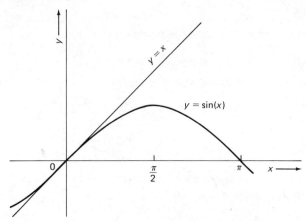

Figure 5-1. The function $\sin(x)$ and the approximation $S_2 = x$.

This gives us a general formula for the coefficients in a Maclaurin series

$$a_n = \frac{1}{n!} \left(\frac{d^n f}{dx^n} \right)_0 \qquad (n = 1, 2, 3, \ldots) \qquad (5.24)$$

where $f(x)$ is any analytic function to be represented by a Maclaurin series.

Problem 5-7

Show that Eq. (5.22) is correct. •

Example 5-7

Find all the coefficients for the series representing $\sin(x)$.

Solution

The derivatives of $\sin(x)$ follow a repeating pattern:

$$f(x) = \sin(x)$$

$$\frac{df}{dx} = \cos(x)$$

$$\frac{d^2 f}{dx^2} = -\sin(x)$$

$$\frac{d^3 f}{dx^3} = -\cos(x)$$

$$\frac{d^4 f}{dx^4} = \sin(x)$$

etc.

When these are evaluated at $x = 0$, we have only odd powers of x in the series:

$$\sin(x) = x - \frac{1}{3!}x^3 + \frac{1}{5!}x^5 - \frac{1}{7!}x^7 + \cdots \qquad \bullet \quad (5.25)$$

Problem 5-8

a. Show that the Maclaurin series for e^x is

$$e^x = 1 + \frac{1}{1!}x + \frac{1}{2!}x^2 + \frac{1}{3!}x^3 + \frac{1}{4!}x^4 + \cdots \qquad (5.26)$$

b. Find the Maclaurin series for $\cos(x)$. $\qquad \bullet$

There are a number of important functions that are not analytic at $x = 0$, and these cannot be represented by a Maclaurin series. Such a function is $\ln(x)$. The first derivative of this is $1/x$, which becomes infinite as $x \to 0$. Although there is no Maclaurin series for $\ln(x)$, you can find a Taylor series for a positive value of h in Eq. (5.5).

In order to find the coefficients for a Taylor series, we require the function and the series to have the same derivatives of all orders at $x = h$. This gives

$$a_n = \frac{1}{n!}\left(\frac{d^n f}{dx^n}\right)_h \qquad (5.27)$$

where $f(x)$ is the function to be represented. We say that h is the value of x about which the function is expanded.

Example 5-8

Find the Taylor series for $\ln(x)$, expanding about $x = 1$.

Solution

$$\ln(x) = (x - 1) - \tfrac{1}{2}(x - 1)^2 + \tfrac{1}{3}(x - 1)^3 - \tfrac{1}{4}(x - 1)^4 + \cdots \bullet \quad (5.28)$$

Problem 5-9

Find the series for $\ln(1 + x)$, expanding about $x = 0$. Notice that you can save some work by looking at Example 5-8. $\qquad \bullet$

Problem 5-10

Find the Taylor series for $\cos(x)$, expanding about $x = \pi/2$. $\qquad \bullet$

The Convergence of Power Series. If a series is to represent a function, it must be convergent in the entire interval of the representation. For a fixed value of x, the series $s(x)$ is no different from a constant series, and all the

120

MATHEMATICAL SERIES / Ch. 5

tests for convergence of Section 5-2 can be applied. We can then consider different fixed values of x and determine the interval in which the series is uniformly convergent.

Example 5-9

Investigate the convergence of the series in Eq. (5.28).

Solution

Let us consider the two cases separately: case a, $x > 1$, and case b, $x < 1$. If $x = 1$, the entire series vanishes, as does the function, so the series converges to the function for $x = 1$.

Case a. For $x > 1$, the series is an alternating series, and we can apply the alternating series test. The nth term of the series is

$$t_n = a_n(x - 1)^n = \frac{(x - 1)^n(-1)^{n-1}}{n}$$

Look at the limit of this as n becomes large.

$$\lim_{n \to \infty} t_n = \begin{cases} 0 & \text{if } |x - 1| \le 1 \quad \text{or} \quad 1 < x \le 2 \\ \infty & \text{if } |x - 1| > 1 \quad \text{or} \quad x > 2 \end{cases}$$

The interval of convergence for $x > 1$ extends up to and includes $x = 2$.

Case b. For $x < 1$, the series is not alternating. We apply the ratio test.

$$r = \lim_{n \to \infty} \frac{t_n}{t_{n-1}} = \lim_{n \to \infty} \left[-\frac{(x - 1)^n/n}{(x - 1)^{n-1}/(n - 1)} \right] = -(x - 1) = 1 - x$$

This will be less than unity if x lies between 0 and 1, but if $x = 0$, the text fails. However, if $x = 0$, the series is the same as the harmonic series except for the the sign, and thus diverges. The interval of convergence is $0 < x \le 2$. •

The behavior in Example 5-9 is quite typical. There is a point at which the function is not analytic, at $x = 0$. The function is analytic to the right of this point, and the series equals the function right up to $x = 0$, approaching from the right. The *interval of convergence* is centered on the point about which the function is expanded, which is $x = 1$ in this case. That is, the distance from $x = 1$ to the left end of the interval of convergence is 1 unit, and the distance from $x = 1$ to the right end of the interval of convergence is also 1 unit. This distance is called the *radius of convergence*. Even though the function is defined for values of x beyond $x = 2$, the series does not converge and cannot represent the function.

Problem 5-11

Find the series for $\ln(x)$, expanding about $x = 2$, and show that the radius of convergence for this series is equal to 2. •

In general, the radius of convergence is the distance from the point about which we are expanding to the closest point at which the function is not analytic. If a function is defined on both sides of such a point, it is represented on the two sides by different series.

Example 5-10

Find the interval of convergence for the series in Eq. (5.26).

Solution

We apply the ratio test.

$$r = \lim_{n \to \infty} \frac{t_n}{t_{n-1}} = \lim_{n \to \infty} \frac{x^n/n!}{x^{n-1}/(n-1)!} = \lim_{n \to \infty} \frac{x}{n}$$

This limit vanishes for any finite value of x, so the series converges for any finite value of x, and the radius of convergence is infinite.   •

Problem 5-12

a. Find the series for $1/(1 - x)$, expanding about $x = 0$. What is the interval of convergence?
b. Find the interval of convergence for the series for $\sin(x)$ and for $\cos(x)$.   •

Unfortunately, the situation is not always so simple as in the examples we have been discussing. For example, if we wanted to construct a Maclaurin series for the function

$$f(x) = \frac{1}{1 + x^2}$$

the radius of convergence would be determined by discontinuities in the function in the complex plane, at $x = i$ and $x = -i$, even though there are no discontinuities for real values of x. The radius of convergence in this case would be unity, the distance from the origin to $x = \pm i$ in the Argand plane. You can read more about this topic in the book by Churchill listed at the end of the chapter.

In physical chemistry, there are a number of applications of power series, but in most cases, a partial sum is used to approximate the series.

For example, the behavior of a nonideal gas is often described by use of the *virial series*,

$$\frac{PV}{RT} = 1 + \frac{B_2}{V} + \frac{B_3}{V^2} + \frac{B_4}{V^3} + \cdots \tag{5.29}$$

where P is the pressure, V the volume of 1 mol of the gas, T the temperature, and R the gas constant. The coefficients B_2, B_3, etc., are called *virial coefficients*.

If all the virial coefficients were known for a particular gas, the virial series would represent exactly the volumetric behavior of that gas at all values of $1/V$. However, for many purposes, the truncated equation is used:

$$\frac{PV}{RT} = 1 + \frac{B_2}{V} \tag{5.30}$$

This will give adequate accuracy for small values of $1/V$.

There is another commonly used virial equation of state, written as

$$PV = RT + BP + CP^2 + DP^3 + \cdots \tag{5.31}$$

This is a power series in P, expanded about $P = 0$. It is also commonly truncated for practical use.

Example 5-11

Show that the coefficient B in Eq. (5.31) is equal to the coefficient B_2 in Eq. (5-29).

Solution

We multiply Eq. (5.29) on both sides by RT/V to obtain

$$P = \frac{RT}{V} + \frac{RTB_2}{V^2} + \frac{RTB_3}{V^3} + \cdots \tag{5.32}$$

This must be equal to

$$P = \frac{RT}{V} + \frac{BP}{V} + \frac{CP^2}{V} + \cdots \tag{5.33}$$

We convert the second series into a series in $1/V$ by substituting the first series into the right-hand side wherever a P occurs. When the entire series on the right-hand side of Eq. (5.32) is squared, every term will have a least a V^2 in the denominator [see Eq. (11) of Appendix 6 for the square of a series]. Therefore, Eq. (5.33) becomes ·

$$P = \frac{RT}{V} + \frac{B}{V}\left(\frac{RT}{V} + \frac{RTB_2}{V^2} + \cdots\right) + O\left(\frac{1}{V}\right)^3 \tag{5.34}$$

where $O(1/V)^3$ stands for terms of order $1/V^3$ or higher (containing no powers of $1/V$ lower than the third power).

Equation (5.34) is thus

$$P = \frac{RT}{V} + \frac{RTB}{V^2} + O\left(\frac{1}{V}\right)^3 \tag{5.35}$$

We now use a fact about series [Eq. (9) of Appendix 6]: If two series are equal, then any coefficient of one is equal to the corresponding coefficient of the other series. Comparison of Eq. (5.35) with Eq. (5.32) shows that

$$B_2 = B \qquad \bullet$$

Another application of a power series in physical chemistry is in the discussion of colligative properties (freezing-point depression, boiling-point elevation, and osmotic pressure). For example, if X_1 is the mole fraction of solvent, ΔH_v is the molar heat of vaporization, T_0 the pure solvent boiling temperature, and T the boiling temperature of the solution,

$$-\ln(X_1) = \frac{\Delta H_v}{R}\left(\frac{1}{T_0} - \frac{1}{T}\right) \tag{5.36}$$

If there is only one solute (component other than the solvent), then its mole fraction, X_2, is given by

$$X_2 = 1 - X_1 \tag{5.37}$$

The left-hand side of Eq. (5.36) is then (see Example 5-8 and Problem 5-9)

$$-\ln(X_1) = -\ln(1 - X_2) = X_2 + \tfrac{1}{2}X_2^2 + \cdots \tag{5.38}$$

If X_2 is not too large, we can write

$$X_2 \approx \frac{\Delta H_v}{R}\left(\frac{1}{T_0} - \frac{1}{T}\right) \tag{5.39}$$

Problem 5-13

Determine how large X_2 can be before the truncation of Eq. (5.38) which was used in Eq. (5.39) is inaccurate by more than 1%.　●

Section 5-4. FOURIER SERIES AND OTHER FUNCTIONAL SERIES

Powers of x or of $x - h$ are not the only choice for basis functions in series such as that of Eq. (5.4). In fact, if we want to produce a series which will converge rapidly, so that we can truncate it after only a few terms, it is good to choose basis functions that have as much as possible in common with the function to be represented.

For example, Fourier series are series designed to represent periodic functions. A Fourier series that represents a periodic function of period $2L$ is

$$f(x) = a_0 + \sum_{n=1}^{\infty} a_n \cos\left(\frac{n\pi x}{L}\right) + \sum_{n=1}^{\infty} b_n \sin\left(\frac{n\pi x}{L}\right) \tag{5.40}$$

Problem 5-14

Show that the basis functions in the series in Eq. (5.40) are periodic with period $2L$. That is, show, for arbitrary n, that

$$\sin\left[\frac{n\pi(x + 2L)}{L}\right] = \sin\left(\frac{n\pi x}{L}\right)$$

and

$$\cos\left[\frac{n\pi(x + 2L)}{L}\right] = \cos\left(\frac{n\pi x}{L}\right)$$

●

There are some interesting mathematical questions about Fourier series, which we do not discuss. One is the question of the convergence of a Fourier series for all value of x. Another is the question of completeness. A set of basis functions is said to be *complete* if a series in these functions can accurately represent any function. The set of sine and cosine functions in eq. (5.40) is a complete set for the represenatation of periodic functions of period $2L$. Completeness is often assumed without proof when dealing with a new set of basis functions.

Finding the Coefficients of a Fourier Series—Orthogonality. In a power series, we found the coefficients by demanding that the function and the series have equal derivatives at the point about which we were expanding. In a Fourier series, we use a different procedure, utilizing a property of the the basis functions that is called *orthogonality*. This property is expressed by the three equations:

$$\int_{-L}^{L} \cos\left(\frac{m\pi x}{L}\right)\cos\left(\frac{n\pi x}{L}\right)dx = L\delta_{mn} = \begin{cases} L & \text{if } m = n \\ 0 & \text{if } m \neq n \end{cases} \tag{5.41}$$

$$\int_{-L}^{L} \cos\left(\frac{m\pi x}{L}\right)\sin\left(\frac{n\pi x}{L}\right)dx = 0 \tag{5.42}$$

$$\int_{-L}^{L} \sin\left(\frac{m\pi x}{L}\right)\sin\left(\frac{n\pi x}{L}\right)dx = L\delta_{mn} \tag{5.43}$$

The quantity δ_{mn} is called the *Kronecker delta*. It is equal to unity if its two indices are equal, and is equal to zero otherwise.

Equations (5.41) and (5.43) do not apply if m and n are both equal to zero. The integral in Eq. (5.41) is equal to $2L$ if $m = n = 0$, and the integral in Eq. (5.43) is equal to zero if $m = n = 0$.

Two functions that yield zero when multiplied together and integrated are said to be *orthogonal* to each other. Equations (5.41), (5.42), and (5.43) indicate that all the basis functions for the Fourier series of period $2L$ are orthogonal to each other. This terminology is analogous to that used with vectors. If two vectors are at right angles to each other, they are said to be orthogonal to each other, and their scalar product is zero (see Chapter 2). An integral such as those in the three equations above is sometimes called a scalar product of the two functions.

To find a_m, where $m \neq 0$, we multiply both sides of Eq. (5.40) by $\cos(m\pi x/L)$ and integrate from $-L$ to L.

$$\int_{-L}^{L} f(x)\cos\left(\frac{m\pi x}{L}\right)dx = \sum_{n=0}^{\infty} a_n \int_{-L}^{L} \cos\left(\frac{n\pi x}{L}\right)\cos\left(\frac{m\pi x}{L}\right)dx$$
$$+ \sum_{n=1}^{\infty} b_n \int_{-L}^{L} \sin\left(\frac{n\pi x}{L}\right)\cos\left(\frac{m\pi x}{L}\right)dx \tag{5.44}$$

Notice that we have incorporated the a_0 term into the first sum, using the fact that $\cos(0) = 1$. We have also used the fact that the integral of a sum is equal to the sum of the integrals of the terms if the series is uniformly convergent.

We now apply the orthogonality facts, Eqs. (5.41) and (5.42), to find that all of the integrals on the right-hand side of Eq. (5.44) vanish except for the term with two cosines in which $n = m$. The result is

$$\int_{-L}^{L} f(x) \cos\left(\frac{m\pi x}{L}\right) dx = a_m L \qquad (5.45)$$

This is a formula for finding all of the a coefficients except for a_0.

To find a_0, we use the fact that

$$\int_{-L}^{L} \cos(0) \cos(0) \, dx = \int_{-L}^{L} dx = 2L \qquad (5.46)$$

which leads to our working equations for the a coefficients:

$$a_0 = \frac{1}{2L} \int_{-L}^{L} f(x) \, dx \qquad (5.47a)$$

$$a_n = \frac{1}{L} \int_{-L}^{L} f(x) \cos\left(\frac{n\pi x}{L}\right) dx \qquad (5.47b)$$

A similar procedure consisting of multiplication by $\sin(m\pi x/L)$ and integration from $-L$ to L yields

$$b_n = \frac{1}{L} \int_{-L}^{L} f(x) \sin\left(\frac{n\pi x}{L}\right) dx \qquad (5.48)$$

Problem 5-15

Show that Eq. (5.48) is correct. ●

There are a few comments that we can make now about Fourier series. The first is that a function does not have to analytic, or even continuous, in order to be represented by a Fourier series. It is only necessary that the function be integrable. As mentioned in Chapter 4, this permits a function to have discontinuities, as long as the step in the function is finite. At a discontinuity, a Fourier series will converge to a value halfway between the value just to the right of the discontinuity and the value just to the left of the discontinuity.

Another comment is that we can represent a function which is not periodic by a Fourier series if we are only interested in representing the function in the interval $-L < x < L$. The Fourier series will be periodic, and if the function is defined outisde this interval, the series will be equal to the function inside the interval, but not necessarily equal to the function outside the interval.

Another comment is that if the function $f(x)$ is an even function, all of the b_n coefficients will vanish, and only the cosine terms will appear in the series. Such a series is called a *Fourier cosine series*. If $f(x)$ is an odd function, only the sine terms will appear, and the series is called a *Fourier sine series*.

A final comment is that if we want to represent a function only in the interval $0 < x < L$, we can regard it as the right half of an odd function or the right half of an even function, and can therefore represent it either with a sine series or a cosine series. These two series would have the same value in the interval $0 < x < L$, but would be the negatives of each other in the interval $-L < x < 0$.

Example 5-12

Find the Fourier series to represent the function $f(x) = x$ for the interval $-L < x < L$.

Solution

The function is odd in the interval $(-L, L)$, so the series will be a sine series. Although we have the definition of the function only for the interval $(-L, L)$, the series will be periodic, and will be the "sawtooth wave" that is shown in Figure 5-2. The coefficients are obtained from Eq. (5.48). Since the integrand is the product of two odd functions, it is an even function, and the integral is equal to twice the integral from 0 to L:

$$b_n = \frac{2}{L}\int_0^L x \sin\left(\frac{n\pi x}{L}\right)dx = \frac{2}{L}\left(\frac{L}{n\pi}\right)^2 \int_0^{n\pi} y\sin(y)\,dy = \frac{2L}{n\pi}(-1)^{n-1}$$

The series is

$$f(x) = \sum_{n=1}^{\infty} \frac{2L}{n\pi}(-1)^{n-1}\sin\left(\frac{n\pi x}{L}\right)$$

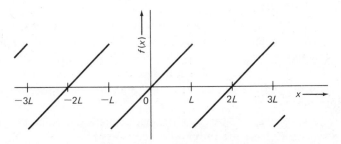

Figure 5-2. The "sawtooth wave" series of Example 5-12.

Problem 5-16

a. Show that the a_n coefficients for the series representing the function in Example 5-12 all vanish.

b. Show that the series equals zero at $x = -L$, $x = L$, $x = 3L$, etc., rather than equaling the function at this point.

c. Find the Fourier cosine series for the even function

$$f(x) = |x| \qquad \text{for } -L < x < L$$

Draw a graph of the periodic function represented by the series. ●

It is a necessary condition for the convergence of Fourier series that the coefficients become smaller and smaller and approach zero as n becomes larger and larger. If a Fourier series is convergent, it will be uniformly convergent for all values of x.

If convergence is fairly rapid, it may be possible to approximate a Fourier series by one of its partial sums. Figure 5-3 shows three different partial sums of the series that represents the "square-wave" function

$$f(x) = \begin{cases} +1 & \text{for } 0 < x < L \\ -1 & \text{for } -L < x < 0 \end{cases}$$

Only the right half of one period is shown. The first partial sum only vaguely resembles the function, but S_{10} is a better approximation. Notice the little "spike" or overshoot near the discontinuity. This is a typical behavior. Even though S_{100} fits the function more closely away from the discontinuity, it

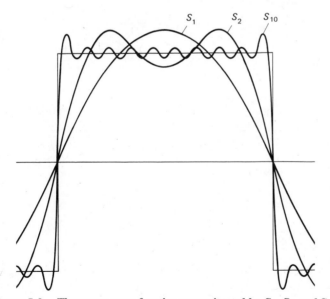

Figure 5-3. The square wave function approximated by S_1, S_2, and S_{10}.

has a spike near the discontinuity which is just as high as that of S_{10}, although much narrower.

Fourier Series with Complex Exponential Basis Functions. The sine and cosine basis functions are closely related to complex exponential functions, as shown in Eqs. (2.124) and (2.125). One can write

$$b_n \sin\left(\frac{n\pi x}{L}\right) + a_n \cos\left(\frac{n\pi x}{L}\right) = \tfrac{1}{2}(a_n - ib_n)e^{in\pi x/L} + \tfrac{1}{2}(a_n + ib_n)e^{-in\pi x/L} \qquad \textbf{(5.49)}$$

It is therefore possible to rewrite Eq. (5.40) as

$$f(x) = \sum_{n=-\infty}^{\infty} c_n e^{in\pi x/L} \qquad \textbf{(5.50)}$$

if we allow c_n to be complex. Note that we have incorporated the terms with negative exponents into the same sum with the other terms by allowing the summation index to take on negative as well as positive values.

Fourier Integrals. We have assumed that L is finite. In the case that $L \to \infty$, the values of $n\pi x/L$ become closer and closer together, and the Fourier series becomes an integral, which is called a *Fourier integral*. We let

$$k = \frac{n\pi}{L} \qquad \textbf{(5.51)}$$

As the limit $L \to \infty$ is taken, k becomes a finite, continuously variable quantity if $n \to \infty$ in the proper way.

The series now becomes

$$f(x) = \frac{1}{\sqrt{2\pi}} \int_{-\infty}^{\infty} c(k)e^{ikx}\, dk \qquad \textbf{(5.52)}$$

where the coefficient c in Eq. (5.50) has become a continuous function of k. We have introduced a factor of $1/\sqrt{2\pi}$ in front of the integral, changing the definition of $c(k)$. The reason for this is that the equation for determining $c(k)$ is now

$$c(k) = \frac{1}{\sqrt{2\pi}} \int_{-\infty}^{\infty} f(x)e^{-ikx}\, dx \qquad \textbf{(5.53)}$$

and it is convenient to have the same factor in front of both integrals.

The function $c(k)$ is called the Fourier transform of $f(x)$, and the function $f(x)$ is called the Fourier transform of $c(k)$. The function $f(x)$ is no longer periodic, because the period has been allowed to become infinite. Since we now have improper integrals, we must have functions $f(x)$ and $c(k)$ such that the integrals converge. For this to happen, $f(x)$ must approach zero as $x \to -\infty$ and $x \to \infty$.

Example 5-13

Find the Fourier transform of the function

$$f(x) = e^{-ax^2}$$

Solution

$$c(k) = \frac{1}{\sqrt{2\pi}} \int_{-\infty}^{\infty} e^{-ax^2} e^{ikx} \, dx$$

From Eq. (2.112),

$$c(k) = \frac{1}{\sqrt{2\pi}} \int_{-\infty}^{\infty} e^{-ax^2} \cos(kx) \, dx + \frac{i}{\sqrt{2\pi}} \int_{-\infty}^{\infty} e^{-ax^2} \sin(kx) \, dx$$

The second integral in this equation is equal to zero because it has an odd integrand. The first integral is twice the integral that is found as Eq. (47) of Appendix 4.

$$c(k) = \frac{2}{\sqrt{2\pi}} \int_{0}^{\infty} e^{-ax^2} \cos(kx) \, dx = \frac{2}{\sqrt{2\pi}} \frac{1}{2} \sqrt{\frac{\pi}{a}} e^{-k^2/4a}$$

$$= \frac{1}{\sqrt{2a}} e^{-k^2/4a} \qquad \bullet$$

Other Functional Series with Orthogonal Basis Sets. Fourier series are just one example of series using orthogonal basis sets. For example, in quantum mechanics it is found that the eigenfunctions of quantum mechanical operators form orthogonal sets of functions, and these can be used as basis functions for series.

Assume that we have a complete set of orthogonal functions, called ϕ_1, ϕ_2, ϕ_3, etc., and that these functions have been normalized. This means that the functions have been multiplied by appropriate constants so that the scalar product of any one of the functions with itself is unity:

$$\int \phi_n^* \phi_m \, dx = \delta_{nm} = \begin{cases} 1 & \text{if } n = m \\ 0 & \text{if } n \neq m \end{cases} \tag{5.54}$$

Notice that the scalar product is defined as the integral of the complex conjugate of the first function times the second function.

Since the set of functions is assumed to be complete, we can expand an arbitrary function, ψ, in terms of the ϕ functions:

$$\psi = \sum_n c_n \phi_n \tag{5.55}$$

The sum in this equation will include one term for each function in the complete set, and there may be an infinite number of them.

In order to find the coefficients c_1, c_2, c_3, etc., we multiply by the complex conjugate of ϕ_m and integrate. With Eq. (5.54), our result is

$$\int \phi_m^* \psi \, dx = \sum_n c_n \int \phi_m^* \phi_n \, dx = \sum_n c_n \delta_{mn} = c_m \tag{5.56}$$

Notice that Eqs. (5.47) and (5.48) are just special cases of this equation.

Section 5-5. MATHEMATICAL OPERATIONS ON SERIES

We close this chapter with a few comments on the mathematical manipulation of series. The principal fact which you whould remember from this section is that *if a series is uniformly convergent, the result of operating on the series is the same as the result of operating on the individual terms and summing the results.* For example, if

$$f(x) = \sum_{n=0}^{\infty} a_n g_n(x) \tag{5.57}$$

and the series is uniformly convergent, then

$$\frac{df}{dx} = \frac{d}{dx} \left[\sum_{n=0}^{\infty} a_n g_n(x) \right] = \sum_{n=0}^{\infty} a_n \frac{dg_n}{dx} \tag{5.58}$$

This amounts to interchange of the operations of summing and differentiating.

Similarly,

$$\int_a^b f(x) \, dx = \int_a^b \left[\sum_{n=0}^{\infty} a_n g_n(x) \right] dx = \sum_{n=0}^{\infty} a_n \int_a^b g_n(x) \, dx \tag{5.59}$$

This amounts to interchange of the operations of summing and integrating.

We have already used Eq. (5.59) in the previous section in deriving the formula for the coefficients in the Fourier series.

Example 5-14

Find the Maclaurin series for $\cos(x)$ from the Maclaurin series for $\sin(x)$, using the fact that $d[\sin(x)]/dx = \cos(x)$.

Solution

From Eq. (5.30), we have

$$\sin(x) = x - \frac{x^3}{3!} + \frac{x^5}{5!} - \frac{x^7}{7!} + \cdots$$

so that

$$\frac{d[\sin(x)]}{dx} = 1 - \frac{x^2}{2!} + \frac{x^4}{4!} - \frac{x^6}{6!} + \cdots$$

●

Problem 5-17

From the Taylor series about $x = 1$ given in Eq. (5.28), find the Taylor series for $1/x$ about $x = 1$, using the fact that

$$\frac{d[\ln(x)]}{dx} = \frac{1}{x}$$

●

ADDITIONAL PROBLEMS

5.18

By use of the Maclaurin series already obtained in this chapter, prove the identity

$$\boxed{e^{ix} = \cos(x) + i\sin(x)} \qquad (5.60)$$

5.19

Show that no Maclaurin series

$$f(x) = a_0 + a_1 x + a_2 x^2 + \cdots$$

can be formed to represent the function $f(x) = +\sqrt{x}$. Why is this?

5.20

If $f(x) = \tan(x)$, find the first few terms of the Taylor series

$$f(x) = a_0 + a_1\left(x - \frac{\pi}{4}\right) + a_2\left(x - \frac{\pi}{4}\right)^2 + \cdots$$

where x is measured in radians. What is the radius of convergence of the series?

5.21

The sine of $\pi/4$ radians (45°) is $\sqrt{2}/2 = 0.70710678 \ldots$. How many terms in the series

$$\sin(x) = x - \frac{x^3}{3!} + \frac{x^5}{5!} - \frac{x^7}{7!} + \cdots$$

must be taken to achieve 1% accuracy at $x = \pi/4$?

5.22

Estimate the largest value of x that allows e^x to be approximated to 1% accuracy by the partial sum

$$e^x \approx 1 + x$$

5.23

Find two different Taylor series to represent the function

$$f(x) = \frac{1}{x^2}$$

One series is

$$f(x) = a_0 + a_1(x - 1) + a_2(x - 1)^2 + \cdots$$

and the other is

$$f(x) = b_0 + b_1(x - 2) + b_3(x - 2)^2 + \cdots$$

Show that $b_n = a_n/2^n$ for any value of n. Find the interval of convergence for each series (the ratio test may be used). Which series must you use in the vicinity of $x = 3$? Why?

5.24

A certain electronic circuit produces the following sawtooth wave.

$$f(t) = \begin{cases} a(-T - t), & -T < t < -T/2 \\ at, & -T/2 < t < T/2 \\ a(T - t), & T/2 < t < T \end{cases}$$

where a and T are constants and t represents the time. The definition of the function given is for only one period of length $2T$, but the function is periodic. Find the Fourier series that represents this function. Sketch the graph of the function and of the first partial sum.

5.25

Find the Fourier series that represents the square wave

$$A(t) = \begin{cases} 0, & -T/2 < t < 0 \\ A_0, & 0 < t < T/2 \end{cases}$$

where A_0 is a constant and T is the period.

5.26

Using the Maclaurin series for e^x, show that the derivative of e^x is equal to e^x.

5.27

Find the Maclaurin series that represents $\tan(x)$. What is its radius of convergence?

5.28

Find the Maclaurin series that represents $\cosh(x)$. What is its radius of convergence?

ADDITIONAL READING

Herbert B. Dwight, *Tables of Integrals and Other Mathematical Data*, 4th ed., Macmillan Publishing Co., New York, 1961.

This little book contains quite a large number of constant series and power series, as well as tables of integrals.

R. V. Churchill, *Introduction to Complex Variables and Applications*, McGraw-Hill Book Company, New York, 1948.

This book is an introduction to the calculus of complex variables. Chapter 6 of the book is devoted to power series of a complex variable, and questions of convergence in the complex plane are discussed. Much of the information about series of complex variables is important for series of a real variable.

L. B. W. Jolley, *Summation of Series*, 2nd ed., Dover Publications, New York, 1961.

This collection of series contains a large number of series which the author has collected from many sources. It is a paperback book which is well worth the small price. There is also a fairly large section of infinite products.

R. V. Churchill, *Fourier Series and Boundary Value Problems*, 2nd ed., McGraw-Hill Book Company, New York, 1963.

This book concerns the use of Fourier series in the solution of partial differential equations occurring in physics. Chapters 4 through 6 concern the principal properties of Fourier series and integrals.

David L. Powers, *Boundary Value Problems*, Academic Press, New York, 1972.

This book covers the same area as the book by Churchill, and includes a 40-page chapter on Fourier series and integrals.

G. P. Tolstov, *Fourier Series*, Prentice-Hall, Englewood Cliffs, N.J., 1962.

This is a good introduction to Fourier series and integrals.

Wilfred Kaplan, *Advanced Calculus*, Addison-Wesley Publishing Co., Reading, Mass., 1952.

This is a text for a calculus course beyond the first year. Chapter 6 concerns infinite series, and Chapter 7 concerns Fourier series.

Angus E. Taylor and Robert W. Mann, *Advanced Calculus*, 2nd ed., Ginn and Company, Lexington, Mass., 1972.

This book is similar to the book by Kaplan.

Konrad Knopp, *Infinite Sequences and Series*, translated by F. Bagemihl, Dover Publications, New York, 1956.

This is a small but authoritative paperback book which discusses most of the important mathematics of sequences and series, including the theory of convergence. A very short chapter on the evaluation of series is included.

In addition to the foregoing works, particular series can be found in various tables of integrals, such as the work by Gradshteyn and Rhyzhik listed in Chapter 4. You can also read about Taylor and Maclaurin series in almost any elementary calculus textbook.

CHAPTER SIX

Calculus with Several Independent Variables

Section 6-1. INTRODUCTION

In physical chemistry, there are a number of applications of functions of several independent variables. For such a function, a value for each of the independent variables must be given in order for the function to provide a value for the dependent variable.

For example, the thermodynamic properties of a simple fluid system of one component and one phase are functions of three independent variables. If we choose temperature, T, volume, V, and number of moles, n, as our independent variables, then the other properties, such as pressure, P, are dependent variables. We write

$$P = P(T, V, n) \tag{6.1}$$

We assume that the functions that represent such actual behavior of physical systems are piecewise continuous with respect to each variable. That is, if we temporarily keep all but one of the independent variables fixed, the function behaves just like a continuous function of one variable. We also assume that the function is piecewise single-valued.

In physical chemistry, we sometimes work with mathematical formulas that represent specific functions, and we sometimes work with identities that allow us to obtain useful results without having a particular function at hand. For example, if the temperature of a gas is fairly high and its volume is large enough, the pressure of a gas is given to a good approximation by

$$P = \frac{nRT}{V} \tag{6.2}$$

We will sometimes use this and other specific mathematical functional relations. However, in other cases, we will write equations involving the pressure without knowing whether the pressure is given by Eq. (6.2) or some other formula.

In this chapter, we will discuss both differential and integral calculus. After studying this chapter, you would be able to do all of the following:

1. **If given a mathematical formula representing a function of several variables, write formulas for the partial derivatives and for the differential of the function, and use these in applications such as the calculation of small changes in a dependent variable.**

2. **Perform a change of independent variables and obtain formulas relating different partial derivatives.**

3. **Use identities involving partial derivatives to eliminate undesirable quantities from thermodynamic formulas.**

4. **Identify an exact differential and an integrating factor.**

5. **Perform a line integral with two independent variables.**

6. **Perform a multiple integral.**

7. **Change independent variables in a multiple integral.**

8. **Use vector derivative operators.**

9. **Find constrained and unconstrained maximum and minimum values of a function of several variables.**

Section 6-2. DIFFERENTIALS AND PARTIAL DERIVATIVES

Representation of Functions.   Functions of several variables can be represented in the same ways as can functions of one variable. Both can be represented by mathematical formulas, as in Eq. (6.2). Both can be represented by graphs and by tables of values. Both can be represented by infinite series. However, graphs, tables, and series become more complicated when used for a function of several variables than with functions of a single variable.

Figure 6-1 shows the dependence of the pressure of a nearly ideal gas as a function of T and $\bar{V}$, defined as V/n. With only two axes on our graph, we have to resort to showing several members of a family of functions of $\bar{V}$, each for a different value of T.

Figure 6-2 attempts to represent the same information. The value of P is given by the height from the horizontal $\bar{V}$–T plane to a surface, which plays the same role as the curve in a two-dimensional graph. As you can see, it is easy to read quantitative information from the two-dimensional graph in Figure 6-1, but the perspective view in Figure 6-2 is more difficult to get numbers from. However, Figure 6-2 communicates more rapidly what the general features of the function are than does Figure 6-1.

If you have more than three variables, a graph is much more difficult to use. Sometimes, attempts are made to show roughly how functions of three variables depend on their independent variables by drawing a perspective view of three axes, one for each of the independent variables, and

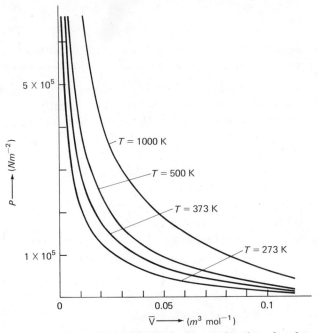

Figure 6-1. The pressure of a nearly ideal gas as a function of molar volume at various fixed temperature.

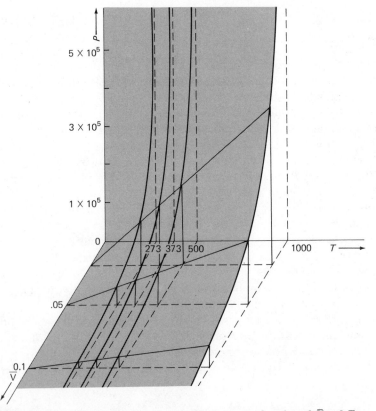

Figure 6-2. The pressure of a nearly ideal gas as a function of $\bar{V}$ and T.

then trying to communicate the approximate value of the dependent variable by a density of shading or a density of dots placed in the diagram.

Tables of values are also cumbersome with two or more independent variables, since a function is now not a set of ordered pairs of numbers but a set of ordered sets of three numbers, four numbers, etc. For two independent variables, we need a rectangular array, with values of one independent variable along the top and values of the other along one side, and values of the dependent variable in the body of the array. For a third independent variable, we would need a different sheet of paper for each value of the third variable.

Changes in a Function of Several Variables. Let us consider a gas contained in a cylinder with a movable piston. Let the cylinder have a valve through which additional gas can be admitted or through which gas can be removed, and let the entire system be immersed in a constant-temperature bath with an adjustable thermoregulator. For the present, we keep the valve closed.

Let us now make an infinitesimal change dV in the volume of the gas, keeping T fixed. Temporarily, P will behave just like a function of V alone, and we discussed in Chapter 3 how to determine the change in a function of one variable. The change in P is

$$dP = \left(\frac{dP}{dV}\right) dV \qquad (n \text{ and } T \text{ fixed}) \tag{6.3}$$

where dP/dV is the derivative of P with respect to V.

We now adopt a new notation, to remind us that P is not just a function of V but is also a function of n and T:

$$dP = \left(\frac{\partial P}{\partial V}\right)_{n,T} dV \qquad (n \text{ and } T \text{ fixed}) \tag{6.4}$$

The quantity $(\partial P/\partial V)_{n,T}$ is called the *partial derivative* of P with respect to V at constant n and T. It is obtained by the techniques of differentiation which we discussed in Chapter 3, with n and T treated like ordinary constants.

If P is given by the ideal gas equation of state, Eq. (6.2), then $(\partial P/\partial V)_{n,T}$ is given by

$$\left(\frac{\partial P}{\partial V}\right)_{n,T} = \left(\frac{\partial}{\partial V}\left[\frac{nRT}{V}\right]\right)_{n,T} = -\frac{nRT}{V^2} \tag{6.5}$$

After making an infinitesimal change in the volume, we can now make an infinitesimal change in the temperature, dT, keeping n and V fixed. P now depends on T in the same way as a function of a single variable, and we can write

$$dP = \left(\frac{\partial P}{\partial T}\right)_{n,V} dT \qquad (n \text{ and } V \text{ fixed}) \tag{6.6}$$

Here $(\partial P/\partial T)_{n,V}$ is called the partial derivative of P with respect to T at constant n and V, and is obtained by the usual techniques of differentiation, treating n and V like ordinary constants. If P is given by the ideal gas equation of state,

$$\left(\frac{\partial P}{\partial T}\right)_{n,V} = \frac{nR}{V} \tag{6.7}$$

If we now make a simultaneous change dV in the volume and dT in the temperature of the gas, we can simply regard this as a change in V with T fixed, followed by a change in T with V fixed, keeping n fixed. The change in P is thus the sum of the expressions in Eqs. (6.4) and (6.6).

$$dP = \left(\frac{\partial P}{\partial V}\right)_{n,T} dV + \left(\frac{\partial P}{\partial T}\right)_{n,V} dT \qquad (n \text{ fixed}) \tag{6.8}$$

Remember that each term of this equation is the change due to the change in only one independent variable, and each partial derivative thus is taken with the other independent variables treated as constants. The combined change in P can be written as a sum of two independent changes because the changes are infinitesimal.

If we make finite but small changes in T and V, we can write as an approximation

$$\Delta P \approx \left(\frac{\partial P}{\partial V}\right)_{n,T} \Delta V + \left(\frac{\partial P}{\partial T}\right)_{n,V} \Delta T \qquad (n \text{ fixed}) \tag{6.9}$$

We now relax our condition that n is held fixed, and consider simultaneous changes dT in T, dV in V, and dn in n. For infinitesimal changes, we can consider these changes to affect P one at a time, and write for the total change in P,

$$dP = \left(\frac{\partial P}{\partial V}\right)_{n,T} dV + \left(\frac{\partial P}{\partial T}\right)_{n,V} dT + \left(\frac{\partial P}{\partial n}\right)_{V,T} dn \tag{6.10}$$

This expression gives the total change in P. The infinitesimal change dP given by this expression is called the *differential* of P, or sometimes the total differential of P. Remember that it is a sum of terms, each of which gives the effect of one variable with the other treated as constants.

In general, if we have a function y that depends on n independent variables, $x_1, x_2, x_3, \ldots, x_n$, its differential is

$$dy = \sum_{i=1}^{n} \left(\frac{\partial y}{\partial x_i}\right)_{x_{j \neq i}} dx_i \tag{6.11}$$

where we have introduced an abbreviation for the subscripts indicating which variables are held fixed. The symbol $x_{j \neq i}$ stands for all of the variables except for x_i. To find each partial derivative, all of the variables except one are treated as constants in the differentiation.

The expression for dP for an ideal gas is

$$dP = -\frac{nRT}{V^2} dV + \frac{nR}{V} dT + \frac{RT}{V} dn \tag{6.12}$$

For small but finite changes, an approximate version of this can be written

$$\Delta P \approx -\frac{nRT}{V^2} \Delta V + \frac{nR}{V} \Delta T + \frac{RT}{V} \Delta n \tag{6.13}$$

Example 6-1

Use Eq. (6.13) to calculate approximately the change in pressure of an ideal gas if the volume is changed from 20.000 liters to 19.800 liters, the temperature is changed from 298.15 K to 299 K, and the number of moles is changed from 1.0000 to 1.0015. Compare with the correct value.

Solution

$$\Delta P \approx -\frac{(1.0000 \text{ mol})(8.3144 \text{ J K}^{-1} \text{ mol}^{-1})(298.15 \text{ K})}{(0.020000 \text{ m}^3)^2}(-0.200 \times 10^{-3} \text{ m}^3)$$

$$+ \frac{(1.0000 \text{ mol})(8.3144 \text{ J K}^{-1} \text{ mol}^{-1})}{0.020000 \text{ m}^3}(0.85 \text{ K})$$

$$+ \frac{(8.3144 \text{ J K}^{-1} \text{ mol}^{-1})(298.15 \text{ K})}{0.020000 \text{ m}^3}(0.0015 \text{ mol})$$

$$\approx 1.779 \times 10^3 \text{ N m}^{-2}$$

where we use the fact that 1 J = 1 N m. To get the correct answer, one must use consistent units. We have used SI units exclusively.

We can calculate the actual change as follows. Let the initial values of n, V, and T be called n_1, V_1, and T_1, and the final values be called n_2, V_2, and T_2.

$$\Delta P = P(n_2, V_2, T_2) - P(n_1, V_1, T_1) = \frac{n_2 R T_2}{V_2} - \frac{n_1 R T_1}{V_1}$$

The result of this calculation is $1.797 \times 10^3 \text{ N m}^{-2}$. Incidentally, the unit of pressure equal to 1 N m^{-2} is also called 1 pascal (1 Pa). •

In Example 6-1, the exact calculation could be made more easily than the approximation. However, in physical chemistry, it is frequently the case that a formula for a function is not known, but values for the partial derivatives have been determined.

Problem 6-1

The differential of the energy can be written as

$$dE = \left(\frac{\partial E}{\partial T}\right)_{P,n} dT + \left(\frac{\partial E}{\partial P}\right)_{T,n} dP + \left(\frac{\partial E}{\partial n}\right)_{P,T} dn \qquad (6.14)$$

for a sample containing only one pure substance.

If the sample is 1.000 mole (0.15384 kg) of carbon tetrachloride, CCl_4, experimental values of the first two partial derivatives are available:

$$\left(\frac{\partial E}{\partial T}\right)_{P,n} = 129.4 \text{ J K}^{-1} \text{ mol}^{-1}$$

$$\left(\frac{\partial E}{\partial P}\right)_{T,n} = 8.51 \times 10^{-4} \text{ J atm}^{-1} \text{ mol}^{-1}$$

where these values are for a temperature of 20 °C and a pressure of 1.000 atm.

Estimate the change in the energy of 1.000 mole of CCl_4 if its temperature is changed from 20 °C to 40 °C and its pressure from 1 atm to 100 atm. ●

Problem 6-2

The volume of a right circular cylinder is given by

$$V = \pi r^2 h$$

where r is the radius and h the height. Calculate the percentage error in the volume if the radius and the height are measured and a 1% error is made in each measurement in the same direction. Use the formula for the differential, and also direct substitution into the formula for the volume, and compare the two answers. ●

Section 6-3. CHANGE OF VARIABLES

In thermodynamics, there is usually the possibility of choosing between different sets of independent variables. For example, we can consider the energy of a one-component, one-phase system to be a function of T, V, and n:

$$E = E(T, V, n) \qquad (6.15)$$

or a function of T, P, and n:

$$E = E(T, P, n) \qquad (6.16)$$

The two choices lead to different expressions for the differential of E:

$$dE = \left(\frac{\partial E}{\partial T}\right)_{V,n} dT + \left(\frac{\partial E}{\partial V}\right)_{T,n} dV + \left(\frac{\partial E}{\partial n}\right)_{T,V} dn \qquad (6.17)$$

and

$$dE = \left(\frac{\partial E}{\partial T}\right)_{P,n} dT + \left(\frac{\partial E}{\partial P}\right)_{T,n} dP + \left(\frac{\partial E}{\partial n}\right)_{T,p} dn \qquad (6.18)$$

You should now be able to see why we are using subscripts to indicate what the other variables are. There are two different derivatives of E with respect to T: $(\partial E/\partial T)_{V,n}$ and $(\partial E/\partial T)_{P,n}$. If we did not use the subscripts, there would be no difference between the symbols for the two derivatives, which are different in value for most systems.

Example 6-2

Express the function $z = x(x, y) = ax^2 + bxy + cy^2$ in terms of x and u, where $u = xy$. Find the two partial derivatives $(\partial z/\partial x)_y$ and $(\partial z/\partial x)_u$.

Solution

$$z = z(x, u) = ax^2 + bu + \frac{cu^2}{x^2}$$

$$\left(\frac{\partial z}{\partial x}\right)_y = 2ax + by$$

$$\left(\frac{\partial z}{\partial x}\right)_u = 2ax - \frac{2cu^2}{x^3} = \left(\frac{\partial z}{\partial x}\right)_y - \frac{bu}{x} - \frac{2cu^2}{x^3}$$

In Example 6-2, there was no difficulty in obtaining an expression for the difference between $(\partial z/\partial x)_y$ and $(\partial z/\partial x)_u$, because we had the formula for the mathematical function. In thermodynamics, it is very unusual to have a functional form in front of us. More commonly we have measured values for partial derivatives, and require a separate means for computing the difference between partial derivatives.

We will obtain a formula

$$\left(\frac{\partial E}{\partial T}\right)_{V,n} = \left(\frac{\partial E}{\partial T}\right)_{P,n} + \, ? \tag{6.19}$$

where the question mark indicates the term which we now find. The procedure that we use is not mathematically rigorous, or even sensible. Regard it merely as a mnemonic device, and do not describe it to any mathematician, because it would probably make him laugh—or worse.

We want to construct the partial derivative on the left-hand side of our equation, so we begin with an expression for dE that contains the derivative on the right-hand side. This is Eq. (6.18). The first thing we do is to divide this differential expression by dT, because the derivative we want on the left-hand side is $(\partial E/\partial T)_{V,n}$. This cannot be done legitimately, because dT is an infinitesimal quantity, but we do it anyway. We get

$$\frac{dE}{dT} = \left(\frac{\partial E}{\partial T}\right)_{P,n} \frac{dT}{dT} + \left(\frac{\partial E}{\partial P}\right)_{T,n} \frac{dP}{dT} + \left(\frac{\partial E}{\partial n}\right)_{P,T} \frac{dn}{dT} \tag{6.20}$$

This equation contains several things that look like ordinary derivatives. However, we must interpret them as partial derivatives, since there are other independent variables besides T. We do this by changing the symbols to the symbols for partial derivatives and adding the appropriate subscripts to indicate the other variables. We choose whatever we want the other variables to be, but they must be the same in all four of the derivatives. The particular choice that we want is for the variables to be T, V, and n, so we write

$$\left(\frac{\partial E}{\partial T}\right)_{V,n} = \left(\frac{\partial E}{\partial T}\right)_{P,n} \left(\frac{\partial T}{\partial T}\right)_{V,n} + \left(\frac{\partial E}{\partial P}\right)_{T,n} \left(\frac{\partial P}{\partial T}\right)_{V,n} + \left(\frac{\partial E}{\partial n}\right)_{P,T} \left(\frac{\partial n}{\partial T}\right)_{V,n} \qquad (6.21)$$

The partial derivative of T with respect to T is equal to unity, no matter what is held constant, and the partial derivative of n with respect to anything is zero if n is held constant, so

$$\boxed{\left(\frac{\partial E}{\partial T}\right)_{V,n} = \left(\frac{\partial E}{\partial T}\right)_{P,n} + \left(\frac{\partial E}{\partial P}\right)_{T,n} \left(\frac{\partial P}{\partial T}\right)_{V,n}} \qquad (6.22)$$

Example 6-3

Apply the foregoing method to the function in Example 6-2, and find the relation between $(\partial z/\partial x)_u$ and $(\partial z/\partial x)_y$.

Solution

$$\left(\frac{\partial z}{\partial x}\right)_u = \left(\frac{\partial z}{\partial x}\right)_y + \left(\frac{\partial z}{\partial y}\right)_x \left(\frac{\partial y}{\partial x}\right)_u$$

$$\left(\frac{\partial z}{\partial y}\right)_x = bx + 2cy = bx + \frac{2cu}{x}$$

$$\left(\frac{\partial y}{dx}\right)_u = \left[\frac{\partial}{\partial x}\left(\frac{u}{x}\right)\right]_u = -\frac{u}{x^2}$$

Thus

$$\left(\frac{\partial z}{\partial x}\right)_u = \left(\frac{\partial z}{\partial x}\right)_y - \frac{bu}{x} - \frac{2cu}{x^3}$$

This agrees with Example 6-2, as it must. ●

Problem 6-3

Complete the following equations.
a. $(\partial H/\partial T)_{P,n} = (\partial H/\partial T)_{V,n} +$?
b. $(\partial S/\partial T)_{E,n} = (\partial S/\partial T)_{E,V} +$?
c. $(\partial z/\partial u)_{x,y} = (\partial z/\partial u)_{x,w} +$?
d. Apply the equation of part c to the case that $z = \cos(x/u) + e^{-y^2/u^2} + 4y/u$ and $w = y/u$. ●

Section 6-4. SOME USEFUL RELATIONS BETWEEN PARTIAL DERIVATIVES

It is fairly common in thermodynamics to have measured values for some partial derivatives, such as $(\partial H/\partial T)_{P,n}$, which is known as the heat capacity at constant pressure. Other partial derivatives are difficult or impossible to measure, and therefore have unknown values for particular systems. It is convenient to be able to eliminate such partial derivatives from equations.

In this section, we present some identities that can be used for this purpose. We have already obtained a method of getting a class of such identities. We will regard Eq. (6.22) and all analogous equations as the first kind of identity.

The next identity is that a derivative is equal to the reciprocal of the derivative with the role of dependent and independent variables reversed:

$$\left(\frac{\partial y}{\partial x}\right)_{z,u} = \frac{1}{(\partial x/\partial y)_{z,u}} \tag{6.23}$$

Notice that the same variables must be held constant in the two derivatives.

Example 6-4

Show that $\left(\dfrac{\partial P}{\partial V}\right)_{n,T} = \dfrac{1}{(\partial V/\partial P)_{n,T}}$ for an ideal gas.

Solution

$$\left(\frac{\partial P}{\partial V}\right)_{n,T} = -\frac{nRT}{V^2}$$

$$\frac{1}{(\partial V/\partial P)_{n,T}} = \frac{1}{-nRT/P^2} = -\frac{P^2}{nRT} = -\frac{(nRT/V)^2}{nRT} = -\frac{nRT}{V^2} \qquad \bullet$$

We will refer to Eq. (6.23) as the *reciprocal identity*.

Problem 6-4

Show that the reciprocal identity is satisfied by $(\partial z/\partial x)_y$ and $(\partial x/\partial z)_y$ if

$$z = \sin\left(\frac{x}{y}\right) \qquad \text{and} \qquad x = y\sin^{-1}(z) \qquad \bullet$$

The next identity is that if $z = z(x, y)$

$$\frac{\partial^2 z}{\partial y\,\partial x} = \frac{\partial^2 z}{\partial x\,\partial y} \tag{6.24}$$

CALCULUS WITH SEVERAL INDEPENDENT VARIABLES / Ch. 6

where the derivative on the left-hand side is the second derivative with respect to x and then with respect to y, and the derivative on the right-hand side is the second derivative with respect to y and then with respect to x.

$$\frac{\partial^2 z}{\partial y \, \partial x} = \left[\frac{\partial}{\partial y} \left(\frac{\partial z}{\partial x} \right)_y \right]_x \tag{6.25}$$

In addition to these mixed second partial derivatives, the second partial derivative with respect to a single variable occurs:

$$\left(\frac{\partial^2 z}{\partial x^2} \right)_y = \left[\frac{\partial}{\partial x} \left(\frac{\partial z}{\partial x} \right)_y \right]_y \tag{6.26}$$

In this case, y is held fixed in both differentiations, but in Eq. (6.25) y is held fixed for the first differentiation and x is held fixed for the second. Since both variables are shown in the symbol, the subscripts are usually omitted. However, if there is a third independent variable, it is listed as a subscript, as in

$$\left(\frac{\partial^2 E}{\partial V \, \partial T} \right)_n = \left[\frac{\partial}{\partial V} \left(\frac{\partial E}{\partial T} \right)_{V,n} \right]_{T,n} \tag{6.27}$$

Example 6-5

Show that

$$\left(\frac{\partial^2 P}{\partial V \, \partial T} \right)_n = \left(\frac{\partial^2 P}{\partial T \, \partial V} \right)_n$$

for an ideal gas.

Solution

$$\left(\frac{\partial^2 P}{\partial V \, \partial T} \right)_n = \left[\frac{\partial}{\partial V} \left(\frac{nR}{V} \right) \right]_{T,n} = -\frac{nR}{V^2}$$

$$\left(\frac{\partial^2 P}{\partial T \, \partial V} \right)_n = \left[\frac{\partial}{\partial T} \left(-\frac{nRT}{V^2} \right) \right]_{V,n} = -\frac{nR}{V^2} \qquad \bullet$$

Problem 6-5

Show that $(\partial^2 z / \partial y \, \partial x) = (\partial^2 z / \partial x \, \partial y)$ if

$$z = e^{-xy^2} \sin(x) \cos(y) \qquad \bullet$$

We will refer to the class of identities represented by Eq. (6.24) as *second-derivative* identities. An important set of such identities is the set of Maxwell relations of thermodynamics.

Another set of identities is the "cycle rule"

$$\boxed{\left(\frac{\partial y}{\partial x} \right)_z \left(\frac{\partial x}{\partial z} \right)_y \left(\frac{\partial z}{\partial y} \right)_x = -1} \tag{6.28}$$

We will "derive" this in the same mnemonic way as was used to obtain Eq. (6.22). We write the differential of y as a function of x and z:

$$dy = \left(\frac{\partial y}{\partial x}\right)_z dx + \left(\frac{\partial y}{\partial z}\right)_x dz \tag{6.29}$$

This equation delivers the value of dy corresponding to arbitrary infinitesimal changes in x and z, so it is still correct if we choose values of dz and dx such that dy vanishes. We now "divide" by dx, and interpret the "quotients" of differentials as partial derivatives, remembering that y is held fixed by our choice:

$$0 = \left(\frac{\partial y}{\partial x}\right)_z \left(\frac{\partial x}{\partial x}\right)_y + \left(\frac{\partial y}{\partial z}\right)_x \left(\frac{\partial z}{\partial x}\right)_y \tag{6.30}$$

Since the partial derivative of x with respect to x is equal to unity, this equation becomes identical with Eq. (6.28) when the reciprocal identity is used.

Problem 6-6

For the particular function $y = x^2/z$, show that Eq. (6.28) is correct. •

The final identity that we present in this section is the partial derivative version of the chain rule

$$\boxed{\left(\frac{\partial z}{\partial y}\right)_{u,v} = \left(\frac{\partial z}{\partial x}\right)_{u,v} \left(\frac{\partial x}{\partial y}\right)_{u,v}} \tag{6.31}$$

This is very similar to Eq. (3.26). Notice that the same variables must be held fixed in all three derivatives.

Example 6-6

Show that if

$$z = ax^2 + bwx$$

and

$$x = uy$$

then Eq. (6.31) is correct.

Solution

$$\left(\frac{\partial z}{\partial x}\right)_{u,w} \left(\frac{\partial x}{\partial y}\right)_{u,w} = (2ax + bw)(u) = 2au^2 y + buw$$

$$\left(\frac{\partial z}{\partial y}\right)_{u,w} = \left[\frac{\partial}{\partial y}(au^2 y^2 + bwuy)\right]_{u,w} = 2au^2 y + buw \qquad •$$

Example 6-7

The following are commonly measured quantities:

Heat capacity at constant pressure $= \left(\dfrac{\partial H}{\partial T}\right)_{P,n} = T\left(\dfrac{\partial S}{\partial T}\right)_{P,n} = C_P$

Heat capacity at constant volume $= \left(\dfrac{\partial E}{\partial T}\right)_{V,n} = T\left(\dfrac{\partial S}{\partial T}\right)_{V,n} = C_V$

Isothermal compressibility $= -\dfrac{1}{V}\left(\dfrac{\partial V}{\partial P}\right)_{T,n} = K_T$

Adiabatic compressibility $= -\dfrac{1}{V}\left(\dfrac{\partial V}{\partial P}\right)_{S,n} = K_S$

Prove that

$$\frac{C_P}{C_V} = \frac{K_T}{K_S}$$

Solution

$$\frac{C_P}{C_V} = \frac{(\partial S/\partial T)_{P,n}}{(\partial S/\partial T)_{V,n}} = \frac{-(\partial S/\partial P)_{T,n}(\partial P/\partial T)_{S,n}}{-(\partial S/\partial V)_{T,n}(\partial V/\partial T)_{S,n}}$$

where we have used the cycle rule on the numerator and the denominator. Next we use the reciprocal identity to write

$$\frac{C_P}{C_V} = \frac{(\partial V/\partial S)_{T,n}(\partial S/\partial P)_{T,n}}{(\partial V/\partial T)_{S,n}(\partial T/\partial P)_{S,n}}$$

Next we use the chain rule to write

$$\frac{C_P}{C_V} = \frac{(\partial V/\partial P)_{T,n}}{(\partial V/\partial P)_{S,n}} = \frac{-(1/V)(\partial V/\partial P)_{T,n}}{-(1/V)(\partial V/\partial P)_{S,n}} = \frac{K_T}{K_S} \qquad \bullet$$

Section 6-5.   EXACT AND INEXACT DIFFERENTIALS

In the earlier sections of this chapter, we have discussed the differential of a function. Such a differential is called an *exact differential*. A general differential form can be written

$$du = M(x, y)\,dx + N(x, y)\,dy \qquad (6.32)$$

If this is the differential of a function, then M and N will be the derivatives of that function. However, if M and N are not the derivatives of the same function, then du is not the differential of a function, and is called an *inexact differential*. It is an infinitesimal quantity, but it is not the change in any function of x and y.

In order to tell whether some differential form is an exact differential or not, we must have a way to tell whether the coefficients are the derivatives of the same function or not. The second derivative identity furnishes a means to do this. If there exists a function

$$u = u(x, y)$$

such that

$$M(x, y) = \left(\frac{\partial u}{\partial x}\right)_y$$

and

$$N(x, y) = \left(\frac{\partial u}{\partial y}\right)_x$$

then from the second derivative identity

$$\frac{\partial^2 u}{\partial x \, \partial y} = \frac{\partial^2 u}{\partial y \, \partial x}$$

which means that

$$\boxed{\left(\frac{\partial N}{\partial x}\right)_y = \left(\frac{\partial M}{\partial y}\right)_x} \qquad \textbf{(6.33)}$$

Equation (6.33) represents a necessary and sufficient condition for the differential of Eq. (6.32) to be exact. That is, if the differential is exact, Eq. (6.33) will be obeyed, and if Eq. (6.33) is obeyed, the differential will be exact.

Example 6-8

Show that the differential

$$du = \left(2xy + \frac{9x^2}{y}\right)dx + \left(x^2 - \frac{3x^2}{y^2}\right)dy$$

is exact.

Solution

$$\left[\frac{\partial}{\partial y}\left(2xy + \frac{9x^2}{y}\right)\right]_x = 2x - \frac{9x^2}{y^2}$$

$$\left[\frac{\partial}{\partial x}\left(x^2 - \frac{3x^3}{y^2}\right)\right]_y = 2x - \frac{9x^2}{y^2}$$

so the differential must be exact. ●

Problem 6-7 .

Determine whether each of the following is an exact differential.
a. $du = (2ax + by^2) dx + (bxy) dy$
b. $du = (x + y) dx + (x + y) dy$
c. $du = (x^2 + 2x + 1) dx + (y^2 + 25y + 24) dy$ ●

Differentials with three or more terms can also either be exact or inexact, and the second derivative identity also provides a test for such differentials. For example, if

$$du = M(x, y, z)\,dx + N(x, y, z)\,dy + P(x, y, z)\,dz \qquad (6.34)$$

then in order for this to be an exact differential, we must have all of the equations obeyed:

$$\left(\frac{\partial M}{\partial y}\right)_{x,z} = \left(\frac{\partial N}{\partial x}\right)_{y,z} \qquad (6.35a)$$

$$\left(\frac{\partial N}{\partial z}\right)_{x,y} = \left(\frac{\partial P}{\partial y}\right)_{x,z} \qquad (6.35b)$$

and

$$\left(\frac{\partial M}{\partial z}\right)_{x,y} = \left(\frac{\partial P}{\partial x}\right)_{y,z} \qquad (6.35c)$$

Problem 6-8

Show that the following is not an exact differential

$$du = (2y)\,dx + (x)\,dy + \cos(z)\,dz \qquad \bullet$$

There are two important inexact differentials in thermodynamics. If a system undergoes an infinitesimal process (one in which the independent variables specifying the state of the system change infinitesimally), dq is the amount of heat transferred to the system, and dw is the amount of work done on the system. For a simple system and reversible processes,

$$dw = -P\,dV \qquad (6.36)$$

Example 6-9

Show that for an ideal gas undergoing reversible processes with n fixed, dw is inexact.

Solution

We choose T and V as our independent variables.

$$dw = M(T, V, n)\,dT + N(T, V, n)\,dV \qquad (n \text{ fixed}) \qquad (6.37)$$

Comparison with Eq. (6.36) shows that $M = 0$ and $N = P = nRT/V$. We apply the test for exactness, Eq. (6.33)

$$\left(\frac{\partial M}{\partial V}\right)_{T,n} = 0$$

$$\left(\frac{\partial N}{\partial T}\right)_{V,n} = \left[\frac{\partial}{\partial T}\left(\frac{nRT}{V}\right)\right]_{V,n} = \frac{nR}{V} \neq 0 \qquad \bullet$$

Problem 6-9

To an excellent approximation, the energy of a monatomic ideal gas is given by

$$E = \frac{3nRT}{2} \tag{6.38}$$

Find the partial derivatives and write the expression for dE using V, T, and n as independent variables. Show that the partial derivatives obey Eqs. (6.35). ●

Integrating Factors. Some inexact differentials are related to exact differentials in that an exact differential is produced if the inexact differential is multiplied by a particular function called an *integrating factor* for that differential.

Example 6-10

Show that the differential

$$du = (2ax^2 + bxy)\,dx + (bx^2 + 2cxy)\,dy$$

is inexact, but that $1/x$ is an integrating factor, so that du/x is exact.

Solution

$$\left[\frac{\partial}{\partial y}(2ax^2 + bxy)\right]_x = bx$$

$$\left[\frac{\partial}{\partial x}(bx^2 + 2cxy)\right]_y = 2bx + 2cy \neq bx$$

so du is inexact.

$$\left[\frac{\partial}{\partial y}(2ax + by)\right]_x = b$$

$$\left[\frac{\partial}{\partial x}(bx + 2cy)\right]_y = b$$

so $(1/x)\,du$ is exact. ●

Problem 6-10

Show that the differential

$$(1 + x)\,dx + \left[\frac{x\ln(x)}{y} + \frac{x^2}{y}\right]dy$$

is inexact, and that y/x is an integrating factor. ●

There is no general method for finding an integrating factor, although we will discuss a method that will work for a particular class of differential forms in Chapter 7, when we discuss differential equations. However, it is true that if a differential possesses one integrating factor, there are infinitely many integrating factors.

Section 6-6. LINE INTEGRALS

In Chapter 4, we found that a finite increment in a function could be constructed by the process of integration. If x_1 and x_0 were particular values of the independent variable x, then

$$F(x_1) - F(x_0) = \int_{x_0}^{x_1} f(x)\,dx = \int_{x_0}^{x_1} dF \tag{6.39}$$

where

$$f(x) = \frac{dF}{dx} \tag{6.40}$$

We can think of the integral of Eq. (6.39) as being a sum of infinitesimal increments equal to $f(x)\,dx$.

We now consider the analogous process for a differential with two or more independent variables. For two independent variables, x and y, we might try to define

$$\int_{x_0,y_0}^{x_1,y_1} du = \int_{x_0,y_0}^{x_1,y_1} [M(x, y)\,dx + N(x, y)]\,dy \tag{6.41}$$

However, this integral is not yet well defined. The situation is shown schematically in Figure 6-3. For a pair of variables, (x_0, y_0) represents one point in the x–y plane, and (x_1, y_1) represents another point in the plane, and many paths join the two points.

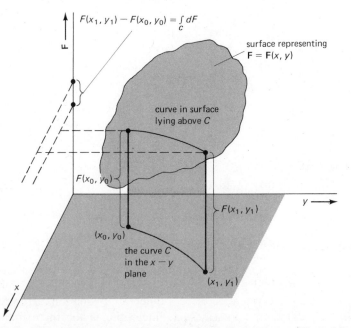

Figure 6-3.   Diagram illustrating the line integral of an exact differential.

In order to complete the definition of the integral in Eq. (6.41), we must specify the path in the x–y plane which joins the point (x_0, y_0) and the point (x_1, y_1). We introduce the notation

$$\int_C du = \int_C [M(x, y)\,dx + N(x, y)\,dy] \qquad (6.42)$$

where the letter C stands for the piece of a curve joining the two points. The integral is called a *line integral*.

We can think of Eq. (6.41) as representing a sum of many infinitesimal contributions, each one given by the appropriate infinitesimal value of du resulting when x is changed by dx and y is changed by dy. However, these changes dx and dy are not independent. They must be related so that we remain on the chosen curve during the integration process.

A curve in the x–y plane specifies y as a function of x, or x as a function of y. For a given curve, we can write

$$y = y(x) \qquad (6.43)$$

or

$$x = x(y) \qquad (6.44)$$

In order to calculate a line integral such as that of Eq. (6.42), we replace y in $M(x, y)$ by the function given in Eq. (6.43), and we replace x in $N(x, y)$ by the function given in Eq. (6.44). With this replacement, M is a function of x only, and N is a function of y only, and each term becomes an ordinary one-variable integral:

$$\int_C du = \int_{x_0}^{x_1} M[x, y(x)]\,dx + \int_{y_0}^{y_1} N[y, x(y)]\,dy \qquad (6.45)$$

In these integrals, specification of the curve C determines not only what the beginning point (x_0, y_0) and the final point (x_1, y_1) are, but also what the functions are which replace y in the dx integral and x in the dy integral.

Example 6-11

Find the value of the line integral

$$\int_C dF = \int_C [(2x + 3y)\,dx + (3x + 4y)]\,dy$$

where C is the straight-line segment given by $y = 2x + 3$ from $(0, 3)$ to $(2, 7)$.

Solution

In the first term, y must be replaced by $2x + 3$, and in the second term x must be replaced by $(1/2)(y - 3)$.

$$\int_C dF = \int_0^2 [2x + 3(2x + 3)]\,dx + \int_3^7 \left[\frac{3}{2}(y - 3) + 4y\right] dy$$

$$= \left(\frac{8x^2}{2} + 9x\right)\Bigg|_0^2 + \left(\frac{\frac{11}{2}y^2}{2} - \frac{9}{2}y\right)\Bigg|_3^7 = 126 \qquad \bullet$$

There is an important theorem, which we now state: *If du is an exact differential, then $\int_C du$ depends only on the initial and final points, and not on the choice of curve joining these points. Further, since the exact differential is the differential of a function, the line integral will have a value equal to the value of the function at the final point minus the value of the function at the beginning point:*

$$\int_C du = \int_C \left[\left(\frac{\partial u}{\partial x} \right) dx + \left(\frac{\partial u}{\partial y} \right) dy \right] = u(x_1, y_1) - u(x_0, y_0) \qquad (6.46)$$

Notice that in this case, we have written M and N as the partial derivatives which they must be equal to in order for du to be exact (see Section 6-5). If du is not an exact differential, there is no such things as a function u, and the line integral will depend not only on the beginning and ending points, but also on the curve of integration joining these points.

Example 6-12

Show that the line integral of Example 6-11 has the same value as the line integral of the same differential on the rectangular path from $(0, 3)$ to $(2, 3)$ and then to $(2, 7)$.

Solution

The path of this integration is not a single curve but two line segments, so we must carry out the integration separately for each segment. This is actually a simplification, because on the first line segment, y is constant, so $dy = 0$ and the dy integral vanishes. On the second line segment, x is constant, so $dx = 0$ and the dx integral vanishes. Therefore,

$$\int_C dF = \int_0^2 (2x + 9) \, dx + \int_3^7 (6 + 4y) \, dy$$

This follows from the fact that $y = 3$ on the first line segment, and from the fact that $x = 2$ on the second line segment. Performing the integrals yields

$$\int_C dF = \left(\frac{2x^2}{2} + 9x \right) \Big|_0^2 + \left(6y + \frac{4y^2}{2} \right) \Big|_3^7 = 126 \qquad \bullet$$

Problem 6-11

a. Show that the following differential is exact:
$$dz = (ye^{xy}) \, dx + (xe^{xy}) \, dy$$

b. Calculate the line integral $\int_C dz$ on the line segment from $(0,0)$ to $(2,2)$. On this line segment, $y = x$.

c. Calculate the line integral $\int_C dz$ on the path going from $(0,0)$ to $(0,2)$ and then to $(2,2)$ (a rectangular path). $\qquad \bullet$

Example 6-13

Show that the differential

$$du = dx + x\,dy$$

is inexact, and carry out the line integral from $(0, 0)$ to $(2, 2)$ by the two different paths: path 1: The straight-line segment from $(0, 0)$ to $(2, 2)$, and path 2: the rectangular path from $(0, 0)$ to $(2, 0)$ and then to $(2, 2)$.

Solution

Test for exactness:

$$\left[\frac{\partial}{\partial y}(1)\right]_x = 0$$

$$\left[\frac{\partial}{\partial x}(x)\right]_y = 1 \neq 0$$

Path 1:

$$\int_{C_1} du = \int_{C_1} dx + \int_{C_1} x\,dy = \int_0^2 dx + \int_0^2 y\,dy$$

where we obtained the second integral by using the fact that $y = x$ on the straight-line segment of path 1.

$$\int_{C_1} du = x\Big|_0^2 + \frac{y^2}{2}\Big|_0^2 = 4$$

Path 2:

$$\int_{C_1} du = x\Big|_0^2 + \frac{y^2}{2}\Big|_0^2 = 2 + 2 = 4$$

$$\int_{C_2} du = \int_{C_2} dx + \int_{C_2} x\,dy = \int_0^2 dx + \int_0^2 2\,dy$$

$$= x\Big|_0^2 + 2y\Big|_0^2 = 2 + 4 = 6$$

The two line integrals have the same beginning point and the same ending point, but are not equal, because the differential is not an exact differential. •

Problem 6-12

Carry out the two line integrals of du from Example 6-13 from $(0, 0)$ to (x_1, y_1):

On path 1:   Rectangular path from $(0, 0)$ to $(0, y_1)$ and then to (x_1, y_1)

On path 2:   Rectangular path from $(0, 0)$ to $(x_1, 0)$ and then to (x_1, y_1) •

In thermodynamics, line integrals sometimes occur that begin and end at the same point. These are used to represent a cyclic process, which begins and ends at the same state of a system. Such a line integral is sometimes denoted by the symbol

$$\oint du$$

Since the beginning and final points are the same, such an integral must vanish if du is an exact differential:

$$\oint du = 0 \qquad \text{(if } du \text{ is exact)} \tag{6.47}$$

If du is inexact, the line integral that begins and ends at the same point will not generally be equal to zero.

There are also line integrals of functions of three independent variables. If $u = u(x, y, z)$, the line integral of the exact differential du is

$$\int_C du = \int_C [M(x, y, z)\, dx + N(x, y, z)\, dy + P(x, y, z)\, dz] \tag{6.48}$$

where C specifies a curve that gives y and z as functions of x, or x and y as functions of z, or x and z as functions of y. If the beginning point of the curve C is (x_0, y_0, z_0) and the ending point is (x_1, y_1, z_1), the line integral is

$$\int_C du = u(x_1, y_1, z_1) - u(x_0, y_0, z_0)$$
$$= \int_{x_0}^{x_1} M[x, y(x), z(x)]\, dx + \int_{y_0}^{y_1} N[x(y), y, z(y)]\, dy$$
$$+ \int_{z_0}^{z_1} P[x(z), y(z), z]\, dz \tag{6.49}$$

The line integral of an inexact differential is given by a similar formula, but is of course not equal to the increment of a function.

Section 6-7. MULTIPLE INTEGRALS

A *double integral*, which is the simplest kind of multiple integral, is written in the form

$$I = \int_{a_1}^{a_2} \int_{b_1}^{b_2} f(x, y)\, dy\, dx \tag{6.50}$$

where $f(x, y)$ is the integrand function, a_1 and a_2 are the limits of the x integration, and b_1 and b_2 are the limits of the y integration.

The double integral is carried out as follows: The "inside" integration is done first. This is the integration over the values of the variable whose differential and limits are written closest to the integrand function. During this integration, the other independent variable is treated as a constant. The result of the first integration is a function of the remaining variable, which is the integrand for the remaining integration.

Example 6-14

Evaluate the double integral

$$\int_0^a \int_0^b (x^2 + 4xy)\, dy\, dx$$

Solution

The inside integration gives, treating x as a constant,

$$\int_0^b (x^2 + 4xy)\, dy = \left(x^2 y + \frac{4xy^2}{2}\right)\Big|_0^b = bx^2 + 2b^2 x$$

This becomes the integrand or the final integration,

$$I = \int_0^a (bx^2 + 2b^2 x)\, dx = \left(\frac{bx^3}{3} + \frac{2b^2 x^2}{2}\right)\Big|_0^a$$

$$= \frac{ba^3}{3} + b^2 a^2 \qquad \bullet$$

Since the result of the inside integration is the integrand for the other integration, the limits of the inside integration can be functions of the variable of the other integration.

Example 6-15

Evaluate the double integral

$$\int_0^a \int_0^{3x} (x^2 + 2xy + y^2)\, dy\, dx$$

Solution

The result of the inside integration is

$$\int_0^{3x} (x^2 + 2xy + y^2)\, dy = \left(x^2 y + \frac{2xy^2}{2} + \frac{y^3}{3}\right)\Big|_0^{3x}$$

$$= 3x^3 + 9x^3 + 9x^3 = 21x^3$$

The x integration thus gives

$$\int_0^a 21x^3\, dx = \frac{21x^4}{4}\Big|_0^a = \frac{21a^4}{4} \qquad \bullet$$

Problem 6-13

Evaluate the double integral

$$\int_2^4 \int_0^\pi x \sin(y)\, dy\, dx \qquad \bullet$$

In Section 4-3, we discussed the fact that a definite integral with one independent variable is equal to the area between the axis and the integrand curve and between the two limits of integration.

A double integral is equal to a volume in an analogous way. This is illustrated in Figure 6-4, which is drawn to correspond to Example 6-16. In the x–y plane, we have an infinitesimal element of area $dx\, dy$, drawn in the figure as though it were finite in size. The vertical distance from the x–y plane to the surface representing the integrand function is the value of the integrand function, so that the volume of the small box shown is $f(x)\, dx\, dy$.

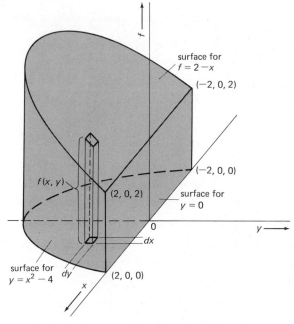

Figure 6-4. The diagram for Example 6-16.

The double integral is the sum of the volume of all such infinitesimal boxes, and thus the volume of the solid bounded by the x–y plane, the surface representing the integrand function, and the limits of integration. If the integrand function is negative in part of the region of integration, we must take the volume above the x–y plane minus the volume below the plane as equal to the integral.

Example 6-16

Calculate the volume of the solid shown in Figure 6-4. The bottom of the solid is the x–y plane. The flat surface corresponds to $y = 0$, the curved vertical surface corresponds to $y = x^2 - 4$, and the top of the solid corresponds to $f = 2 - x$.

Solution

We carry out a double integral with $f = 2 - x$ as the integrand:

$$V = \int_{-2}^{2} \int_{x^2-4}^{0} (2 - x)\, dx\, dy$$

The inside integral is

$$\int_{x^2-4}^{0} (2 - x)\, dy = x^3 - 2x^2 - 4x + 8$$

so that

$$V = \int_{-2}^{2} (x^3 - 2x^2 - 4x + 8)\, dx = \tfrac{64}{3}$$

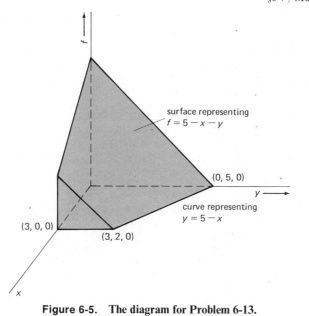

Figure 6-5. The diagram for Problem 6-13.

Problem 6-14

Find the volume of the solid object shown in Figure 6-5. The top of the object corresponds to $f = 5 - x - y$, the bottom of the object is the x–y plane, the trapezoidal face is the x–f plane, and the large triangular face is the y–f plane. The small triangular face corresponds to $x = 3$.   •

Multiple integrals with three or more independent variables also occur. For example, if we have an integrand function depending on x, y, and z, we might have the *triple integral*

$$I = \int_{a_1}^{a_2} \int_{b_1}^{b_2} \int_{c_1}^{c_2} f(x, y, z)\, dz\, dy\, dx \tag{6.51}$$

To evaluate the integral, we first integrate z from c_1 to c_2 and take the result as the integrand for the double integral over y and x. Then we integrate y from b_1 to b_2, and take the result as the integrand for the integral over x from a_1 to a_2.

The limits c_1 and c_2 can depend on y and x, and the limits b_1 and b_2 can depend on x, but a_1 and a_2 cannot depend on x or on y or on z.

Example 6-17

Find the triple integral

$$I = A^2 \int_0^a \int_0^b \int_0^c \sin^2\left(\frac{n\pi x}{a}\right) \sin^2\left(\frac{m\pi y}{b}\right) \sin^2\left(\frac{k\pi z}{c}\right) dz\, dy\, dx$$

This is a normalization integral from quantum mechanics. The integral is equal to the total probability of finding a particle confined to a box whose boundaries are the limits of integration, and it is customary to choose the value of the constant A so that the integral equals unity. The quantities m, n, and k are integral quantum numbers specifying the state of the particle.

Solution

This is a particularly simple triple integral, because the integrand function is a product of three factors, each of which depends on only one variable. The entire integral can therefore be written in factorized form:

$$I = A^2 \left[\int_0^a \sin^2\left(\frac{n\pi x}{a}\right) dx \right]\left[\int_0^b \sin^2\left(\frac{m\pi y}{b}\right) dy \right]\left[\int_0^c \sin^2\left(\frac{k\pi z}{c}\right) dz \right]$$

We first carry out the z integration, using the substitution $u = k\pi z/c$.

$$\int_0^c \sin^2\left(\frac{k\pi z}{c}\right) dz = \frac{c}{k\pi} \int_0^{k\pi} \sin^2(u)\, du$$

The integrand is a periodic function, so that the integral from 0 to $k\pi$ is just k times the integral from 0 to π, which is given as Eq. (8) of Appendix 4:

$$\int_0^c \sin^2\left(\frac{k\pi z}{c}\right) dz = \frac{c}{k\pi}\,(k)\,\frac{\pi}{2} = \frac{c}{2}$$

The other integrals are similar, except for having a or b instead of c, so that

$$I = A^2 \frac{abc}{8} \qquad \bullet$$

You should look carefully at how the integral in Example 6-17 was factorized, because many triple integrals in quantum mechanics are factorized in this way.

Problem 6-15

Find the value of the constant A so that the following integral equals unity.

$$A \int_{-\infty}^{\infty} \int_{-\infty}^{\infty} e^{-x^2 - y^2}\, dy\, dx \qquad \bullet$$

Changing Variables in Multiple Integrals.   We have discussed several multiple integrals whose independent variables were cartesian coordinates. Sometimes it is convenient to take a multiple integral over an area or over a volume using polar coordinates, or spherical polar coordinates, etc.

We first discuss how this is done in polar coordinates, as shown in Figure 2-6. We now require an infinitesimal element of area dA given in terms of the coordinates ρ and ϕ. This element of area is shown in Figure 6-6. Notice that one dimension of the element of area is $d\rho$, but the other dimension is $\rho\, d\phi$, from the fact that an arc length is the radius of the circle times the angle subtended by the arc. Thus, the element of area is $\rho\, d\phi\, d\rho$. Notice that if the element of area were finite, it would not quite be rectangular, and this formula would not be exact.

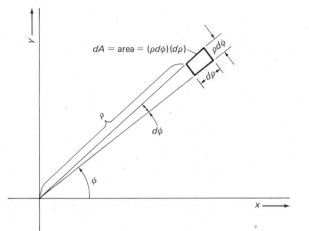

Figure 6-6. An infinitesimal element of area in plane polar coordinates.

We can think of the plane as being covered completely by infinitely many such elements of area, and a double integral over some region of the plane is just the sum of the value of the integrand function at each element of area times the area of the element.

Example 6-18

In cartesian coordinates, the wave function for the ground state of a two-dimensional harmonic oscillator is

$$\psi = B \exp[-a(x^2 + y^2)]$$

Transform this to plane polar coordinates and find the value of B such that the integral of ψ^2 over the entire x–y plane is equal to unity.

Solution

$$\psi = Be^{-a\rho^2}$$

The integral that is to equal unity is

$$B^2 \int_0^\infty \int_0^{2\pi} e^{-2a\rho^2} \rho \, d\phi \, d\rho$$

The integral can be factorized.

$$1 = B^2 \int_0^\infty e^{-2a\rho^2} \rho \, d\rho \int_0^{2\pi} d\phi = 2\pi B^2 \int_0^\infty e^{-2a\rho^2} \rho \, d\rho$$

The ρ integral is done by the method of substitution, letting $u = 2a\rho^2$ and $du = 4a\rho \, d\rho$. We obtain

$$1 = 2\pi B^2 \left(\frac{1}{4a}\right) \int_0^\infty e^{-u} \, du = \frac{B^2 \pi}{2a}$$

$$B = \sqrt{\frac{2a}{\pi}}$$

●

Problem 6-16

Use a double integral to find the volume of a cone of height h and radius a at the base. If the cone is standing with its point upward and with its base centered at the origin, the equation giving the surface of the cone is, in polar coordinates

$$f = h\left(1 - \frac{\rho}{a}\right) \qquad \bullet$$

In transforming from cartesian to plane polar coordinates, the factor ρ which is used with the product of the differentials $d\phi\, d\rho$ is called a *jacobian*, and the symbol $\partial(x, y)/\partial(\rho, \phi)$ is used for this particular jacobian:

$$\iint f(x, y)\, dx\, dy = \iint f(\rho, \phi)\rho\, d\phi\, d\rho = \iint f(\rho, \phi)\frac{\partial(x, y)}{\partial(\rho, \phi)}\, d\rho\, d\phi \quad (6.52)$$

We will not discuss the mathematical theory, but this jacobian is given by

$$\frac{\partial(x, y)}{\partial(\rho, \phi)} = \begin{vmatrix} \partial x/\partial \rho & \partial x/\partial \phi \\ \partial y/\partial \rho & \partial y/\partial \phi \end{vmatrix} \qquad (6.53)$$

The right-hand side of Eq. (6.53) is called a *determinant*. Determinants are discussed in Chapter 8. All we require now is Eq. (8.57a) for a 2 by 2 determinant:

$$\begin{vmatrix} \partial x/\partial \rho & \partial x/\partial \phi \\ \partial y/\partial \rho & \partial y/\partial \phi \end{vmatrix} = \begin{vmatrix} \cos(\phi) & -\rho\sin(\phi) \\ \sin(\phi) & \rho\cos(\phi) \end{vmatrix}$$

$$= \rho\cos^2(\phi) + \rho\sin^2(\phi) = \rho \qquad (6.54)$$

where we have also used Eq. (7) of Appendix 5.

Equation (6.54) gives us the same result as we had before.

$$dA = \text{element of area} = \rho\, d\phi\, d\rho \qquad (6.55)$$

The jacobian for transformation of coordinates in three dimensions is quite similar. If u, v, and w are some set of coordinates such that

$$x = x(u, v, w)$$
$$y = y(u, v, w)$$
$$z = z(u, v, w)$$

then the jacobian for the transformation of a multiple integral from cartesian coordinates to the coordinates u, v, and w is

$$\frac{\partial(x, y, z)}{\partial(u, v, w)} = \begin{vmatrix} \partial x/\partial u & \partial x/\partial v & \partial x/\partial w \\ \partial y/\partial u & \partial y/\partial v & \partial y/\partial w \\ \partial z/\partial u & \partial z/\partial v & \partial z/\partial w \end{vmatrix} \qquad (6.56)$$

Example 6-19

Obtain the jacobian for the transformation from cartesian coordinates to spherical polar coordinates.

Solution

The equations relating the coordinates are Eqs. (2.81), (2.82), and (2.83). From these

$$\frac{\partial(x, y, z)}{\partial(r, \theta, \phi)} = \begin{vmatrix} \sin(\theta)\cos(\phi) & r\cos(\theta)\cos(\phi) & -r\sin(\theta)\sin(\phi) \\ \sin(\theta)\sin(\phi) & r\cos(\theta)\sin(\phi) & r\sin(\theta)\cos(\phi) \\ \cos(\theta) & -r\sin(\theta) & 0 \end{vmatrix}$$

We now have Eq. (8.57b) for a 3 by 3 determinant, giving us

$$\frac{\partial(x, y, z)}{\partial(r, \theta, \phi)} = \cos(\theta)[r^2\cos(\theta)\sin(\theta)\cos^2(\phi) + r^2\sin(\theta)\cos(\theta)\sin^2(\phi)]$$

$$+ r\sin(\theta)[r\sin^2(\theta)\cos^2(\phi) + r\sin^2(\theta)\sin^2(\phi)]$$

$$= r^2\sin(\theta)\cos^2(\theta) + r^2\sin^3(\theta) = r^2\sin(\theta) \qquad (6.57)$$

where we have used Eq. (7) of Appendix 5 several times. •

Problem 6-17

Find the jacobian for the transformation from cartesian to cylindrical polar coordinates. •

A triple integral in cartesian coordinates is transformed into a triple integral in spherical polar coordinates by

$$\iiint f(x, y, z)\,dx\,dy\,dz = \iiint f(r, \theta, \phi)r^2\sin(\theta)\,d\phi\,d\theta\,dr \qquad (6.58)$$

or, equivalently, an element of volume is given by

$$dV = dx\,dy\,dz = r^2\sin(\theta)\,d\phi\,d\theta\,dr \qquad (6.59)$$

To complete the transformation, the limits on r, θ, and ϕ must be found so that they correspond to the limits on x, y, and z. Sometimes the purpose of transforming to spherical polar coordinates is to avoid the task of finding the limits in cartesian coordinates when they can be expressed easily in spherical polar coordinates.

Section 6-8.   VECTOR DERIVATIVE OPERATORS

An operator is a symbol for carrying out a mathematical operation (see Chapter 8). The gradient operator is defined by

$$\nabla = \mathbf{i}\frac{\partial}{\partial x} + \mathbf{j}\frac{\partial}{\partial y} + \mathbf{k}\frac{\partial}{\partial z} \qquad (6.60)$$

where **i**, **j**, and **k** are the unit vectors defined in Chapter 2. The symbol ∇, which is an upside-down capital Greek delta, is called "del." If f is some scalar function of x, y, and z, the gradient of f is

$$\nabla f = \mathbf{i}\left(\frac{\partial f}{\partial x}\right) + \mathbf{j}\left(\frac{\partial f}{\partial y}\right) + \mathbf{k}\left(\frac{\partial f}{\partial z}\right) \tag{6.61}$$

The gradient of f is sometimes denoted by grad f instead of ∇f.

Example 6-20

Find the gradient of the function

$$f = x^2 + 3xy + z^2 \sin\left(\frac{x}{y}\right)$$

Solution

$$\nabla f = \mathbf{i}\left[2x + 3y + \frac{z^2}{y}\cos\left(\frac{x}{y}\right)\right] + \mathbf{j}\left[3x - \frac{xz^2}{y^2}\cos\left(\frac{x}{y}\right)\right] + \mathbf{k}2z\sin\left(\frac{x}{y}\right) \quad \bullet$$

Problem 6-18

Find the gradient of the function

$$g = ax^3 + ye^{bz}$$

where a and b are constants. $\qquad\qquad\bullet$

As you can see, the gradient of a scalar function is a vector. Its direction is the direction in which the function is increasing most rapidly, and its magnitude is the rate of change of the function in that direction.

A common example of gradient is found in mechanics. In a conservative system, the force on a particle is given by

$$F = -\nabla V \tag{6.62}$$

where V is the potential energy of the entire system. The gradient is taken with respect to the coordinates of the particle being considered. The coordinates of any other particles are held fixed in the differentiations.

Example 6-21

The potential energy of an object of mass m near the surface of the earth is

$$V = mgz$$

where g is the acceleration due to gravity. Find the gravitational force on the object.

Solution

$$\mathbf{F} = -\mathbf{k}mg \qquad\qquad\bullet$$

Problem 6-19

Neglecting the attractions of all other celestial bodies, the gravitational potential energy of the earth and the sun is

$$V = -\frac{Gm_sm_e}{r}$$

where G is the universal gravitational constant (see Appendix 1), m_s the mass of the sun, m_e the mass of the earth, and r the distance from the sun to the earth,

$$r = (x^2 + y^2 + z^2)^{1/2}$$

with the center of the sun taken as the origin.

Find the force on the earth in cartesian coordinates. That is, find the force in terms of the unit vectors $\mathbf{i}$, $\mathbf{j}$, and $\mathbf{k}$ with the components expressed in terms of x, y, and z. Find the magnitude of the force. ●

The operator ∇ can operate on vector functions as well as on scalar functions. If $\mathbf{F}$ is a vector function, it has magnitude as well as direction, and depends on some independent variables. An example of a vector function is the velocity of a compressible flowing fluid

$$\mathbf{v} = \mathbf{v}(x, y, z) \tag{6.63}$$

which is the same as the expression

$$\mathbf{v} = \mathbf{i}v_x(x, y, z) + \mathbf{j}v_y(x, y, z) + \mathbf{k}v_z(x, y, z) \tag{6.64}$$

In Chapter 2, we had two different products of two vectors. The derivative that is analogous to the scalar product is

$$\nabla \cdot \mathbf{F} = \left(\frac{\partial F_x}{\partial x}\right) + \left(\frac{\partial F_y}{\partial y}\right) + \left(\frac{\partial F_z}{\partial z}\right) \tag{6.65}$$

where $\mathbf{F}$ is a vector function with cartesian components F_x, F_y, and F_z. This derivative is called the *divergence* of $\mathbf{F}$, and is sometimes denoted by div $\mathbf{F}$ instead of $\nabla \cdot \mathbf{F}$. The divergence of a vector function is a scalar.

One way to visualize the divergence of a function is to consider the divergence of the velocity of a compressible fluid. Curves that are followed by small portions of the fluid are called. *stream lines*. In a region where the stream lines become further from each as the flow is followed, the fluid will become less dense, and in such a region the divergence of the velocity is positive. The divergence provides a measure of the spreading of the stream lines. The equation of continuity of a compressible fluid is

$$\nabla \cdot (\rho\mathbf{v}) = -\left(\frac{\partial \rho}{\partial t}\right) \tag{6.66}$$

where ρ is the density of the fluid and t the time.

Example 6-22

Find the divergence of the function

$$\mathbf{F} = \mathbf{i}x^2 + \mathbf{j}yz + \frac{kxz^2}{y}$$

Solution

$$\nabla \cdot F = 2x + z + \frac{2xz}{y}$$

Problem 6-20

Find the divergence of

$$\mathbf{r} = \mathbf{i}x + \mathbf{j}y + \mathbf{k}z$$

The vector derivative of a vector function is called the *curl* of the vector function and is analogous to the cross product just as the divergence is analogous to the dot product. If you have trouble remembering which is which, just remember that dot and divergence both begin with the letter "d" and that cross and curl both begin with the letter "c." In cartesian coordinates, the curl of the vector function **F** is given by

$$\nabla \times \mathbf{F} = \mathbf{i}\left(\frac{\partial F_z}{\partial y} - \frac{\partial F_y}{\partial z}\right) + \mathbf{j}\left(\frac{\partial F_x}{\partial z} - \frac{\partial F_z}{\partial x}\right) + \mathbf{k}\left(\frac{\partial F_y}{\partial x} - \frac{\partial F_x}{\partial y}\right) \quad (6.67)$$

Note the close analogy of this equation with Eq. (2.95). Some books use the symbol curl **F** or rot **F** instead of $\nabla \times \mathbf{F}$.

The curl of a vector function is more difficult to visualize than is the divergence. In fluid flow, the curl of the velocity gives the *vorticity* of the flow, or the rate of turning of the velocity vector.

Example 6-23

Find the curl of the vector function

$$\mathbf{F} = \mathbf{i}y + \mathbf{j}z + \mathbf{k}x$$

Solution

$$\nabla \times \mathbf{F} = \mathbf{i}(0 - 1) + \mathbf{j}(0 - 1) + \mathbf{k}(0 - 1)$$
$$= -\mathbf{i} - \mathbf{j} - \mathbf{k}$$

Problem 6-21

Find the curl of

$$\mathbf{r} = \mathbf{i}x + \mathbf{j}y + \mathbf{k}z$$

We can define derivatives corresponding to successive application of the del operator. The first such operator is the divergence of the gradient and is an operator that occurs in the Schrödinger equation of quantum mechanics, and in electrostatics. If f is a scalar function, the divergence of the gradient of f is

$$\nabla \cdot \nabla f = \left(\frac{\partial^2 f}{\partial x^2}\right) + \left(\frac{\partial^2 f}{\partial y^2}\right) + \left(\frac{\partial^2 f}{\partial z^2}\right) \qquad (6.68)$$

The operator $\nabla \cdot \nabla$ occurs so commonly that it has its own name, the *laplacian operator*, after Laplace, a famous French mathematician. It is usually given the symbol ∇^2.

Example 6-24

Find the laplacian of the function

$$f(x, y, z) = A \sin(ax) \sin(by) \sin(cz)$$

Solution

$$
\begin{aligned}
\nabla^2 f &= -Aa^2 \sin(ax) \sin(by) \sin(cz) - Ab^2 \sin(ax) \sin(by) \sin(cz) \\
&\quad - Ac^2 \sin(ax) \sin(by) \sin(cz) \\
&= -(a^2 + b^2 + c^2)f
\end{aligned}
$$

●

Problem 6-22

Find $\nabla^2 f$ if

$$f = \exp(x^2 + y^2 + z^2) = e^{x^2} e^{y^2} e^{z^2}$$

●

The other possibilities for successive operation of the del operator are the curl of the gradient and the gradient of the divergence. The curl of the gradient of any differentiable function always vanishes. We will show this for the x component:

$$\frac{\partial}{\partial y}\frac{\partial f}{\partial z} - \frac{\partial}{\partial z}\frac{\partial f}{\partial y} = 0$$

The last equality comes from the second derivative identity.

Problem 6-23

Write the expression for the gradient of the divergence of a vector function **F**.

●

Vector Derivatives in Other Coordinate Systems. It is sometimes necessary or convenient to work in coordinate systems other than cartesian coordinates. For example, in the Schrödinger equation for the motion of the electron in a hydrogen atom, the potential energy is a simple function of r, the distance from the nucleus, but a complicated function of x, y, and z. The complications produced by expressing the laplacian in the Schrödinger equation in spherical polar coordinates are more than outweighed by the simplifications produced by having a simple expression for the potential energy.

Coordinate systems such as spherical polar or cylindrical polar coordinates are called *orthogonal coordinates*, because an infinitesimal displacement produced by changing only one of the coordinates is perpendicular, or orthogonal, to a displacement produced by an infinitesimal change in any one of the other coordinates.

Figure 6-7 shows displacements, drawn as though they were finite, produced by infinitesimal changes in r, θ, and ϕ. These displacements are lengths

$$ds_r = \text{displacement in } r \text{ direction} = dr$$

$$ds_\theta = \text{displacement in } \theta \text{ direction} = r\,d\theta$$

$$ds_\phi = \text{displacement in } \phi \text{ direction} = r\sin(\theta)\,d\phi$$

We define three vectors of unit length, whose directions are those of the infinitesimal displacements in Figure 6-7, called $\mathbf{e}_r$, $\mathbf{e}_\theta$, and $\mathbf{e}_\phi$.

An infinitesimal vector displacement is the sum of displacements in the three orthogonal directions. In spherical polar coordinates,

$$\boxed{d\mathbf{r} = \mathbf{e}_r\,dr + \mathbf{e}_\theta r\,d\theta + \mathbf{e}_\phi r\sin(\theta)\,d\phi} \tag{6.69}$$

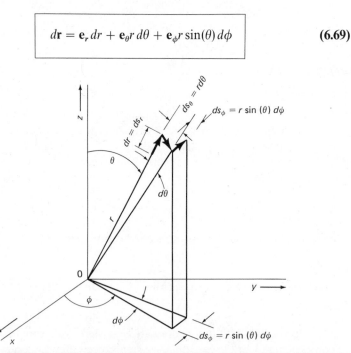

Figure 6-7. Infinitesimal displacements ds_r, ds_θ, and ds_ϕ produced by infinitesimal increments dr, $d\theta$, and $d\phi$.

In cartesian coordinates,

$$dr = \mathbf{i}\,dx + \mathbf{j}\,dy + \mathbf{k}\,dz \qquad\qquad (6.70)$$

Let the three coordinates of an orthogonal system be called q_1, q_2, and q_3. Let the displacements due to the infinitesimal increments be called ds_1, ds_2, and ds_3. Let the unit vectors in the directions of the displacements be called $\mathbf{e}_1$, $\mathbf{e}_2$, and $\mathbf{e}_3$. The equation analogous to Eq. (6.69) is

$$dr = \mathbf{e}_1\,ds_1 + \mathbf{e}_2\,ds_2 + \mathbf{e}_3\,ds_3 \qquad (6.71\text{a})$$
$$= \mathbf{e}_1 h_1\,dq_1 + \mathbf{e}_2 h_2\,dq_2 + \mathbf{e}_3 h_3\,dq_3 \qquad (6.71\text{b})$$

For cartesian coordinates, all three of the h's are equal to unity. For spherical polar coordinates, $h_r = 1$, $h_\theta = r$, and $h_\phi = r\sin(\theta)$.

The gradient of a scalar function f is written in terms of components in the direction of ds_1, ds_2, and ds_3 as follows:

$$\nabla f = \mathbf{e}_1 \frac{\partial f}{\partial s_1} + \mathbf{e}_2 \frac{\partial f}{\partial s_2} + \mathbf{e}_3 \frac{\partial f}{\partial s_3}$$

or

$$\nabla f = \mathbf{e}_1 \frac{1}{h_1}\frac{\partial f}{\partial q_1} + \mathbf{e}_2 \frac{1}{h_2}\frac{\partial f}{\partial q_2} + \mathbf{e}_3 \frac{1}{h_3}\frac{\partial f}{\partial q_3} \qquad (6.72)$$

Example 6-25

Find the expression for the gradient of a function $f = f(r, \theta, \phi)$.

Solution

$$\nabla f = \mathbf{e}_r \frac{\partial f}{\partial r} + \mathbf{e}_\theta \frac{1}{r}\frac{\partial f}{\partial \theta} + \mathbf{e}_\phi \frac{1}{r\sin(\theta)}\frac{\partial f}{\partial \phi} \qquad \bullet\ (6.73)$$

Problem 6-24

Find the expression for the gradient of a function of cylindrical polar coordinates, $f = f(\rho, \phi, z)$. Find the gradient of the function

$$f = e^{-(\rho^2 + z^2)/a^2}\sin(\phi) \qquad\qquad \bullet$$

The divergence of a vector function can similarly be expressed in orthogonal coordinates other than cartesian. If $\mathbf{F}$ is a vector function, it must be expressed in terms of the unit vectors of the coordinate system in which we are to differentiate.

$$\mathbf{F} = \mathbf{e}_1 F_1 + \mathbf{e}_2 F_2 + \mathbf{e}_3 F_3 \tag{6.74}$$

Remember that the components F_1, F_2, and F_3 are not necessarily the cartesian components.

The divergence of the vector function $\mathbf{F}$ is given by

$$\nabla \cdot \mathbf{F} = \frac{1}{h_1 h_2 h_3} \left[\frac{\partial}{\partial q_1} (F_1 h_2 h_3) + \frac{\partial}{\partial q_2} (F_2 h_1 h_3) + \frac{\partial}{\partial q_3} (F_3 h_1 h_2) \right]$$

$$\tag{6.75}$$

Example 6-26

Write the expression for the divergence of a function $\mathbf{F}$ expressed in terms of spherical polar coordinates. Find the divergence of the position vector, which in spherical polar coordinates is

$$\mathbf{r} = \mathbf{e}_r r$$

Solution

$$\nabla \cdot \mathbf{F} = \frac{1}{r^2 \sin(\theta)} \left[\frac{\partial}{\partial r} [F_r r^2 \sin(\theta)] + \frac{\partial}{\partial \theta} [F_\theta r \sin(\theta)] + \frac{\partial}{\partial \phi} (F_\phi r) \right]$$

$$= \frac{1}{r^2} \frac{\partial}{\partial r} (r^2 F_r) + \frac{1}{r \sin(\theta)} \frac{\partial}{\partial \theta} [\sin(\theta) F_\theta] + \frac{1}{r \sin(\theta)} \frac{\partial F_\phi}{\partial \phi} \tag{6.76}$$

The divergence of $\mathbf{r}$ is

$$\nabla \cdot \mathbf{r} = \frac{1}{r^2} 3r^2 + 0 + 0 = 3 \qquad \bullet$$

Problem 6-25

Write the formula for the divergence of a function in cylindrical polar coordinates. $\qquad \bullet$

The curl of a vector function is

$$\nabla \times \mathbf{F} = \mathbf{e}_1 \frac{1}{h_2 h_3} \left[\frac{\partial}{\partial q_2} (h_3 F_3) - \frac{\partial}{\partial q_3} (h_2 F_2) \right]$$

$$+ \mathbf{e}_2 \frac{1}{h_1 h_3} \left[\frac{\partial}{\partial q_3} (h_1 F_1) - \frac{\partial}{\partial q_1} (h_3 F_3) \right]$$

$$+ \mathbf{e}_3 \frac{1}{h_1 h_2} \left[\frac{\partial}{\partial q_1} (h_2 F_2) - \frac{\partial}{\partial q_2} (h_1 F_1) \right] \tag{6.77}$$

The expression for the laplacian of a scalar function, f, is

$$\nabla^2 f = \frac{1}{h_1 h_2 h_3} \left[\frac{\partial}{\partial q_1} \left(\frac{h_2 h_3}{h_1} \frac{\partial f}{\partial q_1} \right) + \frac{\partial}{\partial q_2} \left(\frac{h_1 h_3}{h_2} \frac{\partial f}{\partial q_2} \right) + \frac{\partial}{\partial q_3} \left(\frac{h_1 h_2}{h_3} \frac{\partial f}{\partial q_3} \right) \right]$$

(6.78)

Example 6-27

Write the expression for the laplacian in spherical polar coordinates.

Solution

$$\nabla^2 f = \frac{1}{r^2} \frac{\partial}{\partial r} \left(r^2 \frac{\partial f}{\partial r} \right) + \frac{1}{r^2 \sin(\theta)} \frac{\partial}{\partial \theta} \left[\sin(\theta) \frac{\partial f}{\partial \theta} \right] + \frac{1}{r^2 \sin^2(\theta)} \frac{\partial^2 f}{\partial \phi^2}$$

(6.79)

Section 6-9. MAXIMUM AND MINIMUM VALUES OF FUNCTIONS OF SEVERAL VARIABLES

Just as in the case of a function of one variable, it is sometimes necessary to find the point at which a function of several variables has its minimum or maximum value. In addition, it is sometimes necessary to find the maximum or minimum value of a function subject to some constraint on the independent variables.

Figure 6-8 shows a graph of the function $f = e^{-x^2 - y^2}$. The surface representing the function has a "peak" at which the function attains its maximum value. Shown also in the figure is a curve at which the surface intersects with a plane representing a constraint, that $y = 1 - x$. On this curve there is also a maximum, which is of course a smaller value than the "unconstrained" maximum at the peak. This is the maximum subject to the constraint.

We first discuss the finding of the unconstrained maximum. We use the fact that the plane which is tangent to the surface will be horizontal at any local maximum or minimum. Therefore, the curve representing the intersection of any vertical plane with the surface will have a maximum or a minimum at the same place. The partial derivative with respect to one variable gives the slope of the curve with the other variable constant, so we can find a local maximum or minimum by finding the place where all the partial derivatives of the function vanish simultaneously.

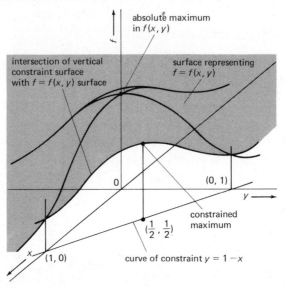

Figure 6-8. The surface representing a function of x and y with the absolute maximum and a constrained maximum shown.

Our method for two variables is therefore:

1. Solve the simultaneous equations

$$\left(\frac{\partial f}{dx}\right)_y = 0 \tag{6.80}$$

$$\left(\frac{\partial f}{\partial y}\right)_x = 0 \tag{6.81}$$

2. Calculate the value of the function at all points satisfying these equations, and also at the boundaries of the region being considered.

The maximum or minimum value in the region being considered must be in this set of values. For more than two independent variables, the method is similar, except that there is one equation in the set of simultaneous equations for each independent variable.

Example 6-28

Find the maximum value of the function that is shown in Figure 6-8:

$$f = e^{-x^2 - y^2}$$

Solution

At the maximum

$$\left(\frac{\partial f}{\partial x}\right)_y = e^{-x^2 - y^2}(-2x) = 0$$

$$\left(\frac{\partial f}{\partial d}\right)_x = e^{-x^2 - y^2}(-2y) = 0$$

The only solution is

$$x = 0, \qquad y = 0$$

No region was specified, so we must consider all values of x and y. For very large magnitudes of x and y, the function vanishes, so we have found the desired absolute maximum. •

In the case of one independent variable, a local maximum could be distinguished from a local minimum or an inflection point by determining the sign of the second derivative. For two independent variables, the following quantity is calculated:

$$D = \left(\frac{\partial^2 f}{\partial x^2}\right)\left(\frac{\partial^2 f}{\partial y^2}\right) - \left(\frac{\partial^2 f}{\partial x\, \partial y}\right)^2 \qquad (6.82)$$

The different cases are as follows:

If $D > 0$ and $\left(\dfrac{\partial^2 f}{\partial x^2}\right) > 0$, then we have a relative minimum.

If $D > 0$ and $\left(\dfrac{\partial^2 f}{\partial x^2}\right) < 0$, then we have a relative maximum.

If $D < 0$, then we have neither a relative maximum nor a relative minimum.

If $D = 0$, the test fails, and we cannot tell what we have.

Problem 6-26

Evaluate D at the point $(0, 0)$ for the function of Example 6-28 and establish that the point is a relative maximum. •

Constrained Maximum–Minimum Problems. It sometimes happens that we must find a maximum or a minimum value of a function subject to some condition. This nearly always means that the maximum which we find, which is called a *constrained maximum*, is smaller than the absolute maximum of the function, or that the minimum which we find is larger than the absolute minimum of the function, in order to satisfy the condition. Let us try to make the situation clear with a simple example:

Example 6-29

Find the maximum value of the function in Example 6-28 subject to the constraint

$$x + y = 1$$

Solution

The situation is shown in Figure 6-8. The constraint corresponds to the specification of y as a function of x by

$$y = 1 - x \qquad (6.83)$$

This function is given by the line in the figure. We are now looking for the place along this curve at which the function has a larger value than at any other place on the curve.

Since y is no longer an independent variable on the curve of the constraint, the direct way to proceed is to replace y by use of Eq. (6.83):

$$f(x, 1 - x) = f(x) = e^{-x^2-(1-x)^2} = e^{-2x^2+2x-1} \tag{6.84}$$

The relative maximum is now at the point where df/dx vanishes:

$$\frac{df}{dx} = e^{-2x^2+2x-1}(-4x + 2) = 0 \tag{6.85}$$

The solution to this is $x = \frac{1}{2}$, which corresponds to $y = \frac{1}{2}$. At this point

$$f(\tfrac{1}{2}, \tfrac{1}{2}) = \exp[-(\tfrac{1}{2})^2 - (\tfrac{1}{2})^2] = e^{-1/2} = 0.6065 \ldots \qquad \bullet$$

Problem 6-27

  a. Find the minimum in the function
$$f(x, y) = x^2 + y^2 + 2x$$
  b. Find the constrained minimum subject to the constraint
$$x + y = 0 \qquad \bullet$$

If we have a constrained maximum–minimum problem with more than two variables, the direct method of substituting the constraint relation into the function is usually not practical.

Lagrange's Method of Undetermined Multipliers. This is a method for finding a constrained maximum or minimum without substituting the constraint relation into the function. If the constraint is written in the form $g(x, y) = 0$, the method for finding the constrained maximum or minimum in $f(x, y)$ is as follows:

  1. Form the new function

$$\boxed{u(x, y) = f(x, y) + \lambda g(x, y)} \tag{6.86}$$

  where λ is a constant called an *undetermined multiplier*,
  2. Form the partial derivatives of u, and set them equal to zero.

$$\left(\frac{\partial u}{\partial x}\right)_y = \left(\frac{\partial f}{\partial x}\right)_y + \lambda\left(\frac{\partial g}{\partial x}\right)_y = 0 \tag{6.87a}$$

$$\left(\frac{\partial u}{\partial y}\right)_x = \left(\frac{\partial f}{\partial y}\right)_x + \lambda\left(\frac{\partial g}{\partial y}\right)_x = 0 \tag{6.87b}$$

  3. Solve the set of equations consisting of $g = 0$ and the two equations of Eq. (6.87) as a set of simultaneous equations for the value of x, the

value of y, and the value of λ that correspond to the relative maximum or minimum.

We will not present a proof of the validity of this method, but you can find such a proof in the book by Taylor and Mann listed at the end of the chapter. Note that the method is equivalent to finding the ordinary maximum of u as a function of x, y, and λ.

Example 6-30

Find the constrained maximum of Example 6.29 by the method of Lagrange.

Solution

The constraining equation is written

$$g(x, y) = x + y - 1 = 0 \tag{6.88}$$

The function u is

$$u(x, y) = e^{-x^2 - y^2} + \lambda(x + y - 1)$$

so that the equations to be solved are Eq. (6.88) and

$$\left(\frac{\partial u}{\partial x}\right)_y = (-2x)e^{-x^2 - y^2} + \lambda = 0 \tag{6.89a}$$

$$\left(\frac{\partial u}{\partial y}\right)_x = (-2y)e^{-x^2 - y^2} + \lambda = 0 \tag{6.89b}$$

Let us begin by solving for λ in terms of x and y. Multiply Eq. (6.89a) by y and Eq. (6.89b) by x and add the two equations. The result can be solved to give

$$\lambda = \frac{4xy}{x + y}e^{-x^2 - y^2} \tag{6.90}$$

Substitute this into Eq. (6.89a) to get

$$(-2x)e^{-x^2 - y^2} + \frac{4xy}{x + y}e^{-x^2 - y^2} = 0 \tag{6.91}$$

The exponential factor is not zero for any finite values of x and y, so

$$-2x + \frac{4xy}{x + y} = 0 \tag{6.92a}$$

When Eq. (6.90) is substituted into Eq. (6.89b) in the same way, the result is

$$-2y + \frac{4xy}{x + y} = 0 \tag{6.92b}$$

When Eq. (6.92b) is subtracted from Eq. (6.92a), the result is

$$-2x + 2y = 0$$

which is solved for y in terms of x to obtain

$$y = x$$

This is substituted into Eq. (6.88) to obtain

$$x + x - 1 = 0$$

which gives

$$x = \tfrac{1}{2}, \qquad y = \tfrac{1}{2}$$

This is the same result as in Example 6-29. •

In this case, the method of Lagrange was more work than the direct method. In more complicated problems, the method of Lagrange will usually be easier.

The method of Lagrange also works if there is more than one constraint. If we desire the relative maximum or minimum of the function

$$f = f(x, y, z) \tag{6.93}$$

subject to the two constraints

$$g_1(x, y, z) = 0 \tag{6.94}$$

and

$$g_2(x, y, z) = 0 \tag{6.95}$$

the procedure is similar, except that two undetermined multipliers are used.

One forms the function

$$u = u(x, y, z) = f(x, y, z) + \lambda_1 g_1(x, y, z) + \lambda_2 g(x, y, z)$$

and solves the set of simultaneous equations consisting of Eqs. (6.94), (6.95), and

$$\left(\frac{\partial u}{\partial x}\right)_{y,z} = 0 \tag{6.96a}$$

$$\left(\frac{\partial u}{\partial y}\right)_{x,z} = 0 \tag{6.96b}$$

$$\left(\frac{\partial u}{\partial z}\right)_{x,y} = 0 \tag{6.96c}$$

The result is a value for λ_1, a value for λ_2, and values for x, y, and z which locate the constrained relative maximum or minimum.

Problem 6-28

Find the constrained minimum of Problem 6-27 using the method of Lagrange. •

ADDITIONAL PROBLEMS

6.29

A certain nonideal gas has an equation of state

$$\frac{P\bar{V}}{RT} = 1 + \frac{B_2}{\bar{V}}$$

where B_2 is given as a function of T by

$$B_2 = [-1.00 \times 10^{-4} - (2.148 \times 10^{-6})e^{1986/T}] \text{ m}^3 \text{ mol}^{-1}$$

where T is the temperature on the Kelvin scale, $\bar{V}$ the volume of 1 mol, P the pressure, and R the gas constant,

Find $(\partial P/\partial \bar{V})_{n,T}$ and $(\partial P/\partial T)_{n,\bar{V}}$ and an expression for dP.

6.30

For a certain system, E is given as a function of S, V, and n by

$$E = E(S, V, n) = Kn^{5/3}V^{-2/3}e^{2S/3nR}$$

where S is the entropy, V the volume, n the number of moles, K a constant, and R the gas constant. Find dE in terms of dS, dV, and dn.

6.31

Find $(\partial f/\partial x)_y$ and $(\partial f/\partial y)_x$ for each of the following functions. a, b, and c are constants.

a. $f = axy \ln(y) + bx \cos(x + y)$
b. $f = ae^{-b(x^2+y^2)} + c \sin(x^2y)$
c. $f = a(x + by)/(c + xy)$
d. $f = (x + y)^{-3}$

6.32

Find $(\partial^2 f/\partial x^2)_y$, $(\partial^2 f/\partial x \partial y)$, and $(\partial^2 f/\partial y^2)_x$, for each of the following functions.

a. $f = (x + y)^{-1}$
b. $f = \cos(x/y)$
c. $f = e^{(ax^2+by^2)}$

6.33

a. Find the area of the semicircle of radius a given by

$$y = +(a^2 - x^2)^{1/2}$$

by doing the double integral

$$\int_{-a}^{a} \int_{0}^{(a^2-x^2)^{1/2}} 1 \, dy \, dx$$

b. Change to polar coordinates and repeat the calculation.

6.34

Test each of the following differentials for exactness.

a. $du = by \cos(bx) \, dx + \sin(bx) \, dy$
b. $du = ay \sin(xy) \, dx + ax \sin(xy) \, dy$

c. $du = \dfrac{y}{(1 + x^2)} \, dx - \tan^{-1}(x) \, dy$

d. $du = x \, dy + y \, dx$
e. $du = y \ln(x) \, dx + x \ln(y) \, dy$
f. $du = 2xe^{axy} \, dx + 2ye^{axy} \, dy$

6.35

If

$$G = -RT \ln\left(\frac{aT^{3/2}V}{n}\right)$$

find dG in terms of dT, dV, and dn. R and a are constants.

6.36

Perform the line integral

$$\int_C df = \int_C (x^2 y\, dx + xy^2\, dy)$$

where C is the line segment from $(0, 0)$ to $(2, 2)$. Would another path with the same end points yield the same result?

6.37

Find the function $f(x, y)$ whose differential is

$$df = (x + y)^{-1}\, dx + (x + y)^{-1}\, dy$$

and which has the value $f(1, 1) = 0$. Do this by performing a line integral on a rectangular path from $(1, 1)$ to (x_1, y_1). Assume $x_1 > 0$, $y_1 > 0$.

6.38

A wheel of radius R has a mass per unit area given by

$$m(\rho) = a\rho^2 + b$$

where a and b are constants, m the mass per unit area, and ρ the distance from the center. Find the moment of inertia, defined by

$$I = \iint m(\rho)\rho^2\, dA$$

where dA represents the element of area and the integral is a double integral over the entire wheel. Carry out the integral in both cartesian and plane polar coordinates.

6.39

Complete the formula

$$\left(\frac{\partial S}{\partial V}\right)_{P,n} = \left(\frac{\partial S}{\partial V}\right)_{T,n} + ?$$

6.40

Find the location of the minimum in the function

$$f = f(x, y) = x^2 - 6x + 8y + y^2$$

What is the value of the function at the minimum?

6.41

Find the minimum in the function of Problem 6-40 subject to the constraint

$$x + y = 2$$

Do this by substitution and by the method of undetermined multipliers.

ADDITIONAL READING

Angus E. Taylor and Robert W. Mann, *Advanced Calculus*, 2nd ed., Ginn and Company, Lexington, Mass., 1972.

This book is a typical textbook for a calculus course to be taken after the elementary sequence, and contains thorough discussions of most of the topics of this chapter. There are a number of similar books, including one by Kaplan listed at the end of Chapter 5.

Henry Margenau and George Mosely Murphy, *The Mathematics of Physics and Chemistry*, 2nd ed., D. Van Nostrand, Princeton, N.J., 1956.

This reference book contains discussions of many of the mathematical topics needed for physical chemistry at the beginning graduate level.

Louis A. Pipes, *Applied Mathematics for Engineers and Physicists*, McGraw-Hill Book Company, New York, 1946.

This book is at about the same level as that of Margenau and Murphy.

Philip M. Morse and Herman Feshbach, *Methods of Theoretical Physics*, McGraw-Hill Book Company, New York, 1953.

This is a large work which comes in two volumes and contains a lot of information about the topics in this chapter. It is at a somewhat higher level than Margenau and Murphy.

CHAPTER SEVEN

Differential Equations and the Motions of Objects

Section 7-1. INTRODUCTION

In this chapter, we discuss differential equations and some techniques for solving them. *Differential equations* are equations that contain derivatives of unknown functions, and the task of solving them is the task of finding what the functions are.

The process of finding an antiderivative function, which was discussed in Section 4-2, is actually the solution of a simple differential equation. We are given the equation

$$\frac{dF}{dx} = f(x) \tag{7.1}$$

where $f(x)$ is a known function, and want to find the function $F(x)$ that satisfies this equation.

We will discuss several kinds of differential equations, some of which are equations of motion for particles. After studying this chapter, you should be able to do all of the following:

1. Construct a differential equation for the motion of a particle from Newton's second law of motion.

2. Solve a linear homogeneous differential equation with constant coefficients.

3. Solve a differential equation whose variables can be separated.

4. Solve an exact differential equation.

5. Use an integrating factor to solve an inexact differential equation.

6. Solve a simple partial differential equation by separation of variables.

Section 7-2. DIFFERENTIAL EQUATIONS AND NEWTON'S LAWS OF MOTION

The velocity of a particle is defined by

$$\mathbf{v} = \frac{d\mathbf{r}}{dt} = \mathbf{i}\frac{dx}{dt} + \mathbf{j}\frac{dy}{dt} + \mathbf{k}\frac{dz}{dt} = \mathbf{i}v_x + \mathbf{j}v_y + \mathbf{k}v_z \qquad (7.2)$$

where $\mathbf{r}$ is the position vector and $\mathbf{i}$, $\mathbf{j}$, and $\mathbf{k}$ are the unit vectors defined in Chapter 2. The acceleration is defined by

$$\mathbf{a} = \frac{d^2\mathbf{r}}{dt^2} = \mathbf{i}\frac{d^2x}{dt^2} + \mathbf{j}\frac{d^2y}{dt^2} + \mathbf{k}\frac{d^2z}{dt^2} = \mathbf{i}a_x + \mathbf{j}a_y + \mathbf{k}a_z \qquad (7.3)$$

Let us consider a particle that moves in the z direction only, so that v_x, v_y, a_x, and a_y vanish. If a_z is known as a function of time,

$$a_z = a_z(t) \qquad (7.4)$$

then Eq. (7.4) is a differential equation for the velocity:

$$\frac{dv_z}{dt} = a_z(t) \qquad (7.5)$$

This equation is solved by doing an integration. We multiply both sides by dt and perform a definite integration

$$v_z(t_1) - v_z(0) = \int_0^{t_1} \left(\frac{dv_z}{dt}\right) dt = \int_0^{t_1} a_z(t)\, dt \qquad (7.6)$$

The result of this integration gives v_z as a function of time, and a second integration gives the position as a function of time:

$$z(t_2) - z(0) = \int_0^{t_2} \left(\frac{dz}{dt_1}\right) dt_1 = \int_0^{t_2} v_z(t_1)\, dt_1 \qquad (7.7)$$

There are inertial navigation systems used on submarines and space vehicles which measure the acceleration as a function of time and perform two numerical integrations in order to determine the position of the vehicle.

Problem 7-1

At time $t = 0$, a certain particle has $z = 0$ and $v_z = 0$. Its acceleration is given as a function of time by

$$a_z = a_0 e^{-t/b}$$

where a_0 and b are constants. Find z as a function of time. Find the speed and the position of the particle at $t = 30$ s if $a_0 = 10$ m s^{-2} and $b = 5$ s. Find the limiting value of the speed as $t \to \infty$. ●

Unfortunately, the acceleration is very seldom known as a function of time, so the simple solution of an equation like Eq. (7.4) cannot often be used to find how a particle moves. Instead, we must obtain the acceleration of the particle from knowledge of the force on it, using Newton's second law.

Newton's Law of Motion. These laws were deduced by Isaac Newton from his analysis of observations of the motions of actual objects, including the motions of apples and celestial bodies. The laws are

1. A body on which no forces act does not accelerate.
2. A body acted on by a force **F** accelerates according to

$$\boxed{\mathbf{F} = m\mathbf{a}} \tag{7.8}$$

where m is the mass of the object.
3. Two bodies exert forces of equal magnitude and opposite direction on each other.

Classical mechanics is primarily the study of the consequences of these laws. Incidentally, the first law is just a special case of the second, and the third law is primarily used to obtain forces for the second law, so Newton's second law is the central fact of classical mechanics.

The most common problem of classical mechanics is to solve for the position and velocity of each particle in a system, given the positions and velocities at some initial time. If the force on a particle can be written as a function of its position alone, we have a differential equation for the motion of that particle, which is called the *equation of motion* for that particle. If the force on a particle depends on the positions of other particles, the equations of motion of the particles are coupled together, and the problem cannot be solved exactly for a system of more than two particles.

The simplest equation of motion is for a particle which can move in only one direction and which has a force on it which depends only on the position of the one particle. From Newton's second law, the equation of motion is

$$\boxed{F_z(z) = \frac{m\,d^2z}{dt^2}} \tag{7.9}$$

181

§7-3 / The Harmonic Oscillator. Linear Differential Equations with Constant Coefficients

In this case, it is possible to obtain the force from a potential energy function, as in Eq. (6.62). Equation (7.9) becomes

$$-\frac{dV}{dz} = \frac{m\,d^2z}{dt^2} \tag{7.10}$$

In many cases, it is possible to solve this equation.

Section 7-3. THE HARMONIC OSCILLATOR. LINEAR DIFFERENTIAL EQUATIONS WITH CONSTANT COEFFICIENTS

Consider an object of mass m attached to the end of a coil spring whose other end is rigidly fastened. Let the position of the object along a line extending through the fastened end of the spring be given by the coordinate z, defined so that $z = 0$ when the spring has its equilibrium length. To a good approximation, the force on the object due to the spring is given for fairly small magnitudes of z by

$$F_z = -kz \tag{7.11}$$

where k is a constant called the *spring constant*. Equation (7.11) is known as *Hooke's law*.

The harmonic oscillator is an idealized model of this system. That is, it is a hypothetical system (existing only in our minds) which has some properties in common with the real system, but it is enough simpler to allow exact mathematical analysis. Once we have solved mathematically for the behavior of a model system, we say that the real system must behave similarly. In our model, we say that the spring has no mass, and we say that Eq. (7.11) is exactly followed, even if z has large magnitudes. We also neglect the force of gravity.

We now write Newton's second law. From Eq. (7.9),

$$\frac{d^2z}{dt^2} + \left(\frac{k}{m}\right)z = 0 \tag{7.12}$$

This is a differential equation which is described as follows:

1. It is linear, because the dependent variable z and its derivatives enter only to the first power.
2. It is homogeneous, because there are no terms that do not contain z.
3. It is second order, because the highest order derivative in the equation is a second derivative.
4. It has a constant coefficients. That is, the quantities that multiply z and its derivatives are constants.

There are two important facts about linear homogeneous differential equations:

1. *If $z_1(t)$ and $z_2(t)$ are both functions that satisfy the equation, then the quantity $z_3(t)$ is also a solution, where*

$$z_3(t) = c_1 z_1(t) + c_2 z_2(t) \qquad (7.13)$$

   and c_1 and c_2 are constants. The expression in Eq. (7.13) is called a linear combination of z_1 and z_2.
2. *If $z(t)$ satisfies the equation, then $cz(t)$ is also a solution, where c is a constant.*

Problem 7-2

A linear differential equation of the nth order can be written

$$f_n(t)\frac{d^n z}{dt^n} + f_{n-1}(t)\frac{d^{n-1}z}{dt^{n-1}} + \cdots + f_1(t)\frac{dz}{dt} + f_0(t)z(t) = g(t) \qquad (7.14)$$

For the case that $g(t) = 0$ (homogeneous equation), prove the two facts given above.   ●

A linear homogeneous differential equation with constant coefficients can be solved by the following "cookbook" method:

1. Begin with the trial solution

$$z(t) = e^{\lambda t} \qquad (7.15)$$

   A trial solution is what the name implies. We try it by substituting it into the equation.
2. Find the values of λ that cause the trial solution to satisfy the equation. Call these $\lambda_1, \lambda_2, \ldots, \lambda_n$.
3. Use fact (1) to write a solution

$$z(t) = c_1 e^{\lambda_1 t} + c_2 e^{\lambda_2 t} + \cdots + c_n e^{\lambda_n t} \qquad (7.16)$$

Example 7-1

Show that the differential equation

$$a_3\left(\frac{d^3 y}{dx^3}\right) + a_2\left(\frac{d^2 y}{dx^2}\right) + a_1\left(\frac{dy}{dx}\right) + a_0 y = 0 \qquad (7.17)$$

can be satisfied by a trial solution $y = e^{\lambda x}$.

183

§7-3 / The Harmonic Oscillator. Linear Differential Equations with Constant Coefficients

Solution

We substitute the trial solution $y = e^{\lambda x}$ into Eq. (7.17):

$$a_3\lambda^3 e^{\lambda x} + a_2\lambda^2 e^{\lambda x} + a_1\lambda^2 e^{\lambda x} + a_1\lambda e^{\lambda x} + a_0 e^{\lambda x} = 0 \qquad \textbf{(7.18)}$$

If x remains finite, we can divide by $e^{\lambda x}$ to obtain the characteristic equation:

$$a_3\lambda^3 + a_2\lambda^2 + a_1\lambda + a_0 = 0 \qquad \textbf{(7.19)}$$

This is an equation that can be solved for three values of λ which cause the trial solution to satisfy Eq. (7.17) •

Example 7-2

Solve the differential equation

$$\frac{d^2y}{dx^2} + \frac{dy}{dx} - 2y = 0 \qquad \textbf{(7.20)}$$

Solution

Substitution of the trial solution $y = e^{\lambda x}$ gives the characteristic equation

$$\lambda^2 + \lambda - 2 = 0$$

The solutions to this equation are

$$\lambda = 1, \quad \lambda = -2$$

The solution to the differential equation is thus

$$y(x) = c_1 e^x + c_2 e^{-2x} \qquad \bullet \ \ \textbf{(7.21)}$$

The solution in Eq. (7.21) satisfies Eq. (7.20) no matter what values c_1 and c_2 have. They are arbitrary constants. The solution is a family of functions, one function for each set of values for c_1 and c_2.

A solution to a linear differential equation of order n that contains n arbitrary constants is a general solution. A *general solution* is a family of functions which includes almost all of the solutions to the differential equation.

A solution to a differential equation that contains no arbitrary constants is called a *particular solution*. A particular solution may be one of the members of the general solution, or it may be another function. We are not prepared to discuss the question of when particular solutions can occur that are not included in the general solution, but you should be aware that they sometimes occur. In most of the differential equations encountered in physical chemistry, you can assume that the general solution includes all solutions.

In the case of a linear homogeneous differential equation with constant coefficients, there will generally be n values of λ that satisfy the characteristic equation if the differential equation is of order n. Therefore, the solution of Eq. (7.16) is a general solution.

There is only one general solution to a differential equation. If you find two general solutions for the same differential equation, and they appear to be different, there must be some mathematical manipulations that will reduce both to the same form.

In physics and chemistry, we are not finished with a problem when we find a general solution to a differential equation. We usually have additional information that will enable us to pick a particular solution out of the family of solutions. Such information is called the *boundary conditions*.

We will use these facts in the solution of Eq. (7.12).

Problem 7-3

Show that the characteristic equation for Eq. (7.12) is

$$\lambda^2 + \frac{k}{m} = 0 \qquad \bullet \quad (7.22)$$

The solution of Eq. (7.22) is

$$\lambda = \pm i\left(\frac{k}{m}\right)^{1/2} \qquad (7.23)$$

where i is the square root of -1, the imaginary unit.

The general solution to Eq. (7.12) is therefore

$$z = z(t) = c_1 \exp\left[+i\left(\frac{k}{m}\right)^{1/2} t\right] + c_2 \exp\left[-i\left(\frac{k}{m}\right)^{1/2} t\right] \qquad (7.24)$$

where c_1 and c_2 are arbitrary constants.

We now apply any boundary conditions to our solution. The first thing we do is to require that our solution be real, because imaginary and complex numbers cannot represent physically measurable quantities. From Eq. (2-112), we have

$$z = c_1[\cos(\omega t) + i\sin(\omega t)] + c_2[\cos(\omega t) - i\sin(\omega t)] \qquad (7.25)$$

where

$$\omega = \left(\frac{k}{m}\right)^{1/2} \qquad (7.26)$$

If we let $c_1 + c_2 = b_1$ and $i(c_1 - c_2) = b_2$, then

$$z = b_1 \cos(\omega t) + b_2 \sin(\omega t) \qquad (7.27)$$

We now eliminate complex solutions by requiring that b_1 and b_2 be real.

Problem 7-4

Show that the function of Eq. (7.27) satisfies Eq. (7.12). $\qquad \bullet$

Let us now describe not only a particular harmonic oscillator, but a particular case of its motion. Say that we have the boundary conditions at $t = 0$:

$$z(0) = 0 \qquad (7.28a)$$

$$v_z(0) = v_0 = \text{constant} \qquad (7.28b)$$

185

§7-3 / The Harmonic Oscillator. Linear Differential Equations with Constant Coefficients

We require one boundary condition for each arbitrary constant in order to get a particular solution, so these two boundary conditions will enable us get a particular solution for the case at hand. Knowledge of the position at time $t = 0$ without knowledge of the velocity would not have sufficed.

For our boundary conditions, b_1 must vanish:

$$z(0) = b_1 \cos(0) + b_2 \sin(0) = b_1 = 0 \tag{7.29}$$

The velocity is given by

$$v_z(t) = \frac{dz}{dt} = b_2 \omega \cos(\omega t) \tag{7.30}$$

so

$$v_z(0) = b_2 \omega \cos(0) = b_2 \omega$$

which gives

$$b_2 = \frac{v_0}{\omega} \tag{7.31}$$

and gives us our particular solution

$$z(t) = \left(\frac{v_0}{\omega}\right) \sin(\omega t) \tag{7.32}$$

The motion given by this solution is called *uniform harmonic motion*. Figure 7-1 shows the position and the velocity of the suspended mass as a

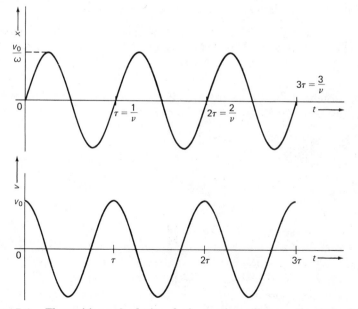

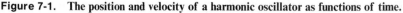

Figure 7-1. The position and velocity of a harmonic oscillator as functions of time.

function of time. The motion is periodic. During one period, the argument of the sine changes by 2π, so that if τ is the length of the period.

$$2\pi = \omega\tau = \left(\frac{k}{m}\right)^{1/2}\tau \tag{7.33}$$

Thus,

$$\tau = 2\pi\left(\frac{m}{k}\right)^{1/2} \tag{7.34}$$

The reciprocal of the period is called the *frequency*, denoted by v:

$$v = \frac{1}{2\pi}\sqrt{\frac{k}{m}} \tag{7.35}$$

Problem 7-5

The vibration of a diatomic molecule resembles that of a harmonic oscillator, except that since both nuclei move, the mass must be replaced by the reduced mass,

$$\mu = \frac{m_1 m_2}{m_1 + m_2}$$

where m_1 is the mass of one nucleus and m_2 the mass of the other nucleus.

Calculate the frequency of vibration of a hydrogen chloride molecule. The force constant is

$$k = 481 \text{ N } m^{-1} = 481 \text{ J m}^{-2}$$

Remember that m_1 and m_2 are masses per atom, not per mole,   •

Let us examine the energy of the harmonic oscillator as a function of time. The total energy is a sum of the kinetic energy

$$K = \tfrac{1}{2}mv^2 = \tfrac{1}{2}mv_z^2 \tag{7.36}$$

and the potential energy. In order for Eq. (6.62) to be satisfied, the potential energy of the harmonic oscillator must be

$$V = \tfrac{1}{2}kz^2 \tag{7.37}$$

The total energy is

$$E = K + V = \tfrac{1}{2}mv_0^2\cos^2(\omega t) + \tfrac{1}{2}k\left(\frac{v_0}{\omega}\right)^2\sin^2(\omega t)$$

$$= \tfrac{1}{2}mv_0^2 \tag{7.38}$$

where we have used Eq. (7) of Appendix 5.

As an undisturbed harmonic oscillator moves, the total energy remains constant. We say that the energy is conserved, and the system is said to be *conservative*. In a conservative system, the force on the particles can be obtained from a potential energy function. As the kinetic energy rises and falls, the potential energy changes so that the sum remains constant.

187

§7-3 / The Harmonic Oscillator. Linear Differential Equations with Constant Coefficients

The Damped Harmonic Oscillator—A Nonconservative System. We now discuss a harmonic oscillator that is subject to an additional force which is proportional to the velocity. An example is the frictional force due to fairly slow motion of an object through a fluid:

$$\mathbf{F}_f = -\zeta\mathbf{v} = -\zeta\frac{d\mathbf{r}}{dt} \tag{7.39}$$

where ζ is the friction constant. Such a force cannot be derived from a potential energy function, and the system is not conservative.

With such a force in addition to the spring force, the equation of motion is

$$F_z = -\zeta\frac{dz}{dt} - kz = m\left(\frac{d^2z}{dt^2}\right) \tag{7.40}$$

A harmonic oscillator governed by this equation of motion is called a *damped harmonic oscillator*. Equation (7.40) is a linear homogeneous equation with constant coefficients, so a trial solution of the form of Eq. (7.15) is guaranteed to work. The characteristic equation for λ is

$$\lambda^2 + \frac{\zeta\lambda}{m} + \frac{k}{m} = 0 \tag{7.41}$$

The solutions of this quadratic equation are

$$\lambda_1 = -\frac{\zeta}{2m} + \frac{\sqrt{(\zeta/m)^2 - 4k/m}}{2} \tag{7.42a}$$

$$\lambda_2 = -\frac{\zeta}{2m} - \frac{\sqrt{(\zeta/m)^2 - 4k/m}}{2} \tag{7.42b}$$

and the general solution to the differential equation is

$$z(t) = c_1 e^{\lambda_1 t} + c_2 e^{\lambda_2 t} \tag{7.43}$$

Problem 7-6

Show that Eq. (7.41) is the correct characteristic equation, that Eq. (7.42) gives the correct solutions to the characteristic equation, and that the function of Eq. (7.43) does satisfy Eq. (7.40). •

There are three distinct cases. In the first case, the quantity inside the square root in Eq. (7.42) is positive, so that λ_1 and λ_2 are both real quantities. This corresponds to a relatively large value of the friction constant, and the case is called *greater than critical damping*. In this case, the mass at the end of the spring does not oscillate, but will return smoothly to its equilibrium position of $z = 0$ if disturbed from this position.

Problem 7-7

From the fact that ζ, k, and m are all positive, show that λ_1 and λ_2 are both negative, and from this fact, show that

$$\lim_{t \to \infty} z(t) = 0 \qquad\qquad • \tag{7.44}$$

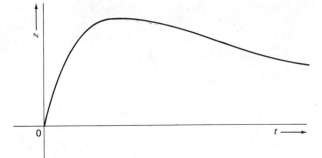

Figure 7-2. The position of a greater than critically damped harmonic oscillator as a function of time.

Figure 7-2 shows the position of a greater than critically damped oscillator as a function of time for a particular set of initial conditions.

The next case is that of relatively small values of ζ, or *less than critical damping*. If

$$\left(\frac{\zeta}{m}\right)^2 < \frac{4k}{m}$$

the quantity inside the square root in Eq. (7.42) is negative, and λ_1 and λ_2 are complex quantities.

$$\lambda_1 = -\frac{\zeta}{2m} + i\omega \tag{7.45a}$$

$$\lambda_2 = -\frac{\zeta}{2m} - i\omega \tag{7.45b}$$

where

$$\omega = \sqrt{\frac{k}{m} - \left(\frac{\zeta}{2m}\right)^2} \tag{7.46}$$

The solution thus becomes

$$z(t) = (c_1 e^{i\omega t} + c_2 e^{i\omega t})e^{-\zeta t/2m} \tag{7.47}$$

which can also be written in the form, similar to Eq. (7.27),

$$z(t) = [b_1 \cos(\omega t) + b_2 \sin(\omega t)]e^{-\zeta t/2m} \tag{7.48}$$

This shows $z(t)$ to be an oscillatory function times an exponentially decreasing function, giving the "ringing" behavior shown in Figure 7-3.

The final case is that of *critical damping*, in which the quantity inside the square root in Eq. (7.42) exactly vanishes. This case is not likely to happen by chance, but it is possible to construct an oscillating object, such as a galvanometer mirror or a two-pan balance beam, which is critically damped. The condition for critical damping is

$$\left(\frac{\zeta}{2m}\right)^2 = \frac{k}{m} \tag{7.49}$$

189

§7-3 / The Harmonic Oscillator. Linear Differential Equations with Constant Coefficients

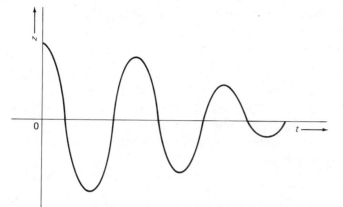

Figure 7-3. The position of a less than critically damped oscillator as a function of time for the initial conditions $z(0) = z_0$, $v_z(0) = 0$.

An interesting thing now happens to the solution of Eq. (7.43). Both values of λ are equal:

$$\lambda_1 = \lambda_2 = -\frac{\zeta}{2m} \tag{7.50}$$

so that Eq. (7.43) becomes

$$z(t) = (c_1 + c_2)e^{\lambda_1 t} = ce^{\lambda_1 t} \tag{7.51}$$

This cannot be a general solution, since a general solution for a second order linear equation must contain two arbitrary constants, and a sum of two constants does not constitute two separate constants.

The problem with our two solutions, $e^{\lambda_1 t}$ and $e^{\lambda_2 t}$, is called *linear dependence*. With two functions, it means that the functions are proportional to each other, so that they are not distinct solutions at all. If we have several solutions, they might be linearly dependent even if all the solutions are different. This would occur if one or more of the solutions happen to equal a linear combination of the others.

Since we do not have a general solution, there must be another family of solutions not included in the solution of Eq. (7.51), and we have to find it. One way to proceed is by using additional trial functions until we find one that works. The one that works is

$$z(t) = te^{\lambda t} \tag{7.52}$$

Problem 7-8

Substitute the trial solution of Eq. (7.52) into Eq. (7.40), using the condition of Eq. (7.49) to restrict the discussion to critical damping, and show that the equation is satisfied.   ●

Our general solution is now

$$z(t) = (c_1 + c_2 t)e^{\lambda_1 t} \tag{7.53}$$

For any particular set of initial conditions, we can find the appropriate values of c_1 and c_2. The behavior of a critically damped oscillator is much the same as that of Figure 7-2.

The Forced Harmonic Oscillator. Inhomogeneous Linear Differential Equations. If the function $g(t)$ in Eq. (7.14) is not zero, the equation is said to be inhomogeneous. We now discuss a harmonic oscillator that has an external force exerted on it in addition to the spring force.

If the external force on the oscillator depends only on the time, the equation of motion is

$$\frac{d^2z}{dt^2} + \left(\frac{k}{m}\right)z = \frac{F(t)}{m} \tag{7.54}$$

where $F(t)$ is the external force. The right-hand side of Eq. (7.54) is the inhomogeneous term. A method for solving such an equation is:

Step 1. Solve the equation obtained by deleting the inhomogeneous term. This equation is called the *complementary equation*, and the general solution to this equation is called the *complementary function*.

Step 2. Find a particular solution to the entire equation by whatever means may be necessary. The general solution to the entire equation is the sum of the complementary function and this particular solution.

Problem 7-9

If $z_c(t)$ is a general solution to the complementary equation and $z_p(t)$ is a particular solution to the inhomogeneous equation, show that $z_c + z_p$ is a solution to the inhomogeneous equation of Eq. (7.14). •

Table 7-1. Particular Trial Solutions for the Variation of Parameters Method*

Inhomogeneous Term	Trial Solution	Forbidden Characteristic Root[†]
1	A	0
t^n	$A_0 + A_1 t + A_2 t^2 + \cdots + A_n t^n$	0
$e^{\alpha t}$	$A e^{\alpha t}$	α
$t^n e^{\alpha t}$	$e^{\alpha t}(A_0 + A_1 t + A_2 t^2 + \cdots + A_n t^n)$	α
$e^{\alpha t} \sin(\beta t)$	$e^{\alpha t}[A \cos(\beta t) + B \sin(\beta t)]$	α, β
$e^{\alpha t} \cos(\beta t)$	$e^{\alpha t}[A \cos(\beta t) + B \sin(\beta t)]$	α, β

* *Source:* M. Morris and O. E. Brown, *Differential Equations*, 3rd ed., Prentice-Hall, Englewood Cliffs, N.J., 1952.

A, B, A_0, A_1, etc., are parameters to be determined. α and β are constants in the differential equation to be solved.

† The trial solution given will not work if the characteristic equation for the complementary differential equation has a root equal to the entry in this column. If such a root occurs with multiplicity k, multiply the trial solution by t^k to obtain a trial solution that will work.

191

§7-3 / The Harmonic Oscillator. Linear Differential Equations with Constant Coefficients

There is a method for finding a particular solution to a linear inhomo-geneous equation, known as the *variation of parameters*. If the inhomogenous term is a power of t, an exponential, a sine, a cosine, or a combination of these functions, this method can be used. One proceeds by taking a suitable trial function that contains parameters (constants whose values need to be determined). This is submitted into the inhomogeneous equation and the values of the parameters are found. Table 7-1 gives a list of suitable trial functions for various inhomogeneous terms.

Let us assume that the external force in Eq. (7.54) is

$$F(t) = F_0 \sin(\alpha t) \tag{7.55}$$

Use of Table 7-1 and determination of the parameters gives the particular solution

$$z_p(t) = \frac{F_0}{m(\omega^2 - \alpha^2)} \sin(\alpha t) \tag{7.56}$$

The complementary function is the solution given in Eq. (7.27), so the general solution is

$$z(t) = b_1 \cos(\omega t) + b_2 \sin(\omega t) + z_p(t) \tag{7.57}$$

Problem 7-10

Verify Eqs. (7.56) and (7.57). •

The motion of the forced harmonic oscillator shows some interesting features. The solution is a linear combination of the natural motion and a motion proportional to the external force. If the frequencies of these are not very different, a motion such as shown in Figure 7-4, known as "beating," can result.

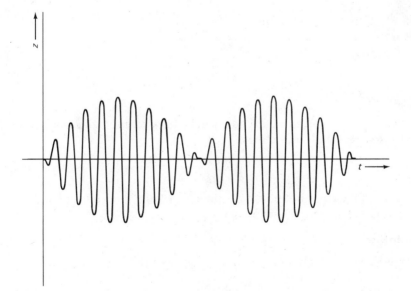

Figure 7-4. The position of a forced harmonic oscillator as a function of time for the case $\alpha = 1.1\omega$.

Section 7-4. DIFFERENTIAL EQUATIONS WITH SEPARABLE VARIABLES

In this section, we discuss equations that can be manipulated algebraically into the form

$$g(y)\left(\frac{dy}{dx}\right) = f(x) \tag{7.58}$$

where $g(y)$ is some integrable function of y and $f(x)$ is some integrable function of x. If we have manipulated the equation into the form of Eq. (7.58), we say that we have separated the variables, because we have no x dependence on the left-hand side of the equation, and no y dependence on the right-hand side.

To solve Eq. (7.58), we multiply both sides of the equation by dx and use

$$\left(\frac{dy}{dx}\right)dx = dy \tag{7.59}$$

We now have

$$g(y)\,dy = f(x)\,dx \tag{7.60}$$

We can integrate both sides of this equation to obtain

$$\int g(y)\,dy = \int f(x)\,dx + C \tag{7.61}$$

where C is a constant of integration, or we can do a definite integral

$$\int_{y_1}^{y_2} g(y)\,dy = \int_{x_1}^{x_2} f(x)\,dx \tag{7.62}$$

where

$$y_1 = y(x_1)$$
$$y_2 = y(x_2)$$

Example 7-3

In a first-order chemical reaction with no back reaction, the concentration of the sole reactant is governed by

$$-\frac{dc}{dt} = kc \tag{7.63}$$

where c is the concentration, t the time, and k the rate constant. Solve the equation to find c as a function of t.

Solution

We divide by c and multiply by dt to separate the variables:

$$\frac{1}{c}\frac{dc}{dt}\,dt = \frac{1}{c}\,dc = -k\,dt$$

We perform an indefinite integration

$$\int \frac{1}{c}\,dc = \ln(c) = -k \int dt + C = -kt + C$$

where C is a constant of integration. We take the exponential of each side of Eq. (7.73) to obtain

$$e^{\ln(c)} = c = e^{C}e^{-kt} = c(0)e^{-kt} \tag{7.64}$$

In the last step, we recognized that e^{C} had to equal the concentration at time $t = 0$. ●

Problem 7-11

In a second-order chemical reaction involving one reactant and having no back reaction,

$$-\frac{dc}{dt} = kc^2$$

Solve this differential equation by separation of variables. Do a definite integration from $t = 0$ to $t = t_1$ instead of an indefinite integration. ●

The separation of variables is a simple method. If you are faced with differential equation, and if you think that there is some chance that the method will work, try it.

Section 7-5. EXACT DIFFERENTIAL EQUATIONS

Sometimes you might be faced with an equation that cannot be manipulated into the form of Eq. (7.68), but which can be manipulated into the somewhat similar *pfaffian form*.

$$M(x, y)\,dx + N(x, y)\,dy = 0 \tag{7.65}$$

In Chapter 6, we discussed differentials like the left-hand side of this equation and found that some such differentials are exact, which means that they are differentials of functions. Others are inexact, which means that they are not differentials of functions.

The test for exactness is the fact that if

$$\boxed{\left(\frac{\partial M}{\partial y}\right)_x = \left(\frac{\partial N}{\partial x}\right)_y} \tag{7.66}$$

then the differential is exact.

If the differential is exact, there is a function $f(x, y)$ such that

$$df = M(x, y)\,dx + N(x, y)\,dy = 0 \tag{7.67}$$

which implies that

$$f(x, y) = C \tag{7.68}$$

where C is a constant, because only a constant has a differential that vanishes. Equation (7.68) is a solution to the differential equation, because this functional relation can be solved for y in terms of x.

In Section 6-6, we discussed the procedure for finding the function in Eq. (7.68), using a line integral:

$$f(x_1, y_1) = f(x_0, y_0) + \int_C df \tag{7.69}$$

where C is a curve beginning at (x_0, y_0) and ending at (x_1, y_1).

The most convenient curve is the rectangular path from (x_0, y_0) to (x_1, y_0) and then to (x_1, y_1). On the first part of this path, y is constant at y_0, so the dy integral vanishes and y is replaced by y_0 in the dx integral. On the second part of the path, x is constant at x_1, so the dx integral vanishes and x is replaced by x_1 in the dy integral:

$$f(x_1, y_1) = f(x_0, y_0) + \int_{x_0}^{x_1} M(x, y_0)\, dx + \int_{y_0}^{y_1} N(x_1, y)\, dy \tag{7.70}$$

Both integrals are now ordinary integrals, so we have a solution if we can perform the integrals. The solution will contain an arbitrary constant, because any constant can be added to both $f(x_1, y_1)$ and $f(x_0, y_0)$ in Eq. (7.70) without changing the equality.

Example 7-4

Solve the differential equation

$$2xy\, dx + x^2\, dy = 0$$

Solution

The equation is exact, because

$$\frac{\partial}{\partial y}(2xy) = 2x \qquad \text{and} \qquad \frac{\partial}{\partial x}(x^2) = 2x$$

We do a line integral from (x_0, y_0) to (x_1, y_0) and then to (x_1, y_1), letting $f(x, y)$ be the function whose differential must vanish:

$$0 = f(x_1, y_1) - f(x_0, y_0) = \int_{x_0}^{x_1} 2xy_0\, dx + \int_{y_0}^{y_1} x_1^2\, dy$$
$$= y_0 x_1^2 - y_0 x_0^2 + x_1^2 y_1 - x_1^2 y_0$$
$$= x_1^2 y_1 - x_0^2 y_0$$

We regard x_0 and y_0 as constant and drop the subscripts on x_1 and y_1:

$$f(x, y) = x^2 y = C$$

where C is a constant. Our solution is

$$y = \frac{C}{x^2}$$

•

Problem 7-12

a.   Show that the solution in Example 7-4 satisfies the equation.
b.   Solve the equation $(4x + y)\,dx + x\,dy = 0$.   •

Section 7-6.   SOLUTION OF INEXACT DIFFERENTIAL EQUATIONS BY THE USE OF INTEGRATING FACTORS

If we have an inexact pfaffian differential equation

$$M(x, y)\,dx + N(x, y)\,dy = 0 \qquad (7.71)$$

we cannot use the method of Section 7-5. However, as mentioned in Section 6-5, some inexact differentials yield an exact differential when multiplied by an integrating factor.

If the function $g(x, y)$ is an integrating factor for the differential in Eq. (7.71), then

$$g(x, y)M(x, y)\,dx + g(x, y)N(x, y)\,dy = 0 \qquad (7.72)$$

is an exact differential equation that can be solved by the method of Section 7-5. A solution for Eq. (7.72) will also be a solution for Eq. (7.71).

Example 7-5

Solve the differential equation

$$\frac{dy}{dx} = \frac{y}{x}$$

Solution

We convert the equation to the pfaffian form:

$$y\,dx - x\,dy = 0$$

Test for exactness:

$$\left(\frac{\partial y}{\partial y}\right)_x = 1$$

$$\left[\frac{\partial(-x)}{\partial x}\right]_y = -1$$

The equation is not exact, but $1/x^2$ is an integrating factor. The equation becomes

$$\left(\frac{y}{x^2}\right)dx - \left(\frac{1}{x}\right)dy = 0 \qquad (7.73)$$

This is exact:

$$\left[\frac{\partial(y/x^2)}{\partial y}\right]_x = \frac{1}{x^2}$$

$$\left[\frac{\partial(-1/x)}{\partial x}\right]_y = \frac{1}{x^2}$$

We can solve Eq. (7.73) by the method of Section 7-5:

$$0 = \int_{x_0}^{x_1} \left(\frac{y_0}{x^2}\right) dx - \int_{y_0}^{y_1} \left(\frac{1}{x_1}\right) dy$$

$$= -y_0\left(\frac{1}{x_1} - \frac{1}{x_0}\right) - \frac{1}{x_1}(y_1 - y_0) = \frac{y_0}{x_0} - \frac{y_1}{x_1}$$

We regard x_0 and y_0 as constants, so that

$$\frac{y}{x} = \frac{y_0}{x_0} = C = \text{constant}$$

or

$$y = Cx$$

This is a general solution, since it contains one arbitrary constant, C, and the original equation was first order. •

If an inexact differential has one integrating factor, it has an infinite number. Therefore, there can be other integrating factors for our equation.

Problem 7-13

Show that $1/y^2$ and $1/(x^2 + y^2)$ are integrating factors for the equation in Example 7-5 and that they lead to the same solution. •

Section 7-7. WAVES IN A STRING.
PARTIAL DIFFERENTIAL EQUATIONS

All the differential equations until now in this chapter have contained only one independent variable. Such differential equations are called *ordinary differential equations*. Differential equations that contain partial derivatives are called *partial differential equations*. Differential equations also occur that contain more than one dependent variable, but you must have a set of such equations and solve them simultaneously. We will not discuss simultaneous differential equations, but you can read about them in some of the books listed at the end of the chapter.

Partial differential equations can be very difficult to solve. However, if you cannot obtain an exact solution, it is nearly always possible to get a good numerical approximation to the solution by using a computer. This is discussed in some of the books listed at the end of the chapter.

We discuss only a rather simple method of solving a partial differential equation, devoting most of the section to an example, the classical equation of motion of a flexible string. You might think that this would be of interest only to a mathematically inclined violinist, but there are similarities between this equation and the Schrödinger equation. In fact, Schrödinger developed his equation from his knowledge of the equation of motion of a string.

197

§7-7 / Waves in a String. Partial Differential Equations

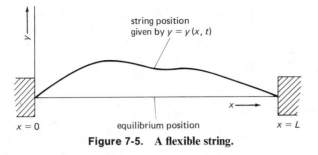

Figure 7-5. A flexible string.

The flexible string that we discuss is a mathematical model. It resembles a real string, but is completely flexible. No force is required to bend the string, so that bending the string in one place puts no force on another part of the string. Our model is also restricted to small vibrations, so that it is not appreciably stretched.

Figure 7-5 shows the string. We choose one end of the string as our origin of coordinates and use the equilibrium position of the string as our x axis. We assume that both ends of the string are fixed in position. The location of the string at a particular value of x is given by a value of y and a value of z for that value of x. Since the string moves, y and z are functions of time as well as of x:

$$y = y(x, t) \tag{7.74a}$$

$$z = z(x, t) \tag{7.74b}$$

The equation of motion of the string is derived by writing Newton's second law for a small segment of the string and taking a mathematical limit as the length of the segment becomes infinitesimal. The result is a partial differential equation

$$\left(\frac{\partial^2 y}{\partial t^2}\right) = \frac{T}{\rho}\left(\frac{\partial^2 y}{\partial x^2}\right) = c^2\left(\frac{\partial^2 y}{\partial x^2}\right) \tag{7.75}$$

and a similar equation for z. In this equation, T is the magnitude of the tension force on the string and ρ the mass of the string per unit length. The quantity c turns out to be the speed of propagation of a wave along the string. Since the two equations are independent of each other, we can solve for y and z separately, and the two solutions will be identical except for the symbol used for the dependent variable.

Solution by Separation of Variables. We do not seek a general solution for Eq. (7.75), but seek only a family of solutions that can be written in the form

$$y(x, t) = \psi(x)\theta(t) \tag{7.76}$$

This is called a *solution with the variables separated*. We regard it as a trial solution, and substitute it into the differential equation to see if it works. Since ψ does not depend on t and θ does not depend on x, the result is

$$\psi(x)\left(\frac{d^2\theta}{dt^2}\right) = c^2\theta(t)\left(\frac{d^2\psi}{dx^2}\right) \tag{7.77}$$

We write ordinary derivatives since we have functions of only one variable.

We now want to separate the variables by manipulating Eq. (7.77) into a form in which one term contains no x dependence and the other term contains no t dependence, just as we manipulated Eq. (7.58) into a form with only one variable in each term. We can do this by dividing both sides of Eq. (7.77) by the product $\psi(x)\theta(t)$. We also divide by c^2, but this is not essential.

$$\frac{1}{\psi(x)}\frac{d^2\psi}{dx^2} = \frac{1}{c^2\theta(t)}\frac{d^2\theta}{dt^2} \tag{7.78}$$

The variables are now separated, and we can use the fact that x and t are both independent variables to extract two ordinary differential equations from Eq. (7.78). We do this by recognizing that if we temporarily keep t fixed at some value, we do not have to keep x fixed. Thus, even if x varies, the left-hand side of Eq. (7.78) must be constant, because it equals a quantity that we can keep fixed.

Therefore,

$$\frac{1}{\psi(x)}\frac{d^2\psi}{dx^2} = \text{constant} = -k^2 \tag{7.79}$$

and

$$\frac{1}{c^2\theta(t)}\frac{d^2\theta}{dt^2} = -k^2 \tag{7.80}$$

We choose to call the constant by the symbol $-k^2$, because it will turn out to be negative.

We now multiply Eq. (7.79) by $\psi(x)$ and multiply Eq. (7.80) by $c^2\theta(t)$:

$$\frac{d^2\psi}{dx^2} + k^2\psi = 0 \tag{7.81}$$

$$\frac{d^2\theta}{dt^2} + k^2c^2\theta = 0 \tag{7.82}$$

The separation of variables is complete, and we have two ordinary differential equations that we can solve.

Except for the symbols used, both of these equations are the same as Eq. (7.12). We can obtain solutions by transcribing the solution to that equation with appropriate changes in symbols:

$$\psi(x) = a_1\cos(kx) + a_2\sin(kx) \tag{7.83}$$

$$\theta(t) = b_1\cos(kct) + b_2\sin(kct) \tag{7.84}$$

199

§7-7 / Waves in a String. Partial Differential Equations

These are general solutions to the ordinary differential equations, but we may not have a general solution to Eq. (7.75), because there may be solutions that are not of the form of Eq. (7.76).

We are now ready to consider a specific case and to find values for the arbitrary constants that make our solution describe the specific case. We consider a string of length L with the ends fixed, as in Figure 7-5. Our first boundary conditions are

$$\psi(0) = 0 \tag{7.85}$$

and

$$\psi(L) = 0 \tag{7.86}$$

Equation (7.85) requires that

$$a_1 = 0 \tag{7.87}$$

since the cosine of zero is unity. Equation (7.86) requires that the argument of the sine function in Eq. (7.83) be equal to some integer times π for $x = L$, because

$$\sin(n\pi) = 0 \qquad (n = 0, 1, 2, \dots) \tag{7.88}$$

Therefore,

$$k = \frac{n\pi}{L} \qquad (n = 1, 2, 3, \dots) \tag{7.89}$$

We exclude $n = 0$, because ψ would then vanish and the string would be stationary at its equilibrium position. We now have

$$\psi(x) = a_2 \sin\left(\frac{n\pi x}{L}\right) \tag{7.90}$$

We have used the boundary conditions required by the nature of the system. Any other boundary conditions will apply to particular cases. For example, let us say that at $t = 0$, the string happens to be passing through its equilibrium position, which is $y = 0$ for all x. If so, then $b_1 = 0$, and we have

$$y(x, t) = A \sin\left(\frac{n\pi x}{L}\right) \sin\left(\frac{n\pi ct}{L}\right) \tag{7.91}$$

where we write $A = a_2 b_2$. The maximum amplitude is A, and another boundary condition is necessary to specify its value.

Figure 7-6 shows the function $\psi(x)$ for several values of n. Each curve represents the shape of the string at an instant when θ equals $+1$. At other times, the string is vibrating between such a position and a position given by $-\psi(x)$. Notice that there are points at which the string is stationary. These points are called *nodes*, and the number of nodes is $n - 1$.

If λ is the wavelength, or the distance for the sine function in ψ to go through a complete period, we have

$$n\lambda = 2L \tag{7.92}$$

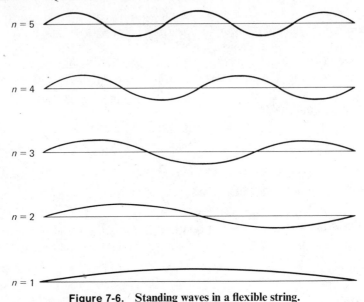

Figure 7-6.   Standing waves in a flexible string.

The period of oscillation is the time required for the sine function in θ to go through a complete period, so if τ is the period,

$$\frac{n\pi c\tau}{L} = 2\pi \tag{7.93}$$

The frequency v is the reciprocal of the period:

$$v = \frac{nc}{2L} = \left(\frac{n}{2L}\right)\left(\frac{T}{\rho}\right)^{1/2} \tag{7.94}$$

In musical acoustic, the pattern of oscillation corresponding to $n = 1$ is called the *fundamental*, that for $n = 2$ is the *first overtone*, etc. The fundamental is also called the *first harmonic*, the first overtone is called the *second harmonic*, etc.

Problem 7-14

A certain violin string has a mass per unit length of 10.00 mg cm^{-1} and a length of 50 cm. Find the tension force necessary to make it produce a fundamental tone of A above middle C (440 oscillations per second).   ●

When a string in a musical instrument is struck or bowed, it will usually not vibrate according to the wave function for a single value of n. The following Fourier series satisfies Eq. (7.75), and can represent any possible

201

§7-7 / Waves in a String. Partial Differential Equations

motion of the string:

$$y(x, t) = \sum_{n=1}^{\infty} \sin\left(\frac{n\pi x}{L}\right)\left[a_n\cos\left(\frac{n\pi ct}{L}\right) + b_n\sin\left(\frac{n\pi ct}{L}\right)\right] \qquad (7.95)$$

The strengths of the different harmonics are represented by the values of the coefficients a_n and b_n. Different musical instruments have different relative strengths of different harmonics.

Problem 7-15

Show that the function in Eq. (7.95) satisfies Eq. (7.75).   ●

Waves in a string with fixed nodes are called *standing waves*. For a string with fixed ends, these are the only waves that occur. For an infinitely long string, *traveling waves* can also occur. The following is a traveling wave:

$$y(x, t) = A\sin[k(x - ct)] \qquad (7.96)$$

Problem 7-16

Show that the function of Eq. (7.96) satisfies Eq. (7.75)   ●

Notice that the function in Eq. (7.96) is not a solution in which the variables are separated. However, using Eq. (14) of Appendix 5, we can show that

$$A\sin k(x - ct) = A[\sin(kx)\cos(kct) - \cos(kx)\sin(kct)] \qquad (7.97)$$

The function of Eq. (7.96) can be shown to be a traveling wave by showing that a particular node in the wave moves along the string. When $t = 0$, there is a node at $x = 0$. After a time of 1 s has passed, this same node has moved along the string a distance equal to c. This follows from the fact that $c \times (1\text{ s})$ is the value of x required to make the argument of the sine vanish.

Problem 7-17

Find the speed of propagation of a traveling wave in an infinite string with the same mass per unit length and the same tension force as the violin string in Problem 7-14.   ●

ADDITIONAL PROBLEMS

7.18

A particle moves in one dimension, say along the x axis. The only force acting on the particle is a frictional force given by

$$F_x = -\zeta\left(\frac{dx}{dt}\right)$$

From Newton's second law, write the equation of motion of the particle. Find the general solution to this equation, and obtain the particular solution that applies if $x(0) = 0$ and $v_x(0) = v_0 = $ constant. Draw a sketch of the position as a function of time.

7.19

A particle moves along the z axis. It is acted upon by a constant gravitational force equal to $-\mathbf{k}mg$, where $\mathbf{k}$ is the unit vector in the z direction. It is also acted on by a frictional force given by

$$\mathbf{F}_f = -\mathbf{k}\left(\frac{dz}{dt}\right)\zeta$$

From Newton's second law, find the equation of motion and solve it. Find z as a function of time if $z(0) = 0$ and $v_z(0) = 0$.

7.20

A body sliding on a solid surface experiences a frictional force that is constant and in the opposite direction to the velocity, if the particle is moving, and zero it is not moving. Find the position of the particle as a function of time if it moves only in the x direction and the initial position is $x(0) = 0$ and the initial velocity is $v_x(0) = v_0 = $ constant.

You can proceed as though the constant force were always present and then "cut the solution off" at the first point at which the velocity vanishes. In other words, after this point, just say that the particle is fixed.

7.21

A certain tank is filled with a solution that is stirred rapidly enough so that it remains uniform at all times. A solution of the same solute is flowing into the tank at a fixed rate of flow, and an overflow pipe allows solution from the tank to flow out at the same rate. If the solution flowing in has a fixed concentration, write and solve the differential equation that governs the number of moles of solute in the tank. The inlet pipe allows A moles per second to flow in and the overflow pipe allows Bn moles per second to flow out, where A and B are constants and n is the number of moles of solute in the tank. Find the value of A and B that correspond to a volume in the tank of 100 liters, an input of 1.000 liter h^{-1} of a solution with 1.000 mol liter^{-1}, and an output of 1.000 liter h^{-1} of the solution in the tank. Find the concentration in the tank after 5.00 h, if the initial concentration is zero.

7.22

An nth-order chemical reaction with one reactant obeys the differential equation

$$\frac{dc}{dt} = -kc^n$$

where c is the concentration of the reactant and k is a constant. Solve this differential equation by separation of variables. If the initial concentration is c_0 moles per liter, find an expression for the time required for half of the reactant to react.

203

§7-7 / Waves in a String. Partial Differential Equations

7.23

Find the solution to the differential equation

$$\left(\frac{d^3y}{dx^3}\right) - 2\left(\frac{d^2y}{dx^2}\right) - \left(\frac{dy}{dx}\right) + 2y = -xe^x$$

7.24

Test the following equations for exactness, and solve if exact:
a. $(x^2 + xy + y^2)dx + (4x^2 - 2xy + 3y^2)dy = 0$
b. $ye^x dx + e^x dy = 0$
c. $[2xy - \cos(x)]dx + (x^2 - 1)dy = 0$

7.25

Solve the differential equation

$$\left(\frac{dy}{dx}\right) + y\cos(x) = e^{-\sin(x)}$$

7.26

Find a particular solution of

$$\left(\frac{d^2y}{dx^2}\right) - 4y = 2e^{3x} + \sin(x)$$

7.27

Radioactive nuclei decay according to the same differential equation which governs first-order chemical reactions, Eq. (7.63). In living matter, the isotope ^{14}C is continually replaced as it decays, but it decays without replacement beginning with the death of the organism. The half-life of the isotope (the time required for half of an initial sample to decay) is 5730 years. If a sample of charcoal from an archaeological specimen exhibits 0.97 disintegrations of ^{14}C per gram of carbon per minute and wood recently taken from a living tree exhibits 15.3 disintegrations of ^{14}C per gram of carbon per minute, estimate the age of the charcoal.

7.28

A pendulum of length L oscillates in a vertical plane. Assuming that the mass of the pendulum is all concentrated at the end of the pendulum, show that it obeys the differential equation

$$L\left(\frac{d^2\theta}{dt^2}\right) = -g\sin(\theta)$$

where g is the acceleration due to gravity and θ the angle between the pendulum and the vertical. This equation cannot be solved exactly. For small oscillations such that

$$\sin(\theta) \approx \theta$$

find the solution to the equation. What is the period of the motion? What is the frequency? Evaluate these quantities if $L = 1.000$ m and if $L = 10.000$ m.

ADDITIONAL READING

Max Morris and Orley E. Brown, *Differential Equations*, 4th ed., Prentice-Hall, Engle-woods Cliffs, N.J., 1964.

This book is a "standard" text for a beginning course in differential equations and is a fairly clear introduction to the subject. It contains a number of examples of differential equations in physical problems.

Earl A. Coddington, *An Introduction to Ordinary Differential Equations*, Prentice-Hall, Englewood Cliffs, N.J., 1961.

A general introductory treatment that assumes only a knowledge of elementary calculus.

Wilfred Kaplan, *Ordinary Differential Equations*, Addison-Wesley, Reading, Mass., 1958.

An introductory book with numerous applications.

Ralph Palmer Agnew, *Differential Equations*, 2nd ed., McGraw-Hill Book Company, New York, 1960.

This book puts some emphasis on underlying theory, and contains some interesting "remarks" which clarify some points.

E. Kamke, *Differentialglechungen—Lösungsmethoden und Lösungen* (Differential Equa-tions—Methods of Solution and Solutions), Vol. I, 2nd ed., Akademische Verlagsgesell-schaft, Leipzig, 1943 (also reprinted by J. W. Edwards, Ann Arbor, Mich., 1945.)

If you read German, you can read a summary of results, but even if you do not, you might be able to use the large compilation of special equations and their solutions which is in-cluded.

Paul W. Berg and James L. McGregor, *Elementary Partial Differential Equations*, Holden-Day, San Francisco, 1966.

This book discusses a number of physical problems, including classical wave equations and heat-flow equations.

David L. Powers, *Boundary Value Problems*, Academic Press, New York, 1972.

This book discusses some heat-flow and wave equations.

Alice B. Dickinson, *Differential Equations, Theory and Use in Time and Motion*, Addison-Wesley Publishing Co., Reading, Mass., 1972.

This book puts special emphasis on the construction of differential equations for models of physical systems. It has a large chapter on solutions which are series.

Lester R. Ford, *Differential Equations*, 2nd ed., McGraw-Hill Book Company, New York, 1955.

This book provides "cookbook" methods for solving quite a few kinds of differential equations.

CHAPTER EIGHT

Operators, Matrices, and Group Theory

Section 8-1. INTRODUCTION

A *mathematical operator* is a symbol standing for the carrying out of a mathematical operation. An example is the derivative operator d/dx, sometimes denoted by $\hat{D}_x$. If f is a differentiable function of x, we write

$$\hat{D}_x f = \frac{df}{dx} \tag{8.1}$$

Another example is the multiplication operator $\hat{g}$. When $\hat{g}$ operates on a function f, the result is the product of the function g with the function f.

A *matrix* is a list of quantities, arranged in rows and columns. If the matrix **A** has four rows and four columns, we call it a 4 by 4 matrix:

$$\mathbf{A} = \begin{bmatrix} a_{11} & a_{12} & a_{13} & a_{14} \\ a_{21} & a_{22} & a_{23} & a_{24} \\ a_{31} & a_{32} & a_{33} & a_{34} \\ a_{41} & a_{42} & a_{43} & a_{44} \end{bmatrix} \tag{8.2}$$

It would appear that matrices and mathematical operators have little in common. However, we will define an algebra involving operators and an algebra involving matrices and will find similarities.

The third topic that we discuss is *group theory*. This is a branch of mathematics that has a useful application to symmetry operators. It can provide useful information about quantum-mechanical wave functions for symmetrical molecules.

After studying this chapter, you should be able to do all of the following:

1. **Perform the elementary operations of operator algebra.**
2. **Identify and use symmetry operators associated with a symmetrical molecule.**

3. Perform the elementary operations of matrix algebra, including finding the inverse of a matrix.

4. Identify a group of symmetry operators and construct a multiplication table for the group.

Section 8-2. OPERATOR ALGEBRA

When an operator operates on a function, the result will generally be another function.

Example 8-1

Let the operator $\hat{A}$ be given by

$$\hat{A} = \hat{x} + \frac{d}{dx}$$

Find $\hat{A}f$ if $f = a\sin(bx)$, where a and b are constants.

Solution

$$\hat{A}a\sin(bx) = xa\sin(bx) + ab\cos(bx) \qquad \bullet$$

If the result of operating on a function with an operator is a function that is proportional to the original function, the function is called an *eigenfunction* of that operator, and the proportionality constant is called an *eigenvalue*. If

$$\boxed{\hat{A}f = af} \qquad (8.3)$$

then f is an eigenfunction of $\hat{A}$ and a is its eigenvalue.

Example 8-2

Find the eigenfunctions and corresponding eigenvalues for the operator d^2/dx^2.

Solution

We need to find a function $f(x)$ and a constant a such that

$$\frac{d^2f}{dx^2} = af$$

This is a differential equation which is solved by the method of Section 7-3. The general solution is

$$f(x) = A\exp(\sqrt{a}\,x) + B\exp(-\sqrt{a}\,x)$$

where A and B are constants. Since no boundary conditions were stated, the eigenvalue a can take on any value, as can the constants A and B. $\quad \bullet$

Problem 8-1

Find the eigenfunctions of the operator $-i\dfrac{d}{dx}$, where $i = \sqrt{-1}$.   ●

Mathematical Operations on Operators.   Since a mathematical operator is a symbol that stands for the carrying out of an operation, it is not obvious that we can define a kind of algebra in which we manipulate these symbols. However, it possible to do so.

If $\hat{A}$ and $\hat{B}$ are two operators, we define the sum of these operators by

$$(\hat{A} + \hat{B})f = \hat{A}f + \hat{B}f \tag{8.4}$$

where f is some function on which $\hat{A}$ and $\hat{B}$ can operate.

The product of two operators is defined as the successive operation of the operators, with the one on the right operating first. That is, if

$$\hat{C} = \hat{A}\hat{B} \tag{8.5}$$

then

$$\hat{C}f = \hat{A}(\hat{B}f) \tag{8.6}$$

The result of $\hat{B}$ operating on f is in turn operated on by $\hat{A}$ and the result is called the *result of operating on f with the product* $\hat{A}\hat{B}$.

Equations such as Eq. (8.5) are called *operator equations*. The two sides are equal in the sense that if each is applied to some function, the two results are the same. Some operator equations are valid for all functions, and others hold only for particular functions.

Example 8-3

Find the operator equal to the operator product $\dfrac{d}{dx}\hat{x}$.

Solution

We take an arbitrary differentiable function $f = f(x)$, and apply the operator product to it:

$$\frac{d}{dx}\hat{x}f = \hat{x}\frac{df}{dx} + f\frac{dx}{dx} = \left(\hat{x}\frac{d}{dx} + \hat{1}\right)f$$

We can write the operator equation

$$\frac{d}{dx}\hat{x} = \hat{x}\frac{d}{dx} + \hat{1}$$

where $\hat{1}$ is the operator for multiplying by unity.   ●

Problem 8-2

Find the operator equal to the operator product $\dfrac{d^2}{dx^2}x$.   ●

The difference of two operators is given by

$$(\hat{A} - \hat{B}) = \hat{A} + (-\hat{1})\hat{B} \tag{8.7}$$

where $-\hat{1}$ is the operator for multiplying by -1.

We now have an operator algebra in which we carry out the operations of addition and multiplication on the operators themselves. These operations have the following properties:

Operator multiplication is associative. This means that, if $\hat{A}$, $\hat{B}$ and $\hat{C}$ are operators, then

$$(\hat{A}\hat{B})\hat{C} = \hat{A}(\hat{B}\hat{C}) \tag{8.8}$$

Operator multiplication and addition are distributive. This means that, if $\hat{A}$, $\hat{B}$, and $\hat{C}$ are operators

$$\hat{A}(\hat{B} + \hat{C}) = \hat{A}\hat{B} + \hat{A}\hat{C} \tag{8.9}$$

Operator multiplication is not necessarily commutative. This means it is possible that

$$\hat{A}\hat{B} \neq \hat{B}\hat{A} \tag{8.10}$$

If the operator $\hat{A}\hat{B}$ is equal to the operator $\hat{B}\hat{A}$, then $\hat{A}$ and $\hat{B}$ are said to commute. The *commutator* of $\hat{A}$ and $\hat{B}$ is denoted by $[\hat{A}, \hat{B}]$ and defined by

$$\boxed{[\hat{A}, \hat{B}] = \hat{A}\hat{B} - \hat{B}\hat{A}} \tag{8.11}$$

If $\hat{A}$ and $\hat{B}$ commute, then $[\hat{A}, \hat{B}] = \hat{0}$, where $\hat{0}$ is the *null operator*.

Example 8-4

Find the commutator $\left[\dfrac{d}{dx}, \hat{x}\right]$.

Solution

We apply the commutator to an arbitrary function $f(x)$:

$$\left[\frac{d}{dx}, \hat{x}\right] f = \frac{d}{dx}(xf) - x\frac{df}{dx} = x\frac{df}{dx} + f - x\frac{df}{dx} = f$$

Therefore,

$$\left[\frac{d}{dx}, \hat{x}\right] = \hat{1} = \hat{E} \qquad \bullet \quad (8.12)$$

The operator $\hat{1}$ which occurred in Example 8-4 is usually denoted by the symbol $\hat{E}$, and is called the *identity operator*.

Problem 8-3

Find the commutator $\left[\hat{x}^2, \dfrac{d^2}{dx^2}\right]$.

$\bullet$

Here are a few facts that will predict in almost all cases whether two operators will commute:

1. An operator containing a multiplication by a function of x and one containing d/dx will not generally commute.
2. Two multiplication operators commute. If g and h are functions of the same or different independent variables, then

$$[\hat{g}, \hat{h}] = 0 \tag{8.13}$$

3. Operators acting on different independent variables commute. For example,

$$\left[\hat{x}\frac{d}{dx}, \frac{d}{dy}\right] = 0 \tag{8.14}$$

4. An operator for multiplication by a constant commutes with any other operator.

Problem 8-4

Show that Eq. (8.14) is correct, and that statement (4) is correct. •

Since we have defined the product of two operators, we have a definition for the powers of an operator. An operator raised to the nth power is the operator for n successive applications of the original operator:

$$\hat{A}^n = \hat{A}\hat{A}\hat{A}\cdots\hat{A} \qquad (n \text{ factors}) \tag{8.15}$$

Example 8-5

If the operator $\hat{A}$ is $\hat{x} + \dfrac{d}{dx}$, find $\hat{A}^3$.

Solution

$$\hat{A}^3 = \left(\hat{x} + \frac{d}{dx}\right)\left(\hat{x} + \frac{d}{dx}\right)\left(\hat{x} + \frac{d}{dx}\right)$$

$$= \left(\hat{x} + \frac{d}{dx}\right)\left(\hat{x}^2 + \frac{d}{dx}\hat{x} + \hat{x}\frac{d}{dx} + \frac{d^2}{dx^2}\right)$$

$$= \hat{x}^3 + \hat{x}\frac{d}{dx}\hat{x} + \hat{x}^2\frac{d}{dx} + \hat{x}\frac{d^2}{dx^2} + \frac{d}{dx}\hat{x}^2 + \frac{d^2}{dx^2}\hat{x} + \frac{d}{dx}\hat{x}\frac{d}{dx} + \frac{d^3}{dx^3}$$

Notice that the order of the individual factors in each term must be maintained, because the two terms in the operator do not commute with each other. •

Problem 8-5

a. For the operator $\hat{A}$ in Example 8-5, find $\hat{A}^3 f$ if $f(x) = \sin(ax)$.

b. Find an expression for $\hat{B}^2$ if $\hat{B} = \hat{x}\dfrac{d^2}{dx^2}$ and find $\hat{B}^2 f$ if $f = bx^4$. •

We do not define the operation of dividing by an operator. However, we define the inverse of an operator as that operator which "undoes" what the first operator does. The inverse of $\hat{A}$ is denoted by $\hat{A}^{-1}$, and

$$\hat{A}^{-1}\hat{A} = \hat{E} \tag{8.16}$$

or

$$\hat{A}^{-1}\hat{A}f = \hat{E}f = f \tag{8.17}$$

Not all operators possess inverses. For example, there is no inverse for $\hat{0}$. The inverse of a multiplication operator is quite simple. It is the operator for multiplication by the reciprocal of the original quantity.

Sometimes operator algebra can be used as a method of solving differential equations.[1] A linear differential equation with constant coefficients can be written in operator notation and solved by operator algebra. For example, the equation

$$\frac{d^2y}{dx^2} - 3\frac{dy}{dx} + 2y = 0 \tag{8.18}$$

can be written as

$$(\hat{D}_x^2 - 3\hat{D}_x + 2)y = 0 \tag{8.19}$$

or as an operator equation

$$(\hat{D}_x^2 - 3\hat{D}_x + 2) = 0 \tag{8.20}$$

Using operator algebra, we manipulate this equation as though it were an ordinary equation. We factorize it to obtain

$$(\hat{D}_x - 2)(\hat{D}_x - 1) = 0 \tag{8.21}$$

The two roots are obtained from

$$\hat{D}_x - 2 = 0 \tag{8.22a}$$

and

$$\hat{D}_x - 1 = 0 \tag{8.22b}$$

Equation (8.22a) is the same as the equation

$$\frac{dy}{dx} - 2y = 0 \tag{8.23a}$$

and Eq. (8.22b) is the same as the equation

$$\frac{dy}{dx} - y = 0 \tag{8.23b}$$

The solution to Eq. (8.23a) is

$$y = e^{2x} \tag{8.24a}$$

---

[1] See, for example, Max Morris and Orley Brown, *Differential Equations*, 3rd ed., Prentice-Hall, Englewood Cliffs, N.J., 1952, pp. 86–89.

and the solution to Eq. (8.23b) is

$$y = e^x \tag{8.24b}$$

Since both of these must be solutions to the original equation, the general solution is

$$y = c_1 e^{2x} + c_2 e^x \tag{8.25}$$

where c_1 and c_2 are arbitrary constants.

Operators in Quantum Mechanics. Operator algebra is important in quantum mechanics, because one of the postulates of the theory is that for every mechanical quantity, there is a mathematical operator. The eigenfunctions and eigenvalues of these operators play a central role in the theory. Operators that commute with each other can have a set of functions which are eigenfunctions of both operators. Since the eigenvalues of an operator are the possible values that measurement of the corresponding mechanical quantity can give, two quantities whose operators commute can simultaneously have well-defined values.

Section 8-3. SYMMETRY OPERATORS

Many common objects are said to be *symmetrical*. The most symmetrical object is a sphere, which looks just the same no matter which way it is turned. A cube, although less symmetrical than a sphere, still has 24 different orientations in which it looks the same. Many biological organisms have approximate *bilateral symmetry*, meaning that the left side looks like a mirror image of the right side.

We define symmetry operators by the action they have on a single point in three-dimensional space. We denote the position of this reference point by its cartesian coordinates, keeping the cartesian coordinate axes fixed as the point moves. We consider only symmetry operators that do not move a reference point if it happens to be at the origin. Such operators are called *point symmetry operators*.

The action of the operator $\hat{O}$ is specified by writing

$$\hat{O}(x_1, y_1, z_1) = (x_2, y_2, z_2) \tag{8.26}$$

where x_1, y_1, z_1 are the coordinates of the original location of the point and x_2, y_2, z_2 are the coordinates of the location to which the operator takes the point.

The first operator that we define is the *identity operator*, $\hat{E}$, which does not move a reference point:

$$\hat{E}(x_1, y_1, z_1) = (x_1, y_1, z_1) \tag{8.27}$$

If $\mathbf{r}_1$ is the vector with components (x_1, y_1, z_1), this can be written

$$\hat{E}\mathbf{r}_1 = \mathbf{r}_1 \tag{8.28}$$

The *inversion operator* is denoted by $\hat{\imath}$. It moves the reference point on a straight line from its original position through the origin to a location at the same distance from the origin as the original position:

$$\hat{i}(x_1, y_1, z_1) = (-x_1, -y_1, -z_1) \tag{8.29}$$

or

$$\boxed{\hat{i}\mathbf{r}_1 = -\mathbf{r}_1} \tag{8.30}$$

For each symmetry operator, we define a *symmetry element*, which is a point, a line, or a plane with respect to which the symmetry operation is performed. For the inversion operator, the symmetry element is the origin. For all point symmetry operators, the symmetry element must include the origin.

A *reflection operator* moves a representative point on a line perpendicular to a specified plane, through the plane to a location at the same distance from the plane as the original point. This motion is called *reflection through the plane*. The specified plane is the symmetry element, and must pass through the origin. There is a different reflection operator for each of the infinitely many planes passing through the origin.

The operator $\hat{\sigma}_h$ corresponds to reflection through the $x-y$ plane (the h subscript stands for "horizontal"). Figure 8-1 shows the action of the $\hat{\sigma}_h$

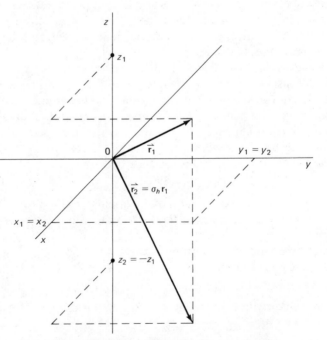

Figure 8-1. The action of the reflection operator, $\hat{\sigma}_h$.

operator. A reflection operator whose symmetry element is a vertical plane is denoted by $\hat{\sigma}_v$, but you must separately specify which vertical plane is the symmetry element. The action of $\hat{\sigma}_h$ corresponds to

$$\hat{\sigma}_h(x_1, y_1, z_1) = (x_1, y_1, -z_1) \qquad \text{(8.31)}$$

Problem 8-6

Write an equation similar to Eq. (8.31) for the $\hat{\sigma}_v$ operator whose symmetry element is the x–z plane, and one for the $\hat{\sigma}_v$ operator whose symmetry element is the y–z plane. •

Next we have a class of *rotation operators*. An ordinary rotation, in which a representative point moves as if it were part of a rigid object being rotated about an axis, is called a *proper rotation*. The axis of rotation is the symmetry element, and the action of the rotation operator is to move the representative point along an arc, staying at a fixed perpendicular distance from the axis. In addition to specifying the axis of rotation, one must specify the direction of rotation and the angle of rotation. The axis of rotation must pass through the origin. We consider only angles of rotation such that n applications of the rotation operator will produce exactly one complete rotation, where n is an integer. Such a rotation operator is denoted by $\hat{C}_n$. The axis of rotation must be specified. For example, the operator for a rotation of 90° about the z axis is sometimes called $\hat{C}_4(z)$. The direction of rotation is taken as counter-clockwise when viewed from the positive end of the axis.

Figure 8-2 shows the action of the $\hat{C}_4(z)$ operator. For this operator,

$$\hat{C}_4(z)(x_1, y_1, z_1) = (-y_1, x_1, z_1) \qquad \text{(8.32)}$$

so that

$$x_2 = -y_1, y_2 = x_1, z_2 = z_1 \qquad \text{(8.33)}$$

Problem 8-7

Find the following.
a. $\hat{C}_4(z)(1, -4, 6)$
b. $C_2(x)(1, 2, -3)$ •

The final class of symmetry operators consists of improper rotations. An *improper rotation* is a proper rotation followed by a reflection through a plane perpendicular to the rotation axis. For this to be a point symmetry

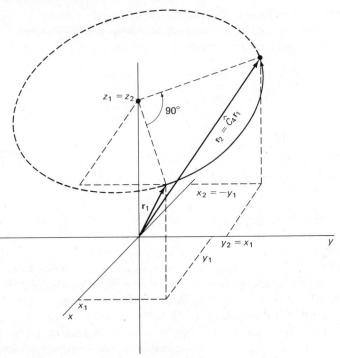

Figure 8-2. The action of a rotation operator, $\hat{C}_4$.

operation, both the rotation axis and the reflection plane must pass through the origin. The symbol for an improper rotation operator is $\hat{S}_n$, where the subscript n has the same meaning as with a proper rotation. The symmetry element for an improper rotation is the axis of rotation, just as with a proper rotation.

The action of the operator for an improper rotation of $90°$ about the z axis is given by

$$\hat{S}_4(z)(x_1, y_1, z_1) = (-y_1, x_1, -z_1) \qquad (8.34)$$

Notice that $\hat{S}_2$ is the same as the inversion operator $\hat{i}$, and that $\hat{S}_1$ is the same as a reflection operator.

Problem 8-8

Find the following.

a. $\hat{S}_4(z)(1, 2, 3)$
b. $\hat{S}_2(y)(3, 4, 5)$

We have defined all of the point symmetry operators. In addition to these, there are other symmetry operators, such as translations, glide planes, screw axes, etc., which are useful in describing crystal lattices but which do not apply to molecules.

The symmetry operators can operate on a set of points, just as on a single point. For example, they can operate simultaneously on the nuclei of a molecule, or on all the particles of a solid object. If the nuclei of a benzene molecule are in their equilibrium positions and the center of mass is at the origin, the inversion operator moves each of the carbon nuclei to the original location of another carbon nucleus, and each of the hydrogen nuclei to the original location of another hydrogen nucleus.

If after such an operation the object is in the same configuration as before except for the exchange of identical particles, we say that the symmetry operators *belongs* to the object. The symmetry properties of the object can be specified by listing all such symmetry operators, or their symmetry elements.

A uniform spherical object is the most highly symmetrical object. If the center of the sphere is at the origin, every mirror plane, every rotation axis, every improper rotation axis, and the inversion center at the origin are symmetry elements of the sphere.

Example 8-6

List the symmetry elements of a uniform cube centered at the origin with its faces parallel to the coordinate planes.

Solution

The symmetry elements are:

The inversion center at the origin.
Three C_4 axes coinciding with the coordinate axes.
Four C_3 axes passing through opposite corners of the cube.
Four S_6 axes coinciding with the C_3 axes.
Six C_2 axes connecting the midpoints of opposite edges.
Three mirror planes in the coordinate planes.
Six mirror planes passing through opposite edges.   •

Notice the standard notation used in this example. A symmetry element is denoted by the symbol for the operator but without the caret (^).

Problem 8-9

List the symmetry elements of a right circular cylinder. The axis of the cylinder is placed on the z axis. Since even an infinitesimal rotation belongs to the object, the z axis is a C_∞ axis.

List the symmetry elements of a uniform regular tetrahedron. It is possible to arrange the object so that its center is at the origin and the four corners are at alternate corners of a cube oriented as in Example 8-6.   •

Example 8-7

List the symmetry elements of the benzene molecule.

Solution

Locate the molecule with its nuclei in their equilibrium configuration as shown in Figure 8-3. The symmetry elements are:

The inversion center at the origin.

The mirror plane containing the nuclei.

A C_6 axis and an S_6 axis on the z axis.

Six vertical mirror planes, three through carbon nuclei and three that pass halfway between adjacent carbon nuclei.

Six C_2 axes located where the mirror planes intersect the x–y plane. These are also S_2 axes.

Some of the elements of symmetry are shown in Figure 8-3. The symbols on the rotation axes identify them, with a hexagon labeling a sixfold axis, a square labeling a fourfold axis, etc.

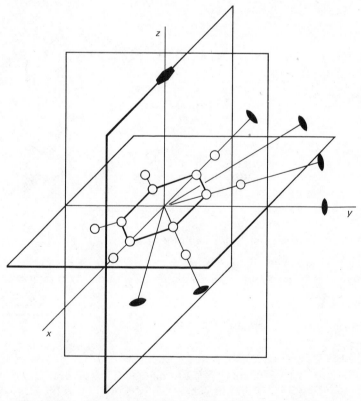

Figure 8-3. The benzene molecule with symmetry elements shown.

Problem 8-10

List the symmetry elements for
a. H_2O
b. CH_4
c. CO_2 (linear)

●

The symmetry of molecules is exploited to gain information about two different kinds of motions: the vibrational motions of the nuclei, and the motions of the electrons. In quantum mechanics, both of these kinds of motions are described through functions called *wave functions*. If the equilibrium nuclear configuration is symmetrical, the electronic and vibrational wave functions will have symmetry properties. Thus far, we have described the action of symmetry operators on points and on particles. We now need to define how they act on functions.

If ψ is some function and $\hat{O}$ is some symmetry operator, then

$$\hat{O}\psi = \phi \tag{8.35}$$

where ϕ is a newly produced function. If $\hat{O}$ is the operator that carries a representative point from (x_1, y_1, z_1) to (x_2, y_2, z_2), we define ϕ to have the same value at (x_2, y_2, z_2) that ψ has at (x_1, y_1, z_1):

$$\phi(x_2, y_2, z_2) = \psi(x_1, y_1, z_1) \tag{8.36}$$

This definition allows us to treat symmetry operators on an equal footing with other mathematical operators.

Example 8-8

The unnormalized $2px$ wave function for the electron in a hydrogen atom is

$$\psi_{2px} = x \exp\left[\frac{-(x^2 + y^2 + z^2)^{1/2}}{2a_0}\right] \tag{8.37}$$

where a_0 is a constant called the *Bohr radius*. Find $\hat{C}_4(z)\psi_{2px}$.

Solution

The effect of this operator is given by Eq. (8.32):

$$\hat{C}_4(z)(x_1, y_1, z_1) = (x_2, y_2, z_2) = (-y_1, x_1, z_1)$$

The transformed function is thus

$$\phi(x_2, y_2, z_2) = \psi_{2px}(x_1, y_1, z_1) = y_2 \exp\left[\frac{-(y_2^2 + x_2^2 + z_2^2)^{1/2}}{2a_0}\right]$$

Thus

$$\hat{C}_4(z)\psi_{2px} = \psi_{2py} \tag{8.38}$$

The symmetry operator has moved the region inside which the wave function is positive in the same way that it moves an object. •

Problem 8-11

Find $\hat{i}\psi_{2px}$. Show that this function is an eigenfunction of the inversion operator, and find its eigenvalue. •

The importance of symmetry operators in the study of electronic wave functions arises from the fact that two commuting operators can have a set of common eignefunctions. The principal operator that is studied in quantum chemistry is the *Hamiltonian operator*, which corresponds to the energy. The electronic Hamiltonian operator has a term in it which is equal to the potential energy as a function of the positions of the nuclei and electrons. A symmetry operator belonging to the nuclear framework will move the electrons in such a way that the potential energy does not change and will thus commute with the Hamiltonian operator. Electronic wave functions can thus occur which are eigenfunctions of the symmetry operators.

Problem 8-12

The potential energy of two charges q_1 and q_2 is, in unrationalized units, $q_1 q_2/r_{12}$, where r_{12} is the distance between the charges. If a hydrogen molecule is placed so that the origin is midway between the two nuclei and the nuclei are on the z axis, show that the inversion operator $\hat{i}$ and the reflection operator $\hat{\sigma}_h$ do not change the potential energy if applied to the electrons but not to the nuclei. •

Full exploitation of the symmetry of electronic wave functions requires a use of group theory, which we briefly introduce in a later section of this chapter. However, we state some facts that can be understood without the use of group theory:

1. If a molecule has a permanent dipole moment, the dipole vector must lie along a proper symmetry axis, and in a plane of symmetry.
2. A molecule with an improper rotation axis cannot be optically active.
3. A given object, such as a nuclear framework of a molecule, cannot possess a completely arbitrary collection of symmetry elements. Some of the elements of symmetry are mutually exclusive. Group theory can tell us which ones can belong together.

Section 8-4. MATRIX ALGEBRA

A matrix is an array or list of numbers. Most of the matrices that you will encounter are two-dimensional arrays. That is, they have rows and columns. If the matrix **A** has m rows and n columns, it is called an *m by n* *matrix* and is written

$$A = \begin{bmatrix} a_{11} & a_{12} & a_{13} & \cdots & a_{1n} \\ a_{21} & a_{22} & a_{23} & \cdots & a_{2n} \\ \cdots & & & & \\ \cdots & & & & \\ a_{m1} & a_{m2} & a_{m3} & \cdots & a_{mn} \end{bmatrix} \qquad (8.39)$$

The quantities that are entries in the two-dimensional list are called *elements* of the matrix. The brackets written on the left and right are included to show where the matrix starts and stops. If $m = n$, we say that the matrix is *square*.

Two matrices are equal to each other if and only if every element of one is equal to the corresponding element of the other. Of course, the two matrices must have the same number of rows and the same number of columns.

The sum of two matrices is defined by

$$\mathbf{C} = \mathbf{A} + \mathbf{B} \quad \text{if and only if} \quad c_{ij} = a_{ij} + b_{ij} \qquad \text{for every } i \text{ and } j \quad (8.40)$$

The product of a matrix and a scalar is defined by

$$\mathbf{B} = c\mathbf{A} \quad \text{if and only if} \quad b_{ij} = ca_{ij} \qquad \text{for every } i \text{ and } j \quad (8.41)$$

The product of two matrices is similar to the scalar product of two vectors. If we rename the components of two vectors F_1, F_2, F_3, G_1, G_2, and G_3 instead of F_x, F_y, F_z, G_x, G_y, and G_z, we can write Eq. (2.89) as

$$\mathbf{F} \cdot \mathbf{G} = F_1 G_1 + F_2 G_2 + F_3 G_3 = \sum_{i=1}^{3} F_i G_i \qquad (8.42)$$

If $\mathbf{C}$ is the product $\mathbf{AB}$, we define

$$\boxed{c_{ij} = \sum_{k=1}^{n} a_{ik} b_{kj}} \qquad (8.43)$$

In this equation, n is the number of columns in $\mathbf{A}$. The matrix $\mathbf{B}$ is required to have as many rows as $\mathbf{A}$ has columns. $\mathbf{C}$ will have as many rows as $\mathbf{A}$ and as many columns as $\mathbf{B}$.

If we wish, we think of the vector $\mathbf{F}$ in Eq. (8.42) as a matrix with one row and three columns (a row vector) and the vector $\mathbf{G}$ as being a matrix with three rows and one column (a column vector). Equation (8.42) is then a special case of Eq. (8.43):

$$\mathbf{F} \cdot \mathbf{G} = \begin{bmatrix} F_1 & F_2 & F_3 \end{bmatrix} \begin{bmatrix} G_1 \\ G_2 \\ G_3 \end{bmatrix}$$

If **A** is a 2 by 3 matrix and **B** is a 3 by 3 matrix, we can write their matrix product as

$$\begin{bmatrix} a_{11} & a_{12} & a_{13} \\ a_{21} & a_{22} & a_{23} \end{bmatrix} \begin{bmatrix} b_{11} & b_{12} & b_{13} \\ b_{21} & b_{22} & b_{23} \\ b_{31} & b_{32} & b_{33} \end{bmatrix} = \begin{bmatrix} c_{11} & c_{12} & c_{13} \\ c_{21} & c_{22} & c_{23} \end{bmatrix} \qquad \textbf{(8.44)}$$

Each element in **C** is obtained in the same way as taking a scalar product of a row from **A** and a column from **B**. For a particular element in **C**, we take the row in **A** which is in the same position as the row in **C** containing the desired element, and the column in **B** which is in the same position as the column in **C** containing the desired element.

Example 8-9

Find the matrix product

$$\begin{bmatrix} 1 & 0 & 2 \\ 0 & -1 & 1 \\ 0 & 0 & 1 \end{bmatrix} \begin{bmatrix} 0 & 0 & 2 \\ 3 & 0 & 1 \\ 1 & 1 & -1 \end{bmatrix}$$

Solution

$$\begin{bmatrix} 1 & 0 & 2 \\ 0 & -1 & 1 \\ 0 & 0 & 1 \end{bmatrix} \begin{bmatrix} 0 & 0 & 2 \\ 3 & 0 & 1 \\ 1 & 1 & -1 \end{bmatrix} = \begin{bmatrix} 2 & 2 & 0 \\ -2 & 1 & -2 \\ 1 & 1 & -1 \end{bmatrix} \qquad \bullet$$

Problem 8-13

Find the two matrix products

$$\begin{bmatrix} 1 & 2 & 3 \\ 3 & 2 & 1 \\ 1 & -1 & 2 \end{bmatrix} \begin{bmatrix} 1 & 3 & 2 \\ 2 & 2 & -1 \\ -2 & 1 & -1 \end{bmatrix} \quad \text{and} \quad \begin{bmatrix} 1 & 3 & 2 \\ 2 & 2 & -1 \\ -2 & 1 & -1 \end{bmatrix} \begin{bmatrix} 1 & 2 & 3 \\ 3 & 2 & 1 \\ 1 & -1 & 2 \end{bmatrix}$$

Notice that the left factor in one product is equal to the right factor in the other product, and vice versa. Are the two products equal? $\qquad \bullet$

If two matrices are square, as in Problem 8-13, they can be multiplied together in either order. In general, the multiplication is not commutative. That is,

$$\textbf{AB} \neq \textbf{BA} \qquad \textbf{(8.45)}$$

except in special cases.

However, matrix multiplication is associative:

$$\textbf{A(BC)} = \textbf{(AB)C} \qquad \textbf{(8.46)}$$

and matrix multiplication and addition are distributive:

$$\textbf{A(B + C)} = \textbf{AB} + \textbf{AC} \qquad \textbf{(8.47)}$$

Problem 8-14

Show that the properties of Eqs. (8.46) and (8.47) are obeyed by the particular matrices

$$A = \begin{bmatrix} 1 & 2 & 3 \\ 4 & 5 & 6 \\ 7 & 8 & 9 \end{bmatrix}, \quad B = \begin{bmatrix} 0 & 2 & 2 \\ -3 & 1 & 2 \\ 1 & -2 & -3 \end{bmatrix}, \quad C = \begin{bmatrix} 1 & 0 & 1 \\ 0 & 3 & -2 \\ 2 & 7 & -7 \end{bmatrix} \quad \bullet$$

The matrix multiplication that we have defined has much in common with the operator multiplication defined in Section 8-2. Both are associative and distributive but not necessarily commutative. In Section 8-2, we defined an identity operator, and we need to define an identity matrix, which we call **E**. We require

$$EA = AE = A$$

The fact that we require **E** to be the identity matrix when multiplied on either side of **A** requires both **A** and **E** to be square. In fact, only with square matrices will we get a strict similarity between operator algebra and matrix algebra. We will require an identity matrix for each number of rows and columns.

The identity matrix **E** has the form

$$E = \begin{bmatrix} 1 & 0 & 0 & \cdots & 0 \\ 0 & 1 & 0 & \cdots & 0 \\ 0 & 0 & 1 & \cdots & 0 \\ & & \cdots \cdots \cdots \cdots & & \\ 0 & 0 & 0 & \cdots & 1 \end{bmatrix} \tag{8.48}$$

The nonzero elements of the identity matrix are those with both indices equal. These are called the *diagonal elements*. The elements of **E** obey

$$e_{ij} = \delta_{ij} = \begin{cases} 1 & \text{if } i = j \\ 0 & \text{if } i \neq j \end{cases}$$

The quantity δ_{ij} is called the *Kronecker delta*.

Problem 8-15

Show by explicit multiplication that

$$\begin{bmatrix} 1 & 0 & 0 & 0 \\ 0 & 1 & 0 & 0 \\ 0 & 0 & 1 & 0 \\ 0 & 0 & 0 & 1 \end{bmatrix} \begin{bmatrix} a_{11} & a_{12} & a_{13} & a_{14} \\ a_{21} & a_{22} & a_{23} & a_{24} \\ a_{31} & a_{32} & a_{33} & a_{34} \\ a_{41} & a_{42} & a_{43} & a_{44} \end{bmatrix} = \begin{bmatrix} a_{11} & a_{12} & a_{13} & a_{14} \\ a_{21} & a_{22} & a_{23} & a_{24} \\ a_{31} & a_{32} & a_{33} & a_{34} \\ a_{41} & a_{42} & a_{43} & a_{44} \end{bmatrix} \quad \bullet$$

In the operator algebra of Section 8-2, we defined an inverse, which undoes the effect of a given operator. We define now the *inverse of a matrix*. Just as in operator algebra, we do not define division by a matrix. If we denote the inverse of **A** by A^{-1}, we must have

$$A^{-1}A = AA^{-1} = E \qquad (8.49)$$

Notice that the matrix product is to be taken in either order. This is because **A** must be the inverse of A^{-1}.

From Eq. (8.43) we can write the second equality in Eq. (8.49) as

$$\sum_{k=1}^{n} a_{ik}(A^{-1})_{kj} = \delta_{ij} \qquad (8.50)$$

where we write $(A^{-1})_{kj}$ for the element of A^{-1}. This is a set of equations, one for each value of i and each value of j, so that there are just enough equations to determine the elements of A^{-1}. Simultaneous equations are discussed in Chapter 9, but we will develop a method of finding A^{-1} now.

Our method is called *Gauss–Jordan elimination*, and consists of a set of operations to be applied to the elements of **A** to turn it into the identity matrix **E** without affecting A^{-1}. These operations will also affect the right-hand side of Eq. (8.50), turning it into a matrix that we temporarily call **D**. Equation (8.50) will be transformed into

$$E(A^{-1}) = D \qquad (8.51)$$

so that A^{-1} will be the same matrix as **D**.

We can see from Eq. (8.43) that the equation will still be valid if we multiply every element in a given row of **A** by some constant and multiply every element in the same row of **C** by the same constant, without changing **B**. This multiplies both sides of some of the equations by the same constant. This is an example of a row operation.

The next row operation that we can apply is to subtract one row of **A** from another row of **A**, element by element, while doing the same thing to **C** but not to **B**. This amounts to subtracting the left-hand sides of pairs of equations and subtracting at the same time the right-hand sides of the equations, which produces valid equations. These two row operations are sufficient to transform the left factor of a matrix product into the identity matrix. If we apply them to Eq. (8.50), we will transform it into Eq. (8.51). We illustrate the procedure in the following example.

Example 8-10

Find the inverse of the matrix

$$A = \begin{bmatrix} 2 & 1 & 0 \\ 1 & 2 & 1 \\ 0 & 1 & 2 \end{bmatrix}$$

Solution

The method of Gauss–Jordan elimination is usually carried out by writing the two matrices on which we operate side by side. The matrix that is not operated on is not written. Our version of Eq. (8.50) is

$$\begin{bmatrix} 2 & 1 & 0 \\ 1 & 2 & 1 \\ 0 & 1 & 2 \end{bmatrix} \begin{bmatrix} (A^{-1})_{11} & (A^{-1})_{12} & (A^{-1})_{13} \\ (A^{-1})_{21} & (A^{-1})_{22} & (A^{-1})_{23} \\ (A^{-1})_{22} & (A^{-1})_{32} & (A^{-1})_{33} \end{bmatrix} = \begin{bmatrix} 1 & 0 & 0 \\ 0 & 1 & 0 \\ 0 & 0 & 1 \end{bmatrix}$$

For our operations we represent this by the "double" matrix

$$\begin{bmatrix} 2 & 1 & 0 & | & 1 & 0 & 0 \\ 1 & 2 & 1 & | & 0 & 1 & 0 \\ 0 & 1 & 2 & | & 0 & 0 & 1 \end{bmatrix}$$

We first want to get a zero in the place of a_{21}. It is usual to clear out the columns from left to right. We multiply the first row by $\frac{1}{2}$, obtaining

$$\begin{bmatrix} 1 & \frac{1}{2} & 0 & | & \frac{1}{2} & 0 & 0 \\ 1 & 2 & 1 & | & 0 & 1 & 0 \\ 0 & 1 & 2 & | & 0 & 0 & 1 \end{bmatrix}$$

We subtract the first row from the second and replace the second by this difference. The result is

$$\begin{bmatrix} 1 & \frac{1}{2} & 0 & | & \frac{1}{2} & 0 & 0 \\ 0 & \frac{3}{2} & 1 & | & -\frac{1}{2} & 1 & 0 \\ 0 & 1 & 2 & | & 0 & 0 & 1 \end{bmatrix}$$

We say that we have used the element a_{11} as the *pivot element* in this procedure. The left column is now cleared except for the unity that we desire in the corner. We now use the a_{22} element as the pivot element to clear the second column. We multiply the second row by $\frac{1}{3}$ and replace the first row by the difference of the first row and the second to obtain

$$\begin{bmatrix} 1 & 0 & -\frac{1}{3} & | & \frac{2}{3} & -\frac{1}{3} & 0 \\ 0 & \frac{1}{2} & \frac{1}{3} & | & -\frac{1}{6} & \frac{1}{3} & 0 \\ 0 & 1 & 2 & | & 0 & 0 & 1 \end{bmatrix}$$

We now want to clear out the unity in the a_{32} position. We do this by multiplying the second row by 2, subtracting this row from the third row and replacing the third row by the difference. The result is

$$\begin{bmatrix} 1 & 0 & -\frac{1}{3} & | & \frac{2}{3} & -\frac{1}{3} & 0 \\ 0 & 1 & \frac{2}{3} & | & -\frac{1}{3} & \frac{2}{3} & 0 \\ 0 & 0 & \frac{4}{3} & | & \frac{1}{3} & -\frac{2}{3} & 1 \end{bmatrix}$$

We now multiply the third row by $\frac{1}{2}$ in order to use the a_{33} element as the pivot element. We subtract the third row from the second and replace the second row by the difference, obtaining

$$\begin{bmatrix} 1 & 0 & -\frac{1}{3} & | & \frac{2}{3} & -\frac{1}{3} & 0 \\ 0 & 1 & 0 & | & -\frac{1}{2} & 1 & -\frac{1}{2} \\ 0 & 0 & \frac{2}{3} & | & \frac{1}{6} & -\frac{1}{3} & \frac{1}{2} \end{bmatrix}$$

We now multiply the third row by $\frac{1}{2}$ and add it to the first row, and replace the first row by the sum. The result is

$$\begin{bmatrix} 1 & 0 & 0 & | & \frac{3}{4} & -\frac{1}{2} & \frac{1}{4} \\ 0 & 1 & 0 & | & -\frac{1}{2} & 1 & -\frac{1}{2} \\ 0 & 0 & \frac{1}{3} & | & \frac{1}{12} & -\frac{1}{6} & \frac{1}{4} \end{bmatrix}$$

The final row operation is multiplication of the third row by 3 to obtain

$$\begin{bmatrix} 1 & 0 & 0 & | & \frac{3}{4} & -\frac{1}{2} & \frac{1}{4} \\ 0 & 1 & 0 & | & -\frac{1}{2} & 1 & -\frac{1}{2} \\ 0 & 0 & 1 & | & \frac{1}{4} & -\frac{1}{2} & \frac{3}{4} \end{bmatrix}$$

The right half of this double matrix is $\mathbf{A}^{-1}$:

$$\mathbf{A}^{-1} = \begin{bmatrix} \frac{3}{4} & -\frac{1}{2} & \frac{1}{4} \\ -\frac{1}{2} & 1 & -\frac{1}{2} \\ \frac{1}{4} & -\frac{1}{2} & \frac{3}{4} \end{bmatrix} \qquad \bullet$$

Problem 8-16

a. Show that $\mathbf{A}\mathbf{A}^{-1} = \mathbf{E}$ and that $\mathbf{A}^{-1}\mathbf{A} = \mathbf{E}$ for the matrices of Example 8-10.

b. Find the inverse of the matrix

$$\mathbf{A} = \begin{bmatrix} 1 & 2 \\ 3 & 4 \end{bmatrix}$$

and show that your result really is the inverse. $\qquad \bullet$

We must make a few comments about matrix inversion. We have already stated that only square matrices have inverses. However, not all square matrices possess inverses. Associated with each square matrix is a determinant, which we define in the next section. If the determinant of a square matrix vanishes, the matrix is said to be singular. A singular matrix has no inverse.

There exist a lot of computer programs that will invert matrices for you if the elements are constants. Even a small computer can quickly find the inverse of a 20 by 20 matrix. In fact, most BASIC compilers have programs for matrix inversion built into them (see Chapter 11).

We conclude this section with the definition of several terms that apply to square matrices.

The *trace* of a matrix is the sum of the diagonal elements of the matrix:

$$\text{Tr}(\mathbf{A}) = \sum_{i=1}^{n} a_{ii} \tag{8.52}$$

The trace is sometimes called the *spur*, from the German word *Spur*, which means track, or trace. For example, the trace of the n by n identity matrix is equal to n.

A matrix in which all the elements below the diagonal elements happen to be zero is called an *upper triangular matrix*. A matrix in which all the elements above the diagonal elements happen to be zero is called a *lower triangular matrix*, and a matrix in which all the elements except the diagonal elements vanish is called a *diagonal matrix*. The matrix in which all of the elements vanish is called the *null matrix* or the *zero matrix*.

The *transpose of a matrix* is obtained by replacing the first column by the first row, the second column by the second row of the original matrix, etc. The transpose of $\mathbf{A}$ is denoted by $\tilde{\mathbf{A}}$ (pronounced "A tilde").

$$(\tilde{\mathbf{A}})_{ij} = \tilde{a}_{ij} = a_{ji} \tag{8.53}$$

If a matrix is equal to its transpose, it is said to be a symmetric matrix. Notice that the matrix in Example 8-10 is symmetric, and that its inverse is also symmetric.

The *hermitian conjugate* of a matrix is obtained by taking the complex conjugate of each element and then taking the transpose of the matrix. If a matrix has only real elements, the hermitian conjugate is the same as the transpose. The hermitian conjugate is also called the *adjoint* (mostly by physicists) and the *associate* (mostly by mathematicians, who use the term adjoint for something else). The hermitian conjugate is denoted by $\mathbf{A}^{\dagger}$.

$$(\mathbf{A}^{\dagger})_{ij} = a_{ji}^{*} \tag{8.54}$$

A matrix that is equal to its hermitian conjugate is said to be hermitian.

An orthogonal matrix is one whose inverse is equal to its transpose. If $\mathbf{A}$ is orthogonal, then

$$\mathbf{A}^{-1} = \mathbf{A} \quad \text{(orthogonal matrix)} \tag{8.55}$$

A unitary matrix is one whose inverse is equal to its hermitian conjugate. If $\mathbf{A}$ is unitary, then

$$\mathbf{A}^{-1} = \mathbf{A}^{\dagger} = \tilde{\mathbf{A}}^{*} \quad \text{(unitary matrix)} \tag{8.56}$$

Problem 8-17

Which of the following matrices are diagonal, symmetric, hermitian, orthogonal, or unitary?

a. $\begin{bmatrix} 1 & 0 \\ 0 & 1 \end{bmatrix}$ b. $\begin{bmatrix} 1 & 0 \\ 1 & 1 \end{bmatrix}$ c. $\begin{bmatrix} i & i \\ 0 & 1 \end{bmatrix}$ d. $\begin{bmatrix} 0 & i \\ -i & 1 \end{bmatrix}$

Section 8-5. DETERMINANTS

Associated with every square matrix is a quantity called a *determinant*. If the elements of the matrix are constants, the determinant is a single constant, defined as a certain sum of products of subsets of the elements. If the matrix has n rows and columns, each term in the sum making up the determinant will have n factors in it.

The determinant of a 2 by 2 matrix is given by

$$\det(A) = \begin{vmatrix} a_{11} & a_{12} \\ a_{21} & a_{22} \end{vmatrix} = a_{11}a_{22} - a_{12}a_{21} \qquad (8.57a)$$

The determinant is written in much the same way as the matrix, except that straight vertical lines are used.

Example 8-11

Find the value of the determinant

$$\begin{vmatrix} 3 & -17 \\ 1 & 5 \end{vmatrix}$$

Solution

$$\begin{vmatrix} 3 & -17 \\ 1 & 5 \end{vmatrix} = (3)(5) - (-17)(1) = 15 + 17 = 32 \qquad \bullet$$

Finding the value of a large determinant is tedious. One way to do it is by *expanding by minors*. This is done as follows:

1. Pick a row or a column of the determinant. Any row or column will do, but one with zeros in it is best.
2. Write a series of terms, one for each element in the row or column. Each term consists of an element of the chosen row or column times the *minor* of that element. The minor of an element is the determinant that is obtained by deleting the row and the column containing that element, so it is a determinant with one less row and one less column than the original determinant. The minor is also called the *cofactor*.
3. Determine the signs of the terms as follows. Count the number of "steps" of one row or one column required to get from the upper left element to the element whose minor is in the term. If the number of steps is odd, the sign is negative. If the number of steps is even, the sign is positive.
4. Repeat the entire process with each determinant in the expansion. If necessary, repeat it again and again until you have a sum of 2 by 2 determinants, which can be evaluated by Eq (8.57a).

Example 8-12

Expand the 3 by 3 determinant of the matrix **A** by minors.

Solution

$$\begin{vmatrix} a_{11} & a_{12} & a_{13} \\ a_{21} & a_{22} & a_{23} \\ a_{31} & a_{32} & a_{33} \end{vmatrix} = a_{11} \begin{vmatrix} a_{22} & a_{23} \\ a_{32} & a_{33} \end{vmatrix} - a_{12} \begin{vmatrix} a_{21} & a_{23} \\ a_{31} & a_{33} \end{vmatrix} + a_{13} \begin{vmatrix} a_{21} & a_{22} \\ a_{31} & a_{32} \end{vmatrix}$$

$$= a_{11}a_{22}a_{33} - a_{11}a_{23}a_{32} - a_{12}a_{21}a_{33} + a_{12}a_{23}a_{31}$$
$$+ a_{13}a_{21}a_{32} - a_{13}a_{22}a_{31}$$

• (8.57b)

Problem 8-18

Expand the following determinant by minors.

$$\begin{vmatrix} 3 & 2 & 0 \\ 7 & -1 & 5 \\ 2 & 3 & 4 \end{vmatrix}$$

•

Problem 8-19

a. Expand the 4 by 4 determinant by minors

$$\begin{vmatrix} a_{11} & a_{12} & a_{13} & a_{14} \\ a_{21} & a_{22} & a_{23} & a_{24} \\ a_{31} & a_{32} & a_{33} & a_{34} \\ a_{41} & a_{42} & a_{43} & a_{44} \end{vmatrix}$$

b. Write a computer program in the BASIC language to evaluate the determinant for constant elements. •

Determinants have a number of important properties. We will state some of them without proof:

Property 1. If two rows of a determinant are interchanged, the result will be a determinant whose value is the negative of the original determinant. The same is true if two columns are interchanged.

Property 2. If two rows or two columns of a determinant are equal, the determinant has value zero.

Property 3. If each element in one row or one column of a determinant is multiplied by the same quantity, say c, the value of the new determinant is c times the value of the original determinant. Therefore, if an n by n determinant has every element multiplied by c, the new determinant is c^n times the original determinant.

Property 4. If every element in some row or in some column of a determinant is zero, the value of the determinant is zero.

Property 5. If any row is replaced, element by element, by that row plus a constant times another row, the value of the determinant is unchanged. The same is true for two columns. For Example,

$$\begin{vmatrix} a_{11} + ca_{12} & a_{12} & a_{13} \\ a_{21} + ca_{22} & a_{22} & a_{13} \\ a_{31} + ca_{32} & a_{32} & a_{33} \end{vmatrix} = \begin{vmatrix} a_{11} & a_{12} & a_{13} \\ a_{21} & a_{22} & a_{23} \\ a_{31} & a_{32} & a_{33} \end{vmatrix} \tag{8.58}$$

Property 6. The determinant of a triangular matrix (a triangular determinant) is equal to the product of the diagonal elements. For example,

$$\begin{vmatrix} a_{11} & 0 & 0 \\ a_{22} & a_{22} & 0 \\ a_{31} & a_{32} & a_{33} \end{vmatrix} = a_{11}a_{22}a_{33} \tag{8.59}$$

Property 7. The determinant of a matrix is equal to the determinant of the transpose of that matrix.

$$\det(\tilde{\mathbf{A}}) = \det(\mathbf{A}) \tag{8.60}$$

For example,

$$\begin{vmatrix} a_{11} & a_{12} & a_{13} \\ a_{21} & a_{22} & a_{23} \\ a_{31} & a_{32} & a_{33} \end{vmatrix} = \begin{vmatrix} a_{11} & a_{21} & a_{31} \\ a_{12} & a_{22} & a_{32} \\ a_{13} & a_{23} & a_{33} \end{vmatrix}$$

These properties can all be proved from the expansion of a determinant by minors. You might want to prove them as an exercise.

Problem 8-20

Find the value of the determinant

$$\begin{vmatrix} 3 & 4 & 5 \\ 2 & 1 & 6 \\ 3 & -5 & 10 \end{vmatrix}$$

Interchange the first and second columns, and find the value of the resulting determinant. Replace the second column by the sum of the first and second columns and find the value of the resulting determinant. Replace the second column by the first, thus making two identical columns, and find the value of the resulting determinant. ●

There is an application of determinants in quantum chemistry which comes from Property 1. The wave function of a system containing two or more electrons must have the property that it changes sign if two of the electrons are interchanged in position. If $\mathbf{r}_1$ and $\mathbf{r}_2$ are the position vectors of two electrons and Ψ is the wave function, then

$$\Psi(\mathbf{r}_1, \mathbf{r}_2) = -\Psi(\mathbf{r}_2, \mathbf{r}_1) \tag{8.61}$$

Many approximate multielectron wave functions are constructed as a product of several one-electron wave functions, or orbitals. If ψ_1, ψ_2, etc., are orbitals, such a wave function might be, for n electrons,

$$\Psi = \psi_1(\mathbf{r}_1)\psi_2(\mathbf{r}_2)\psi_3(\mathbf{r}_3)\psi_4(\mathbf{r}_4) \cdots \psi_n(\mathbf{r}_n) \tag{8.62}$$

This wave function does not obey Eq. (8.61). A wave function that does obey this equation can be written as a determinant called a *Slater determinant* (after John C. Slater, a prominent quantum mechanician). The elements of this determinant are functions, instead of constants, so the determinant is equal to a function, not a constant.

$$\Psi(\mathbf{r}_1, \mathbf{r}_2, \ldots, \mathbf{r}_n) = \frac{1}{\sqrt{n!}} \begin{vmatrix} \psi_1(\mathbf{r}_1) & \psi_1(\mathbf{r}_2) & \psi_1(\mathbf{r}_3) & \cdots & \psi_1(\mathbf{r}_n) \\ \psi_2(\mathbf{r}_1) & \psi_2(\mathbf{r}_2) & \psi_2(\mathbf{r}_3) & \cdots & \psi_2(\mathbf{r}_n) \\ \cdots\cdots\cdots\cdots\cdots\cdots\cdots\cdots\cdots\cdots\cdots\cdots \\ \psi_n(\mathbf{r}_1) & \psi_n(\mathbf{r}_2) & \psi_n(\mathbf{r}_3) & \cdots & \psi_n(\mathbf{r}_n) \end{vmatrix} \tag{8.63}$$

This obeys Eq. (8.61), since interchanging $\mathbf{r}_1$ and $\mathbf{r}_2$, for example, is the same as interchanging two columns, which changes the sign of the determinant. The factor $1/\sqrt{n!}$ in front is a normalizing factor, which is not important to us right now.

Notice that if we attempt to write such a wave function with two electrons in the same orbital (two of the ψ factors identical), then two rows of the determinant are identical, and the entire determinant vanishes by Property 2. This is the *Pauli exclusion principle*, which states that no two electrons in the same atom or molecule can be in the same orbital.

Section 8-6. AN ELEMENTARY INTRODUCTION TO GROUP THEORY

A *group* is a collection of elements with a single operation for combining two elements of the group. We call the operation multiplication. The following requirements must be met:

1. If **A** and **B** are members of the group, and **F** is the product **AB**, then **F** must be a member of the group.
2. The group must contain the identity element, **E**, such that

$$\mathbf{AE} = \mathbf{EA} = \mathbf{A} \tag{8.64}$$

3. The inverse of every element of the group must be a member of the group.
4. The associative law must hold:

$$\mathbf{A(BC)} = \mathbf{(AB)C} \tag{8.65}$$

It is not necessary that the elements of the group commute with each other. That is, it is possible that

$$\mathbf{AB} \neq \mathbf{BA} \tag{8.66}$$

If all the members of the group do commute, the group is called *abelian*, after a mathematician named Abel.

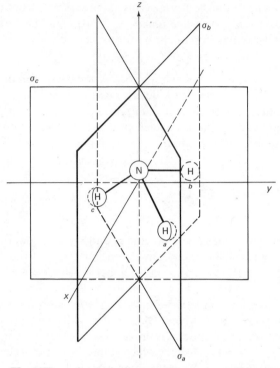

Figure 8-4a. The NH_3 molecule in its coordinate axes, with symmetry elements shown (after Levine).

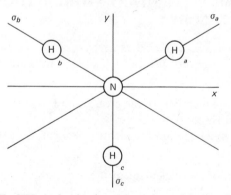

Figure 8-4b. The NH_3 molecule viewed from the positive z axis (after Levine).

It is a fact that the set of symmetry operators which "belong" to a symmetrical object in the sense of Section 8-4 form a group. We will illustrate this fact for a particular molecule, NH_3.[2]

---

[2] This discussion is adapted from Ira N. Levine, *Molecular Spectroscopy*, John Wiley and Sons, New York, 1975, pp. 390ff.

Table 8-1. **Multiplication Table for the Symmetry Operators of the NH$_3$ Molecule**

	$\hat{E}$	$\hat{C}_3$	$\hat{C}_3^2$	$\hat{\sigma}_a$	$\hat{\sigma}_b$	$\hat{\sigma}_c$
$\hat{E}$	$\hat{E}$	$\hat{C}_3$	$\hat{C}_3^2$	$\hat{\sigma}_a$	$\hat{\sigma}_b$	$\hat{\sigma}_c$
$\hat{C}_3$	$\hat{C}_3$	$\hat{C}_3^2$	$\hat{E}$	$\hat{\sigma}_c$	$\hat{\sigma}_a$	$\hat{\sigma}_b$
$\hat{C}_3^2$	$\hat{C}_3^2$	$\hat{E}$	$\hat{C}_3$	$\hat{\sigma}_b$	$\hat{\sigma}_c$	$\hat{\sigma}_a$
$\hat{\sigma}_a$	$\hat{\sigma}_a$	$\hat{\sigma}_b$	$\hat{\sigma}_c$	$\hat{E}$	$\hat{C}_3$	$\hat{C}_3^2$
$\hat{\sigma}_b$	$\hat{\sigma}_b$	$\hat{\sigma}_c$	$\hat{\sigma}_a$	$\hat{C}_3^2$	$\hat{E}$	$\hat{C}_3$
$\hat{\sigma}_c$	$\hat{\sigma}_c$	$\hat{\sigma}_a$	$\hat{\sigma}_b$	$\hat{C}_3$	$\hat{C}_3^2$	$\hat{E}$

The NH$_3$ molecule is a triangular pyramid in its equilibrium configuration. Figure 8-4a shows the nuclear framework as viewed from the first octant of the coordinate system, and Figure 8-4b shows the framework as viewed from the positive end of the z axis. The molecule is placed in the coordinate system with the center of mass at the origin, and the rotation axis of highest order (largest value of n) along the z axis.

The symmetry elements of the molecule are shown. The symmetry operators that belong to the molecule are $\hat{E}$, $\hat{C}_3$, $\hat{C}_2^2$, and the three reflection operators corresponding to vertical mirror planes passing through each of the three hydrogen nuclei. The square of the $\hat{C}_3$ operator is included, because in order to have a group, we must include all possible operators which put the framework into equivalent configurations.

We now satisfy ourselves that the four conditions to have a group are met:

Condition 1. The product of any two members of the group must be a member of the group. We show this by constructing a multiplication table, as shown in Table 8-1. The operators listed in the first column of the table are to be used as the left factor, and the operators listed in the first row of the table are to be used as the right factor in a product. We must specify which factor comes first because the operators do not necessarily commute. The entries in the table are obtained as in the example:

Example 8-13

Find the product $\hat{\sigma}_c\hat{C}_3$.

Solution

Both of these operators leave the nitrogen nucleus where it is. The $\hat{C}_3$ operator moves the hydrogen nucleus originally at the σ_a plane to the σ_b plane, the nucleus originally at the σ_b plane to the σ_c plane, and the nucleus originally at the σ_c plane to the σ_a plane. The $\hat{\sigma}_c$ operator reflects in the σ_c plane, so that it exchanges the nuclei at the σ_a and σ_b planes. It thus returns the nucleus originally at the σ_a plane to its original position and moves the nucleus originally at the σ_c plane to the σ_b plane. This is the same as the effect that the $\hat{\sigma}_a$ operator would have, so

$$\hat{\sigma}_c\hat{C} = \hat{\sigma}_a \qquad \bullet \quad (8.67)$$

Problem 8-21

Verify several of the entries in Table 8-1. •

As you can see, some pairs of operators commute, whereas others do not. For example, $\hat{C}_3\hat{\sigma}_c = \hat{\sigma}_b$, whereas $\hat{\sigma}_c\hat{C}_3 = \hat{\sigma}_a$.

Condition 2. The group does contain the identity operator, $\hat{E}$.
Condition 3. The inverse of every operator is in the group. Each reflection operator is its own inverse, and the inverse of $\hat{C}_3$ is $\hat{C}_3^2$. This condition would not be met if we had omitted $\hat{C}_3^2$.
Condition 4. The multiplication operation is associative, because operator multiplication is always associative.

We have thus shown that the symmetry operators belonging to the NH_3 molecule form a group. Such a group, consisting of point symmetry operators, is called a *point group*.

There is only a limited number of point groups that exist, and each is assigned a symbol, called a *Schoenflies symbol.*For example, the point group of the NH_3 molecule is called the C_{3v} group. This symbol is chosen because the principal rotation axis is a C_3 axis, and because there are vertical mirror planes. To a knowledgeable listener, you can communicate immediately what the symmetry properties of the NH_3 molecule are by saying that it has C_{3v} symmetry.[3]

The H_2O molecule belongs to the C_{2v} point group, which contains the operators $\hat{E}$, $\hat{C}_2$, and two reflection operators, one whose mirror plane is the plane of the nuclei and one whose mirror plane is perpendicular to the first.

Problem 8-22

Obtain the multiplication table for the C_{2v} point group, and show that it satisfies the conditions to be a group. •

Symmetry Operators and Matrices. As we have seen, there is great similarity between operator algebra and matrix algebra. We now see how matrices can be used to represent symmetry operators.

Equation (8.26) represents the action of a symmetry operator, $\hat{A}$, on the location of a point. Let the original location of the point be given by the cartesian coordinates (x, y, z), and the final coordinates be given by (x', y', z'):

$$\hat{A}(x, y, z) = (x', y', z') \qquad (8.68)$$

---

[3] See R. A. Alberty and F. Daniels, *Physical Chemistry*, 5th ed., John Wiley & Sons, New York, 1979, Chap. 9, for a method of assigning Schoenflies symbols.

If we represent the position vectors by 3 by 1 matrices (column vectors) this can be written as a matrix equation:

$$\begin{bmatrix} a_{11} & a_{12} & a_{13} \\ a_{21} & a_{22} & a_{23} \\ a_{31} & a_{32} & a_{33} \end{bmatrix} \begin{bmatrix} x \\ y \\ z \end{bmatrix} = \begin{bmatrix} x' \\ y' \\ z' \end{bmatrix} \qquad \textbf{(8.69)}$$

Equation (8.69) is the same as three ordinary equations:

$$a_{11}x + a_{12}y + a_{13}z = x' \qquad \textbf{(8.70a)}$$
$$a_{21}x + a_{22}y + a_{23}z = y' \qquad \textbf{(8.70b)}$$
$$a_{31}x + a_{32}y + a_{33}z = z' \qquad \textbf{(8.70c)}$$

We can obtain the elements of the matrix for $\hat{A}$ by comparing these equations with the equations obtained in Section 8-3 for various symmetry operators. For example, in the case of the identity operator, $x = x'$, $y = y'$, and $z = z'$, so that the matrix for the identity symmetry operator is the 3 by 3 identity matrix.

$$\hat{E} \leftrightarrow \mathbf{E} = \begin{bmatrix} 1 & 0 & 0 \\ 0 & 1 & 0 \\ 0 & 0 & 1 \end{bmatrix} \qquad \textbf{(8.71)}$$

The double-headed arrow means that the symmetry operator $\hat{E}$ and the matrix $\mathbf{E}$ are equivalent. The matrix product as in Eq. (8.69) and the operator expression as in Eq. (8.68) give the same result.

Example 8-14

Find the matrix equivalent to $\hat{C}_n(z)$.

Solution

Let $\alpha = 2\pi/n$ radians, the angle through which the operator rotates a particle.

$$x' = \cos(\alpha)x - \sin(\alpha)y$$
$$y' = \sin(\alpha)x + \cos(\alpha)y$$
$$z' = z$$

Comparison of this with Eq. (8.70) gives

$$\hat{C}_n(z) \leftrightarrow \begin{bmatrix} \cos(\alpha) & -\sin(\alpha) & 0 \\ \sin(\alpha) & \cos(\alpha) & 0 \\ 0 & 0 & 1 \end{bmatrix} \qquad \bullet \;\; \textbf{(8.72)}$$

Problem 8-23

a. Verify by matrix multiplication that Eqs. (8.71) and (8.72) give the correct results.

b. Use Eq. (8.72) to find the matrix for $\hat{C}_2(z)$.

c. Find the matrices equivalent to $\hat{S}_3(z)$ and $\hat{\sigma}_h$. •

The matrices that are equivalent to a group of symmetry operators have exactly the same effect as the symmetry operators, so they must multiply together in exactly the same way. We can show this by carrying out the matrix multiplications. The matrices for the C_{3v} group are as follows:

$$\hat{E} \leftrightarrow \begin{bmatrix} 1 & 0 & 0 \\ 0 & 1 & 0 \\ 0 & 0 & 1 \end{bmatrix} = \mathbf{E} \qquad \hat{C}_3 \leftrightarrow \begin{bmatrix} -1/2 & -\sqrt{3}/2 & 0 \\ \sqrt{3}/2 & -1/2 & 0 \\ 0 & 0 & 1 \end{bmatrix} = \mathbf{A}$$

$$\hat{C}_3^2 \leftrightarrow \begin{bmatrix} -1/2 & \sqrt{3}/2 & 0 \\ -\sqrt{3}/2 & -1/2 & 0 \\ 0 & 0 & 1 \end{bmatrix} = \mathbf{B} \qquad \hat{\sigma}_a \leftrightarrow \begin{bmatrix} 1/2 & \sqrt{3}/2 & 0 \\ \sqrt{3}/2 & -1/2 & 0 \\ 0 & 0 & 1 \end{bmatrix} = \mathbf{C}$$

$$\hat{\sigma}_b \leftrightarrow \begin{bmatrix} 1/2 & -\sqrt{3}/2 & 0 \\ -\sqrt{3}/2 & -1/2 & 0 \\ 0 & 0 & 1 \end{bmatrix} = \mathbf{D} \qquad \hat{\sigma}_c \leftrightarrow \begin{bmatrix} -1 & 0 & 0 \\ 0 & 1 & 0 \\ 0 & 0 & 1 \end{bmatrix} = \mathbf{F} \qquad (8.73)$$

We have given each matrix a letter symbol which was arbitrarily chosen.

Problem 8-24

By transcribing Table 8-1 with appropriate changes in symbols, generate the multiplication table for the matrices in Eq. (8.73). Verify several of the entries by matrix multiplication. •

Each of the matrices in Eq. (8.73) is equivalent to one of the symmetry operators in the C_{3v} group, but it is not exactly identical to it, being only one possible way to represent the symmetry operator. Therefore, the set of matrices forms another group of the same *order* (same number of members) as the C_{3v} point group. The fact that it obeys the same multiplication table is expressed by saying that it is *isomorphic* with the group of symmetry operators. A group of matrices that is isomorphic with a group of symmetry operators is called a *faithful representation* of the group. Our group of matrices consists of 3 by 3 matrices and is said to be of *dimension* 3.

There can be other representations of a symmetry group besides the group of matrices that are equivalent to the symmetry operators. The matrices do not have to have any physical interpretation. The only requirement is that they multiply in the same way as do the symmetry operators. In fact, in some representations, called *unfaithful* or *homomorphic*, there are fewer matrices than there are symmetry operators, and one matrix occurs in the places in the multiplication table where two or more symmetry operators occur. However, all the matrices in a given representation have to be square, and all must have the same number of rows and columns.

Group representations are divided into two kinds, *reducible* and *irreducible*. In a *reducible representation*, the matrices are "block-diagonal" or can be put into block-diagonal form by a similarity transformation. A *block-diagonal matrix* is one in which all elements are zero except those in square regions along the diagonal. The following matrix has two 2 by 2 blocks and a 1 by 1 block:

$$\begin{bmatrix} 1 & 2 & 0 & 0 & 0 \\ 3 & 2 & 0 & 0 & 0 \\ 0 & 0 & 4 & 3 & 0 \\ 0 & 0 & 3 & 3 & 0 \\ 0 & 0 & 0 & 0 & 2 \end{bmatrix}$$

All the matrices in a reducible representation have to have the same size blocks in the same order.

A similarity transformation on the matrix $\mathbf{A}$ consists of the multiplication of $\mathbf{A}$ on the right by some matrix (not necessarily one of the group) and on the left by the inverse of the same matrix, as in

$$\mathbf{B} = \mathbf{X}^{-1}\mathbf{A}\mathbf{X} \qquad (8.74)$$

The representation of the group C_{3v} given in Eq. (8.73) is reducible, since each matrix has a 2 by 2 block and a 1 by 1 block. When two block-diagonal matrices with the same size blocks are multiplied together, the result is a matrix that is also block-diagonal with the same blocks. This is apparent in the case of the matrices in Eq. (8.73), which produce only each other when multiplied together.

Problem 8-25

Show by matrix multiplication that two matrices with a 2 by 2 block and two 1 by 1 blocks produce another such matrix when multiplied together. •

Because of the way in which block-diagonal matrices multiply, the 2 by 2 blocks in the matrices in Eq. (8.73) if taken alone form another representation of the C_{3v} group. The 1 by 1 blocks also form such a representation, but this representation is an unfaithful or homomorphic one, in which every operator is represented by the 1 by 1 identity matrix. Both the representations obtained from the submatrices are irreducible. The one-dimensional representation is called the *totally symmetric representation*. When a reducible representation is written with its matrices in block-diagonal form, the block submatrices always form *irreducible representations*, and the reducible representation is said to be the *direct sum* of the irreducible representations.

Notice in this particular case that we could not get three one-dimensional representations, because the $\hat{C}_3(z)$ operator mixes the x and y coordinates of a particle, preventing the matrices from being diagonal.

Problem 8-26

Pick a few pairs of 2 by 2 submatrices from Eq. (8.73) and show that they multiply in the same way as the 3 by 3 matrices.   •

In any representation of a symmetry group, the trace of a matrix is called the *character* of the corresponding operator for that representation.

Problem 8-27

Find the characters of the operators in the C_{3v} group for the representation in Eq. (8.73).   ●

The two irreducible representations of C_{3v} that we have obtained thus far are said to be *nonequivalent*, since they have different dimensions. There are several theorems governing irreducible representations for a particular group.[4,5] These theorems can be used, for example, to determine that three irreducible representations of the C_{3v} group occur, and that their dimensions are 2, 1, and 1. The other one-dimensional representation is

$$\hat{E} \leftrightarrow 1 \qquad \hat{C}_3 \leftrightarrow 1 \qquad \hat{C}_3^2 \leftrightarrow 1$$
$$\hat{\sigma}_a \leftrightarrow -1 \qquad \hat{\sigma}_b \leftrightarrow -1 \qquad \hat{\sigma}_c \leftrightarrow -1 \qquad (8.74)$$

Problem 8-28

Show that the 1 by 1 matrices (scalars) in Eq. (8.74) obey the same multiplication table as does the group of symmetry operators.   ●

We will not discuss the application of group theory to molecular quantum mechanics. A considerable amount of information about the energy levels (sets of states of the same energy) for vibration and electronic motion in molecules can be obtained from group theory. For example, there is a theorem which says that there is a correspondence between an energy level and some one of the irreducible representations of the symmetry group of the molecule, and that the degeneracy (number of states in the level) is equal to the dimension of that irreducible representation.

ADDITIONAL PROBLEMS

8.29

Find the following commutators:

a. $[\hat{D}_x, \sin(x)]$ b. $[\hat{D}_x^3, x]$ c. $[\hat{D}_x^2, f(x)]$

8.30

Show that the x and z components of the angular momentum have quantum mechanical operators that do not commute:

$$\hat{L}_x = \frac{\hbar}{i}\left(\hat{y}\frac{d}{dz} - \hat{z}\frac{d}{dy}\right), \qquad \hat{L}_z = \frac{\hbar}{i}\left(\hat{x}\frac{d}{dy} - \hat{y}\frac{d}{dx}\right)$$

8.31

The hamiltonian operator for a one-dimensional harmonic oscillator moving in the x direction is

$$\hat{H} = -\frac{\hbar^2}{2m}\frac{d^2}{dx^2} + \frac{kx^2}{2}$$

[4] M. Hamermesh, *Group Theory*, Addison-Wesley Publishing Co., Reading, Mass., 1962, pp. 101–111.

[5] Ira N. Levine, *Quantum Chemistry*, Vol. II, Allyn and Bacon, Boston, 1970, p. 389.

Find the value of the constant a such that the function e^{-ax^2} is an eigen-function of the hamiltonian operator. The quantity k is a constant called the *force constant*, m is the mass of the oscillating particle, and $\hbar$ is Planck's constant divided by 2π.

8.32

In quantum mechanics, the expectation value for a mechanical quantity is given by

$$\langle A \rangle = \frac{\int \psi^* \hat{A} \psi \, dx}{\int \psi^* \psi \, dx}$$

where $\hat{A}$ is the operator for the mechanical quantity and ψ is the wave function for the state of the system. The integrals are over all permitted values of the coordinates of the system. The expectation value is defined as the prediction of the mean of a large number of measurements of the mechanical quantity, given that the system is in the state corresponding to ψ prior to each measurement.

For a particle moving in the x direction only and confined to a region on the x axis from $x = 0$ to $x = a$, the integrals are single integrals from 0 to a and $\hat{p}$ is given by $(\hbar/i)\, \partial/\partial x$. For this case, find the expectation value of p_x and of p_x^2 if the wave function is

$$\psi = C \sin\left(\frac{\pi x}{a}\right)$$

where C is a constant. Comment on your results.

8.33

If $\hat{A}$ is the operator corresponding to the mechanical quantity A and ϕ_n is an eigenfunction of $\hat{A}$, such that

$$\hat{A}\phi_n = a_n \phi_n$$

show that the expectation value of A is equal to a_n if the state of the system corresponds to ϕ_n. See Problem 8.32 for the formula for the expectation value.

8.34

If x is an ordinary variable, the Maclaurin series for $1/(1 - x)$ is

$$\frac{1}{1 - x} = 1 + x^2 + x^3 + x^4 + \cdots$$

If $\hat{X}$ is some operator, show that the series

$$1 + \hat{X} + \hat{X}^2 + \hat{X}^3 + \hat{X}^4 + \cdots$$

is the inverse of the operator $1 - \hat{X}$.

8.35

Find the final location of a representative point for each initial location and symmetry operator:

  a. $(1, 1, 1)$ $\hat{C}_2(y)$

    b. $(1, 1, 1)$ $\hat{C}_3(z)$
    c. $(1, 1, 1)$ $\hat{S}_4(z)$
    d. $(1, 1, 1)$ $\hat{C}_2(z)\hat{i}\hat{\sigma}_h$
    e. $(1, 1, 0)$ $\hat{S}_2(y)\hat{\sigma}_h$

8.36

Find the 3 by 3 matrix that is equivalent in its action to each of the symmetry operators:

    a. $\hat{S}_2(z)$ b. $\hat{C}_2(x)$ c. $\hat{C}_8(x)$ d. $\hat{S}_6(x)$

8.37

Give the function that results if the given symmetry operator operates on the given function for each of the following:

Operator	Function
a. $\hat{C}_4(z)$	x^2
b. $\hat{\sigma}_h$	$x\cos(x/y)$
c. $\hat{i}$	$(x + y + z^2)$
d. $\hat{S}_4(x)$	$(x + y + z)$

8.38

Find the matrix products:

a.
$$\begin{bmatrix} 0 & 1 & 2 \\ 4 & 3 & 2 \\ 7 & 6 & 1 \end{bmatrix} \begin{bmatrix} 1 & 2 & 3 \\ 6 & 8 & 1 \\ 7 & 4 & 3 \end{bmatrix}$$

b.
$$\begin{bmatrix} 6 & 3 & 2 & -1 \\ -7 & 4 & 3 & 2 \\ 1 & 3 & 2 & -2 \\ 6 & 7 & -1 & -3 \end{bmatrix} \begin{bmatrix} 4 & 7 & -6 & -8 \\ 3 & -6 & 8 & -6 \\ 2 & 3 & -3 & 4 \\ -1 & 4 & 2 & 3 \end{bmatrix}$$

c.
$$\begin{bmatrix} 3 & 2 & 1 & 4 \end{bmatrix} \begin{bmatrix} 1 & 2 & 3 \\ 0 & 3 & -4 \\ 1 & -2 & 1 \\ 3 & 1 & 0 \end{bmatrix}$$

d.
$$\begin{bmatrix} 6 & 3 & 1 \\ 7 & 4 & -2 \end{bmatrix} \begin{bmatrix} 1 & 4 & -7 & 3 \\ 2 & 5 & 8 & -2 \\ 3 & 6 & -9 & 1 \end{bmatrix}$$

8.39

Show that $(\mathbf{AB})\mathbf{C} = \mathbf{A}(\mathbf{BC})$ for the example matrices:

$$\mathbf{A} = \begin{bmatrix} 0 & 1 & 2 \\ 3 & 1 & -4 \\ 2 & 3 & 1 \end{bmatrix} \quad \mathbf{B} = \begin{bmatrix} 3 & 1 & 4 \\ -2 & 0 & 1 \\ 3 & -2 & 1 \end{bmatrix}, \quad \mathbf{C} = \begin{bmatrix} 0 & 3 & 1 \\ -4 & 2 & 3 \\ 3 & 1 & -2 \end{bmatrix}$$

8.40

Show that $A(B + C) = AB + AC$ for the example matrices in Problem 8.39.

8.41

Test the following matrices for singularity. Find the inverses of any that are nonsingular. Multiply the original matrix by its inverse to check your work.

a.
$$\begin{bmatrix} 0 & 1 & 2 \\ 2 & 3 & 1 \\ 2 & 4 & 3 \end{bmatrix}$$

b.
$$\begin{bmatrix} 6 & 8 & 1 \\ 7 & 3 & 2 \\ 4 & 6 & -9 \end{bmatrix}$$

c.
$$\begin{bmatrix} 3 & 2 & -1 \\ -4 & 6 & 3 \\ 7 & 2 & -1 \end{bmatrix}$$

d.
$$\begin{bmatrix} 0 & 2 & 3 \\ 1 & 1 & 1 \\ 2 & 0 & 1 \end{bmatrix}$$

8.42

Find the matrix, P, which results from the similarity transformation $P = X^{-1}QX$, where

$$Q = \begin{bmatrix} 1 & 2 \\ 2 & 1 \end{bmatrix}, \qquad X = \begin{bmatrix} 2 & 3 \\ 4 & 3 \end{bmatrix}$$

8.43

The H_2O molecule belongs to the point group C_{2v}, which contains the symmetry operators $\hat{E}$, $\hat{C}_2$, $\hat{\sigma}_a$ and $\hat{\sigma}_b$, where the C_2 axis passes through the oxygen nucleus and midway between the two hydrogen nuclei, and where the σ_a mirror plane contains the three nuclei and the σ_b mirror plane is perpendicular to the σ_a mirror plane.

   a. Find the 3 by 3 matrix that is equivalent to each symmetry operator.
   b. Show that the matrices obtained in part (b) have the same multiplication table as the symmetry operators, and that they form a group. The multiplication table for the group was to be obtained in Problem 8-22.

8.44

Permutation operators are operators that interchange objects. Three objects can be arranged in $3! = 6$ different ways, or permutations. From a given arrangement, all six permutations can be attained by application of the six operators: $\hat{E}$, the identity operator; $\hat{P}_{12}$, which interchanges objects 1 and 2; $\hat{P}_{23}$, which interchanges objects 2 and 3; $\hat{P}_{13}$, which interchanges objects 1 and 3; $\hat{P}_{23}\hat{P}_{12}$, which interchanges objects 1 and 2 and then interchanges objects 2 and 3, and $\hat{P}_{23}\hat{P}_{13}$, which interchanges objects 1 and 3 and then interchanges objects 2 and 3.

Satisfy yourself that these operators produce different arrangements. Show that the six operators form a group. Construct a multiplication table for the group.

ADDITIONAL READING

H. H. Jaffe and Milton Orchin, *Symmetry in Chemistry*, John Wiley & Sons, New York, 1965.

This is a small book that gives an elementary introduction to the subject of symmetry for chemists.

Milton Orchin and H. H. Jaffe, *Symmetry, Orbitals, and Spectra*, Wiley-Interscience, New York, 1971.

This book contains a discussion of both quantum mechanics and the application of group theory to it. It includes a separate pamphlet with tables of characters for group representations.

Robert H. Dicke and James P. Wittke, *Introduction to Quantum Mechanics*, Addison-Wesley Publishing Co., Reading, Mass., 1960.

This is a quantum mechanics text for physicists. It contains a discussion of operator algebra and the construction of quantum mechanical operators. There are many similar books, and you can read about this subject in almost any of them.

Robert A. Alberty and Farrington Daniels, *Physical Chemistry*, 5th ed., John Wiley & Sons, New York, 1979.

This is a standard physical chemistry textbook. However, it contains a chapter on symmetry, including a procedure for assigning any molecule to a symmetry group.

D. S. Schonland, *Molecular Symmetry*, D. Van Nostrand, Princeton, N.J., 1965.

This is a careful introduction to the subject, written for chemists.

M. Hamermesh, *Group Theory*, Addison-Wesley Publishing Co., Reading, Mass., 1962.

This is an advanced work, carefully done, which is probably better suited to physicists than to chemists.

Ira N. Levine, *Quantum Chemistry*, Allyn and Bacon, Boston, 1970.

This work comes in two volumes. Volume I contains a discussion of operators, and Volume II contains a very good discussion of symmetry, group theory, and group representations. The book is written at the senior /first year graduate level.

Ira N. Levine, *Molecular Spectroscopy*, John Wiley and Sons, New York, 1975.

This book is similar to Vol. II of the above work.

W. W. Bell, *Matrices for Scientists and Engineers*, Van Nostrand Reinhold Company, New York, 1975.

This book is a discussion of matrix algebra and its application to scientific problems.

Robert W. Hornbeck, *Numerical Methods*, Quantum Publishers, New York, 1975.

This book is a clear introduction to numerical analysis. It contains a discussion of matrix inversion, including a FORTRAN computer program to carry out the procedure, as well as a number of other useful things. It is available in an inexpensive paperback version.

CHAPTER NINE

Algebraic Equations

Section 9-1. INTRODUCTION

In this chapter we discuss methods for solving algebraic equations, both exactly and approximately. For example, if an equation is written

$$f(x) = 0 \qquad\qquad (9.1)$$

we seek to find those values of x such that the equation is correct. These values are called *solutions*, or *roots*, of the equation.

If there are two variables in the equation, such as

$$F(x, y) = 0 \qquad\qquad (9.2)$$

then the equation can be solved for y as a function of x or x as a function of y, but in order to solve for specific values of both variables, a second equation, such as

$$G(x, y) = 0 \qquad\qquad (9.3)$$

is required. In general, if there are n variables, n independent and consistent equations are required. A set of such equations is called a *set of simultaneous equations*.

After studying this chapter, you should be able to do the following:

1. Solve any quadratic equation and determine which root is physically meaningful.
2. Obtain an accurate numerical approximation to the roots of any single equation in one unknown.
3. Solve any fairly simple set of two or three simultaneous equations by the method of elimination.
4. Solve a set of linear inhomogeneous simultaneous equations by Cramer's method and by matrix inversion.

Section 9-2. ONE EQUATION AND ONE UNKNOWN

One common kind of equations to be solved is the *polynomial equation*

$$a_0 + a_1x + a_2x^2 + \cdots + a_nx^n = 0 \tag{9.4}$$

where n is some positive integer. If $n = 1$, the equation is a *linear equation*. If $n = 2$, the equation is a *quadratic equation*. If $n = 3$, the equation is a *cubic equation*. If $n = 4$, it is a *quartic*, and so on.

Generally, there are n roots to such an equation, but two or more of the roots may be equal to each other, and some of the roots may be complex numbers. If x is a physical quantity, any complex roots must be discarded, and in some problems, negative roots are also inadmissible. Ordinarily, there will be only one root in a physical problem that is admissible.

A linear equation is simple to solve. If

$$a_0 + a_1x = 0 \tag{9.5}$$

then the single root is

$$x = -\frac{a_0}{a_1} \tag{9.6}$$

Quadratic equations are more interesting. A quadratic equation can be written

$$ax^2 + bx + c = 0 \tag{9.7}$$

Occasionally, it is possible to factorize the quadratic expression, which means that the equation can be written

$$a(x - x_1)(x - x_2) = 0 \tag{9.8}$$

where x_1 and x_2 are constants. If this can be done, then x_1 and x_2 are the two values of x that satisfy the equation.

If a quadratic equation cannot be factorized, you can apply the famous *quadratic formula*

$$\boxed{x = \frac{-b \pm \sqrt{b^2 - 4ac}}{2a}} \tag{9.9}$$

This provides two roots, one when the $+$ sign in front of the square root is chosen and one when the $-$ sign is chosen. There are three cases: (1) The quantity $b^2 - 4ac$, called the *discriminant*, is positive. In this case, the roots will be real and unequal. (2) The discriminant is equal to zero. In this case, the two roots will be equal. (3) The discriminant is negative. In this case, the roots will be unequal and complex, since the square root of a negative quantity is imaginary.

Problem 9-1

Show by substituting Eq. (9.9) into Eq. (9.7) that the quadratic formula provides the roots to a quadratic equation. •

A common application of a quadratic equation in elementary chemistry is to the calculation of the hydrogen-ion concentration in a solution of a weak acid. If activity coefficients are assumed to equal unity, the equilibrium expression is

$$K_a = \frac{C_H C_A}{C_{HA}} \qquad (9.10)$$

where C_H is the hydrogen-ion concentration, C_A the acid-anion concentration, C_{HA} is the concentration of the undissociated acid, and K_a the acid dissociation constant.

Example 9-1

For acetic acid, $K_a = 1.754 \times 10^{-5}$ mol liter^{-1} at 25°C. Find C_H if 0.10000 mole of acetic acid is dissolved in enough water to make 1.000 liter.

Solution

Assuming that no other solutes producing hydrogen ions or acetate ions are present, $C_A = C_H$, which we denote by x.

$$K_a = \frac{x^2}{0.1 - x}$$

or

$$x^2 + K_a x - 0.1 K_a = 0$$

From Eq. (9.9), our solution is

$$x = \frac{-K_a \pm \sqrt{K_a - 0.4 K_a}}{2}$$

$$= 1.316 \times 10^{-3} \text{ mol liter}^{-1} \quad \text{or} \quad -1.333 \times 10^{-3} \text{ mol liter}^{-1}$$

We must discard the negative root, because a concentration cannot be negative. •

Problem 9-2

a. Express the answer to Example 9-1 in terms of pH.
b. Find the pH of a solution formed from 0.075 mol of NH_3 and enough water to make 1 liter of NH_4OH solution. The equilibrium expression is

$$K_b = \frac{C_N C_{OH}}{C_b}$$

where we use the symbol C_N for the concentration of NH_4^+ ion, C_{OH} for the concentration of OH^- ion, and C_b for the concentration of undissociated NH_4OH. K_b equals 1.70×10^{-5} mol liter^{-1} for NH_4OH. •

If you have a cubic or higher-order polynomial equation, it is probably best to find numerical approximations to the roots rather than attempting an exact solution, even though exact methods are available for cubic and at least some quartic equations.[1] The same is true of equations other than polynomial equations (equations containing "transcendental" functions such as sine, cosines, logarithms, etc., are sometimes called *transcendental equations*).

We will discuss three methods for obtaining approximations to roots of equations: (1) making approximations on the equation to be solved, turning it into an equation that can be solved exactly but is itself only approximately correct; (2) graphical solution; and (3) numerical analysis.

As an example of the first method, let us consider an equation for the hydrogen-ion concentration in a solution of a weak acid which is more nearly correct than Eq. (9.10). This equation is[2]

$$C_a = \left(1 + \frac{C_H}{K_a}\right)\left(C_H - \frac{K_w}{C_H}\right) \tag{9.11}$$

The difference between this equation and Eq. (9.10) is that this equation includes the hydrogen ions produced by the ionization of the water. In the equation, C_a is the gross acid concentration (the concentration of acid that would occur if no ionization occurred), and K_w is the ionization constant of water, equal to 1.0×10^{-14} mol^2 liter^{-2} near 25°C. If we let $C_H = x$ and multiply this equation out, we obtain the cubic equation

$$\frac{x^3}{K_a} + x^2 - \left(\frac{C_a + K_w}{K_a}\right)x - K_w = 0 \tag{9.12}$$

In some cases, it may not be necessary to solve this cubic equation to obtain adequate accuracy.

Example 9-2

Convert Eq. (9.11) to a simpler approximate equation by discarding any terms that are small enough to be negligible. Consider the case of acetic acid with a gross acid concentration of 0.100 mol/liter.

[1] Older editions of *The Handbook of Chemistry and Physics* give methods for cubic equations. See, for example, the 33rd edition, Chemical Rubber Publishing Co., Cleveland, 1951–1952, pp. 272–273.

[2] F. Daniels and R. A. Alberty, *Physical Chemistry*, 4th ed., John Wiley & Sons, New York, 1975, p. 224.

Solution

Equation (9.11) is a product of two binomials. If one term in a binomial is much smaller than the other term, it may be possible to neglect this term. In the second binomial, we know that C_H will lie somewhere between 0.1 and 10^{-7} mol liter^{-1}, the value for pure water. In fact, we know from our approximate solution in Example 9-1 that C_H is near 10^{-3} mol liter^{-1}. Since K_w is 10^{-14} mol^2 liter^{-2}, the second term must be near 10^{-11} mol liter^{-1}, which is smaller than the first term by a factor of 10^8. We therefore drop the term K_w/C_H and obtain the equation

$$C_a = \frac{C_H + C_H^2}{K_a} \tag{9.13}$$

which is the same as Eq. (9.10). •

Problem 9-3

Show that Eq. (9.13) is the same as Eq. (9.10) •

It is possible in some cases to make a further approximation on Eq. (9.11). The first term, unity, in the first binomial, is somewhat smaller than the second if the gross acid concentration is fairly large so that C_H is also fairly large. In the case of acetic acid and a gross acid concentration of 0.1 mol liter^{-1}, C_H is approximately equal to 10^{-3} mol liter^{-1}, so that C_H/K_a is approximately equal to 100. If we neglect the unity compared with this term, we would expect an error of about 1%. If we can tolerate this, we can further approximate Eq. (9.11) as

$$C_a = \frac{C_H^2}{K_a} \tag{9.14}$$

which has the solution

$$C_H = \sqrt{C_a K_a} \tag{9.15}$$

However, as C_a is made smaller, Eq. (9.15) quickly becomes a poor approximation, and for very small acid concentrations, Eq. (9.13) also becomes inaccurate. Table 9-1 shows the results from the three equations at different acid concentrations. You can see that Eq. (9.13), the quadratic equation, remains fairly accurate down to $C_a = 10^{-5}$ mol liter^{-1}, but that Eq. (9.15) is wrong by about 7% at 10^{-3} mol liter^{-1}, and much worse than that at lower concentrations.

Table 9-1. Results for the Hydrogen-Ion Concentration in Acetic Acid Solutions at 25°C from Different Equations at Different Concentrations

	C_H (mol liter^{-1})		
C_a (mol liter^{-1})	by Eq. (9.11)	by Eq. (9.13)	by Eq. (9.15)
0.1000	1.31565×10^{-3}	1.31565×10^{-3}	1.324×10^{-3}
1.000×10^{-3}	1.23959×10^{-4}	1.23959×10^{-4}	1.324×10^{-4}
1.000×10^{-5}	0.711545×10^{-5}	0.711436×10^{-5}	1.324×10^{-5}
1.000×10^{-7}	0.161145×10^{-6}	0.099435×10^{-6}	1.324×10^{-6}

In the next example, we see another way in which an equation can be made into a tractable approximate equation, by linearizing a function.

Example 9-3

The ratio of the equilibrium populations of two states whose energies differ by ΔE is given by the *Boltzmann formula*

$$\frac{p_i}{p_j} = \exp\left(-\frac{E_i - E_j}{kT}\right) = \exp\left(-\frac{\Delta E}{kT}\right) \tag{9.16}$$

where k is Boltzmann's constant and T is the absolute temperature. The difference in energy between the lowest rotational energy level of the HCl molecule and the next rotational energy level is 8.88×10^{-22} J.

Find approximately the temperature at which the population of a state of the first excited level is 96% of that of the lowest level.

Solution

We need to find T in the equation

$$0.96 = \exp\left[-\frac{8.88 \times 10^{-22}\ \text{J}}{(1.3807 \times 10^{-23}\ \text{J K}^{-1})T}\right] = e^{-a/T}$$

where we use the symbol a for the constant equal to 64.32 K. We expand the exponential function in a Maclaurin series, using Eq. (5.26).

$$0.96 = 1 - \frac{a}{T} + \frac{1}{2!}\left(\frac{a}{T}\right)^2 + \cdots \tag{9.17}$$

Inspection of the series shows that a/T must be approximately equal to 0.04, and that $(a/T)^2$ therefore must be approximately equal to $(0.04)^2$ or 0.0016, which is considerably smaller than a/T. We therefore discard all of the terms past the a/T term and write

$$0.96 = 1 - \frac{a}{T}$$

We say that we have linearized the equation, since we have kept only the constant term and the term linear in a/T. The solution is

$$T = \frac{a}{0.04} = \frac{64.32\ \text{K}}{0.04} = 1608\ \text{K}$$

This is only an approximate result, since it satisfies an approximate equation. The correct answer is 1576 K. •

Problem 9-4

Convert Eq. (9.17) into a quadratic equation by discarding all terms in the series beyond the $(a/T)^2$ term and solve this equation for the temperature at which the population ratio is 0.96. Compare your answer with the solution of Example 9-3 and with the correct answer given in that solution •

The second method that we discuss for finding numerical approximations to roots of an equation is the *graphical method*. If x is our variable, the equation can always be written as

$$f(x) = 0 \qquad\qquad (9.18)$$

If an accurate graph of the function f is drawn, any real roots to the equation can be obtained by reading off the values of x where the curve crosses the x axis.

Example 9-4

By graphing, find a root of the equation

$$2\sin(x) - x = 0$$

Solution

A graph of the function for the interval $1.8 < x < 2.0$ is shown in Figure 9-1. From the graph, it appears that the root is near $x = 1.895$. •

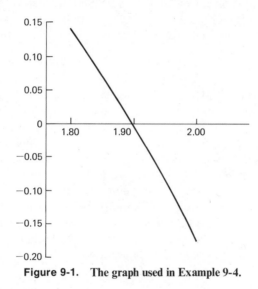

Figure 9-1.   The graph used in Example 9-4.

Problem 9-5

Find approximately the smallest positive root of the equation

$$\tan(x) - x = 0 \qquad\qquad •$$

Section 9-3.   NUMERICAL SOLUTION OF EQUATIONS

Strictly speaking, one does not solve an equation numerically. One only obtains an approximation to a root. However, with a computer or with a hand calculator, it is easy to obtain enough significant digits for almost any purpose.

The first numerical method is by trial and error. One repeatedly evaluates the function that is to vanish, choosing different values of the independent variable, until the function nearly vanishes. When using this method with a hand calculator, it is usually possible to adopt a strategy of finding two values of the independent variable that produce function values of different sign, and then to choose values carefully within this interval.

Example 9-5

Use the method of trial and error to solve the equation of Example 9-4.

Solution

We let $f(x) = 2\sin(x) - x$. We find quickly that $f(1) = 0.68294$. We find that $f(2) = -0.1814$, so that there must be a root between $x = 1$ and $x = 2$. We find that $f(1.5) = 0.49499$, so the root lies between 1.5 and 2. We find that $f(1.75) = 0.21798$, so the root is larger than 1.75. However, $f(1.9) = -0.00740$, so the root is smaller than 1.9. We find that $f(1.89) = 0.00897$, so the root is between 1.89 and 1.90. We find that $f(1.895) = 0.000809$ and that $f(1.896) = -0.000829$. To five significant digits, the root is $x = 1.8955$. •

Problem 9-6

Use the method of trial and error to solve the equation of Problem 9-5. •

The Method of Bisection. This method is just a variation of the method of trial and error, but it is made systematic so that a computer program can be written to carry it out. The method consists of starting with two values of x for which the function $f(x)$ has opposite signs, and then evaluating the function for the midpoint of the interval. If the function has the same sign at the midpoint as at the left end of the interval, the root is in the right half of the interval, and vice versa. The midpoint of the half of the original interval containing the root is taken, and it is determined which half of this new interval contains the root. The method is continued, repeating the process until the interval known to contain the root is as small as twice the error you are willing to tolerate. The middle of the last interval is then taken as the approximation to the root.

Occasionally, a root will occur at which the graph of the function does not cross the x axis, but is tangent to it at the root. The method of bisection will not work for this kind of a root, but the method can be used on the first derivative of the function, which must change sign at such a root.

Problem 9-7

The equation

$$x^3 - 2x^2 + x - 1 = 0$$

has one positive real root. Find it to five significant digits, using the method of bisection. •

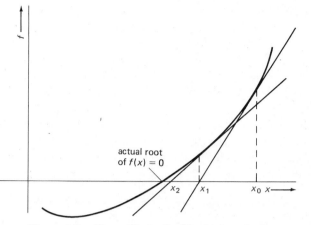

Figure 9-2.   Figure illustrating Newton's method.

The Method of Newton.   This method, which is also called the *Newton–Raphson method*, is an iterative procedure, as was the method of bisection. This means that a procedure is repeated again and again until the desired degree of accuracy is attained.

The procedure is illustrated in Figure 9-2. We assume that we have an equation written as in Eq. (9.18). The process is as follows:

Step 1.   Guess at a value, x_0, which is "not too far" from the actual root. A rough graph of the function $f(x)$ can help to choose a good value for x_0.

Step 2.   Find the value of $f(x)$ at $x = x_0$, and the value of df/dx at $x = x_0$.

Step 3.   Using the value of $f(x)$ and df/dx, find the value of x at which the tangent line to the curve at $x = x_0$ crosses the axis. This value of x, which we call x_1 is our next approximation to the root. It is given by

$$x_1 = x_0 - \frac{f(x_0)}{f'(x_0)} \tag{9.19}$$

where we use the notation

$$f'(x_0) = \frac{df}{dx}\bigg|_{x=x_0} \tag{9.20}$$

Step 4.   Repeat the process until you are satisfied with the accuracy obtained. The nth approximation is given by

$$x_n = x_{n-1} - \frac{f(x_{n-1})}{f'(x_{n-1})} \tag{9.21}$$

Problem 9-8

Using the definition of the derivative, show that Eqs. (9.19) and (9.21) are correct. ●

You must decide when to stop your iteration. If the graph of the function $f(x)$ crosses the x axis at the root, you can probably stop when the difference between x_n and x_{n+1} is smaller than the error you can tolerate. However, if the curve becomes tangent to the x axis at the root, the method may converge very slowly near the root. It may be necessary to pick another trial root on the other side of the root and to compare the results from the two iterations.

```
ACIDEQ  04:41 PM        04-Mar-80
100 PRINT "THIS PROGRAM SOLVES FOR THE HYDROGEN ION CONCENTRATION IN A "
110 PRINT "SOLUTION OF A WEAK ACID, USING NEWTON'S METHOD ON THE CUBIC"
120 PRINT "EQUATION WHICH INCLUDES HYDROGEN IONS FROM THE DISSOCIATION"
130 PRINT "OF THE WATER.  ACTIVITY COEFFICIENTS ARE ASSUME TO EQUAL UNITY."
140 PRINT "ITERATION IS DISCONTINUED WHEN THE NEXT ITERATION CHANGES THE"
150 PRINT "VALUE OF THE APPROXIMATION TO THE ROOT BY LESS THAN ONE MILLIONTH"
160 PRINT "OF THE VALUE OF THE ROOT."
200 PRINT "VALUE OF THE ACID DISSOCIATION CONSTANT?"
210 INPUT K1
220 PRINT "GROSS ACID CONCENTRATION IN MOLES/LITER?"
230 INPUT CO
240 PRINT "VALUE OF THE TRIAL ROOT?"
250 INPUT X
255 K0=1.00E-14
260 FOR I=1 TO 100
270 Y=X^3/K1+X^2-(CO+K0/K1)*X-K0
280 Y1=3*X^2/K1+2*X-CO-K0/K1
290 Z=Y/Y1
300 IF ABS(Z/X) < 1E-6 THEN 400
310 X=X-Z
320 NEXT I
400 PRINT "NUMBER OF ITERATIONS ="I
410 PRINT "HYDROGEN ION CONCENTRATION ="X
420 P= - LOG(X)/2.303
430 PRINT "PH ="P
999 END

Ready

RUN
ACIDEQ  04:42 PM        04-Mar-80
THIS PROGRAM SOLVES FOR THE HYDROGEN ION CONCENTRATION IN A
SOLUTION OF A WEAK ACID, USING NEWTON'S METHOD ON THE CUBIC
EQUATION WHICH INCLUDES HYDROGEN IONS FROM THE DISSOCIATION
OF THE WATER.  ACTIVITY COEFFICIENTS ARE ASSUME TO EQUAL UNITY.
ITERATION IS DISCONTINUED WHEN THE NEXT ITERATION CHANGES THE
VALUE OF THE APPROXIMATION TO THE ROOT BY LESS THAN ONE MILLIONTH
OF THE VALUE OF THE ROOT.
VALUE OF THE ACID DISSOCIATION CONSTANT?
? 1.754E-5
GROSS ACID CONCENTRATION IN MOLES/LITER?
? 1E-3
VALUE OF THE TRIAL ROOT?
? 1E-4
NUMBER OF ITERATIONS = 5
HYDROGEN ION CONCENTRATION = .123959E-3
PH = 3.90602

Ready
```

Figure 9-3. The computer program in BASIC to carry out Newton's method for Example 9-6.

251

§9-4 / Simultaneous Equations. Two Equations and Two Unknowns

You should always try to make sure that you do not demand too much of the method. A poor choice of x_0 can make the method converge to the wrong root, especially if the function is oscillatory. If your choice of x_0 is near a local maximum or a local minimum, the first application of the procedure might give a value of x_1 that is nowhere near the desired root. These kinds of problems can be especially troublesome when you are using a computer to do the iterations, because you will not see the values of x_1, x_2, etc., until the program has finished execution, in most cases. Remember the first law of computing: "Garbage in, garbage out."

Example 9-6

Write a computer program in BASIC to carry out Newton's method on Eq. (9.11) and use it to obtain one of the results in Table 9-1.

Solution

The program is shown in Figure 9-3, along with the output. Notice that the program is written so that iteration stops if the difference between two approximations is less than one part in 10^6 of the value of the root. If the root were at $x = 0$, this could not be used. In the present case, this choice should give about five significant digits in the answer.   •

Newton's method can be modified to work on simultaneous equations.[3]

Section 9-4. SIMULTANEOUS EQUATIONS. TWO EQUATIONS AND TWO UNKNOWNS

The simplest set of simultaneous equations is

$$a_{11}x + a_{12}y = c_1 \tag{9.22a}$$
$$a_{21}x + a_{22}y = c_2 \tag{9.22b}$$

where the a's and the c's are constants. If certain conditions are met, such a set of equations can be solved for a single value of x and a single value of y. These two roots constitute the solution to the set of equations. The set of equations in Eq. (9.22) is called *linear*, because the unknowns x and y enter only to the first power, and is called *inhomogeneous*, because there are constant terms, not containing x or y.

The first method of solution that we discuss is the *method of substitution*. One of the equations is solved to give one variable as a function of the other. This function is substituted into the other equation to give an equation in one unknown which can be solved. The result is substituted into either of the original equations, which is then solved for the second variable.

---

[3] S. D. Conte and Carl de Boor, *Elementary Numerical Analysis*, 2nd. ed., McGraw-Hill Book Company, New York, 1972, pp. 86ff.

Example 9-7

Use the method of substitution on Eq. (9.22).

Solution

We solve Eq. (9.22a) for y in terms of x:

$$y = \frac{-a_{11}x}{a_{12}} + \frac{c_1}{a_{12}} \tag{9.23}$$

We substitute this into Eq. (9-22b):

$$a_{21}x + a_{22}\left(\frac{c_1}{a_{12}} - \frac{a_{11}x}{a_{12}}\right) = c_2 \tag{9.24}$$

This contains only x and not y, so it can be solved for x to give the root

$$x = \frac{c_1 a_{22} - c_2 a_{12}}{a_{11}a_{22} - a_{12}a_{21}} \tag{9.25}$$

Equation (9.25) can be substituted into Eq. (9.22a) or (9.22b) to yield

$$y = \frac{c_2 a_{11} - c_1 a_{21}}{a_{11}a_{22} - a_{12}a_{21}} \qquad \bullet \;\; (9.26)$$

Problem 9-9

Do the algebraic manipulations to obtain Eq. (9.26).   •

The method of substitution is usually not practical for more than two equations in two unknowns. However, most of the other methods that we are going to discuss are limited to linear equations. The method of substitution is not limited to linear equations.

Problem 9-10

Solve the pair of simultaneous equations by the method of substitution:

$$x^2 - 2xy - x = 0$$

$$\frac{1}{x} + \frac{1}{y} = 2$$

(*Hint:* Multiply the second equation by xy before proceeding.)   •

Notice that in Problem 9-10 there are two solutions, since the first equation is quadratic in x. When it is solved after substituting to eliminate y, two values of x are found to satisfy the equation, and there is a root in y for each of these.

The *method of elimination* uses the process of subtracting one equation from another to obtain a simpler equation. That is, the left-hand side of one equation is subtracted from the left-hand side of another equation and the right-hand side of the first equation is subtracted from the right-hand side of the second. If both original equations are valid, the new equation will also be valid.

253

§9-4 / Simultaneous Equations. Two Equations and Two Unknowns

Example 9-8

Solve the following pair of equations.

$$x + y = 3$$
$$2x + y = 0$$

Solution

We subtract the first equation from the second to obtain

$$x = -3$$

This is substituted into either of the original equations to obtain

$$y = 6 \qquad \bullet$$

It is also possible to multiply one of the equations by a constant before taking the difference, and it is possible to add the equations instead of subtracting.

Problem 9-11

Solve the set of equations

$$3x + 2y = 40$$
$$2x - y = 10 \qquad \bullet$$

We can now consider the two common difficulties that can arise with pairs of simultaneous equations. These are (1) that the equations might be inconsistent, and (2) that the equations might not be independent. If two equations are inconsistent, there is no solution that can satisfy both of them, and if the equations are not independent, they express the same information, so that there is really only one equation, which cannot be solved for values of two variables.

Example 9-9

Show that the pair of equations is inconsistent.

$$2x + 3y = 15$$
$$4x + 6y = 45$$

Solution

We attempt solution by elimination. We multiply the first equation by 2 and subtract the second from the first, obtaining

$$0 = -15$$

which is obviously not correct. The equations are therefore inconsistent. $\bullet$

Example 9-10

Show that the equations are not independent.

$$3x + 4y = 7$$
$$6x + 8y = 14$$

Solution

We attempt a solution by elimination, multiplying the first equation by 2. However, this makes the two equations identical, so that if we subtract one from the other, we obtain

$$0 = 0$$

which is obviously correct, but not of any use to us. We really have just one equation instead of two, so that we could solve for x in terms of y or for y in terms of x, but not for values of either x or y. •

We can understand consistency and independence in simultaneous equations by looking at the graphs of the equations. Each of the equations represents y as a function of x. In the linear equations that are now being considered, these are represented by straight lines. Figure 9-4 shows the two lines representing the two equations of Example 9-8. The two equations are consistent and independent, and the lines cross at the point whose coordinates are the roots.

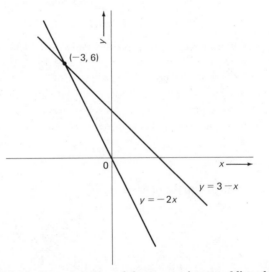

Figure 9-4. **Graphical representation of the two consistent and linearly independent equations of Example 9-8.**

Figure 9-5 shows the lines representing the two equations of Example 9-9. As you can see, the lines do not cross. There is no solution to this pair of inconsistent equations.

A single line represents both of the equations of Example 9-10. Any point on the line satisfies both equations, which are said to be *linearly dependent*.

255

§9-4 / Simultaneous Equations. Two Equations and Two Unknowns

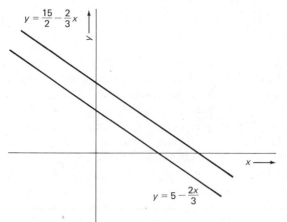

$$y = \frac{15}{2} - \frac{2}{3}x$$

$$y = 5 - \frac{2x}{3}$$

Figure 9-5. Graphical representation of two inconsistent equations of Example 9-9.

The pairs of linear equations that we have seen so far have been inhomogeneous. A pair of homogeneous linear equations is similar to Eq. (9.22) except that both c_1 and c_2 vanish.

Consider the set of homogeneous linear equations

$$a_{11}x + a_{12}y = 0$$
$$a_{21}x + a_{22}y = 0 \qquad (9.27)$$

Solve both of these equations for y in terms of x:

$$y = \frac{-a_{11}x}{a_{12}} \qquad (9.28a)$$

$$y = \frac{-a_{21}x}{a_{22}} \qquad (9.28b)$$

Both of these are represented by straight lines with zero intercept. There are two possibilities: Either the lines cross at the origin, or they coincide everywhere. In other words, either $x = 0$, $y = 0$ is the solution, or else the equations are linearly dependent. The solution $x = 0$, $y = 0$ is sometimes called a *trivial solution*, and the two equations must be linearly dependent in order for a nontrivial solution to exist. A nontrivial solution consists of specifying y as a function of x, not in finding unique roots for both x and y.

Problem 9-12

Determine whether the set of equations has a nontrivial solution, and find the solution if it exists.

$$7x + 15y = 0$$
$$101x + 195y = 0$$

•

Section 9-5. SIMULTANEOUS EQUATIONS WITH MORE THAN TWO UNKNOWNS

Systems of several equations are similar to pairs of equations for two unknowns. For a unique solution, we must have n independent and consistent equations to solve for n unknowns. Sometimes, in practical calculations, you will have more equations than you have unknowns. If all of them are consistent, you can simply pick out n independent equations to solve for n unknowns. If the equations are not all consistent, you have what is called an *overdetermined system of equations*, which has no solution.

If the equations arise from experimental measurements, the only source of inconsistency may be experimental error. In this case, you can pick various sets of n equations and solve them separately, presumably getting slightly different answers because of experimental error. The variation can be used to get an idea of the effects of the errors.

With sets of several equations, the methods of substitution and elimination can still be used. However, it is convenient to have systematic ways of applying these methods. Two such methods are known as Cramer's rule and Gauss–Jordan elimination. We will also discuss solution by matrix inversion.

Cramer's Rule. This method is equivalent to the method of substitution. The solutions are written as quotients of determinants. Equation (9.25) can be written

$$x = \frac{\begin{vmatrix} c_1 & a_{12} \\ c_2 & a_{22} \end{vmatrix}}{\begin{vmatrix} a_{11} & a_{12} \\ a_{21} & a_{22} \end{vmatrix}} \tag{9.29}$$

and Eq. (9.26) can be written

$$y = \frac{\begin{vmatrix} a_{11} & c_1 \\ a_{21} & c_2 \end{vmatrix}}{\begin{vmatrix} a_{11} & a_{12} \\ a_{21} & a_{22} \end{vmatrix}} \tag{9.30}$$

Notice the pattern: The denominator in each expression is the determinant of the matrix **A**, and the numerator is the determinant of this matrix with one of the columns replaced by the column vector $\begin{pmatrix} c_1 \\ c_2 \end{pmatrix}$. In the expression for x, the column of coefficients for x is replaced, and in the expression for y, the column of coefficients for y is replaced.

Problem 9-13

Use Cramer's rule to solve

$$4x + y = 13$$
$$2x - 3y = 3$$

If there are more than two variables and more than two linear inhomogeneous equations, the pattern is the same. For the set

$$a_{11}x_1 + a_{12}x_2 + a_{13}x_3 = c_3$$
$$a_{21}x_1 + a_{22}x_2 + a_{23}x_3 = c_2 \qquad \text{(9.31)}$$
$$a_{31}x_1 + a_{32}x_2 + a_{33}x_3 = c_3$$

the value of x_1 is given by

$$x_1 = \frac{\begin{vmatrix} c_1 & a_{12} & a_{13} \\ c_2 & a_{22} & a_{23} \\ c_3 & a_{32} & a_{33} \end{vmatrix}}{\det(\mathbf{A})} \qquad \text{(9.32)}$$

where $\det(\mathbf{A})$ is the determinant of the 3 by 3 matrix of the a coefficients. The values of x_2 and x_3 are given by similar expressions, with the second column in the determinant in the numerator replaced by the constants c_1, c_2, and c_3 in the expression for x_2, and with the third column replaced in the expression for x_3.

Equation (9.31) can be written in matrix notation as

$$\begin{bmatrix} a_{11} & a_{12} & a_{13} \\ a_{21} & a_{22} & a_{23} \\ a_{31} & a_{32} & a_{33} \end{bmatrix} \begin{bmatrix} x_1 \\ x_2 \\ x_3 \end{bmatrix} = \begin{bmatrix} c_1 \\ c_2 \\ c_3 \end{bmatrix} \qquad \text{(9.33)}$$

or

$$\mathbf{AX} = \mathbf{C} \qquad \text{(9.34)}$$

where $\mathbf{X}$ and $\mathbf{C}$ are the column vectors shown.

Let $\mathbf{A}_n$ be the matrix which is obtained from $\mathbf{A}$ by replacing the nth column by the column vector $\mathbf{C}$. Cramer's rule is now written

$$x_n = \frac{\det(\mathbf{A}_n)}{\det(\mathbf{A})} \qquad \text{(9.35)}$$

Example 9-11

Use Cramer's rule to find the value of x_1 that satisfies

$$\begin{bmatrix} 2 & 4 & 1 \\ 1 & -1 & 1 \\ 1 & 1 & 1 \end{bmatrix} \begin{bmatrix} x_1 \\ x_2 \\ x_3 \end{bmatrix} = \begin{bmatrix} 21 \\ 4 \\ 10 \end{bmatrix} \qquad \text{(9.36)}$$

Solution

$$x^1 = \frac{\begin{vmatrix} 21 & 4 & 1 \\ 4 & -1 & 1 \\ 10 & 1 & 1 \end{vmatrix}}{\begin{vmatrix} 2 & 4 & 1 \\ 1 & -1 & 1 \\ 1 & 1 & 1 \end{vmatrix}} = \frac{21\begin{vmatrix} -1 & 1 \\ 1 & 1 \end{vmatrix} - 4\begin{vmatrix} 4 & 1 \\ 1 & 1 \end{vmatrix} + 10\begin{vmatrix} 4 & 1 \\ -1 & 1 \end{vmatrix}}{2\begin{vmatrix} -1 & 1 \\ 1 & 1 \end{vmatrix} - 1\begin{vmatrix} 4 & 1 \\ 1 & 1 \end{vmatrix} + 1\begin{vmatrix} 4 & 1 \\ -1 & 1 \end{vmatrix}}$$

$$= \frac{21(-1-1) - 4(4-1) + 10(4+1)}{2(-1-1) - (4-1) + (4+1)} = \frac{-4}{-2} = 2$$

Problem 9-14

Find the values of x_2 and x_3 that satisfy Eq. (9.36).

We will not discuss completely the questions of consistency and independence for sets of more than two equations, but we will make the following comments, which apply to sets of linear inhomogeneous equations:

1.   A set of n equations is said to be *linearly dependent* if a set of constants $b_1, b_2, \ldots, b_n$, not all equal to zero, can be found such that if the first equation is multiplied by b_1, the second equation by b_2, the third equation by b_3, etc., the equations add to zero for all values of the variables. A simple example of linear dependence is for two of the equations to be identical. In this case, we could multiply one of these equations by $+1$ and the other by -1 and all of the remaining equations by 0 and have them sum to zero. If two equations are identical, one has of course only $n - 1$ usable equations, and cannot solve for n variables. If other, more complicated types of linear dependence occur, one still has only $n - 1$ usable equations (or perhaps even fewer).

2.   In the case of two identical equations, the determinant of the matrix **A** vanishes, from property 2 of determinants, described in Section 8-5. The determinant of **A** will also vanish in more complicated types of linear dependence. If det(**A**) vanishes, you are in trouble. Either the equations are linearly dependent or they are inconsistent.

3.   It is possible for a set of equations to appear to be overdetermined, and not be, if some of the set of equations are linearly dependent.

Problem 9-15

See if the set of four equations in three unknowns can be solved.

$$\begin{aligned} x_1 + x_2 + x_3 &= 6 \\ x_1 + x_2 - x_3 &= 0 \\ 3x_1 + 3x_2 + x_3 &= 12 \\ 2x_1 + x_2 + 4x_3 &= 16 \end{aligned}$$

Solution by Matrix Inversion. We have written a set of three linear inhomogeneous equations in matrix notation in Eq. (9.33), and the same notation can be used for n such equations in n unknowns. We write

$$\boxed{AX = C} \tag{9.37}$$

where A is now an n by n square matrix, X is an n by 1 column matrix, and C is another n by 1 column matrix.

If we possess the inverse of A, we can multiply both sides of Eq. (9.37) on the left by A^{-1} to get

$$\boxed{A^{-1}AX = X = A^{-1}C} \tag{9.38}$$

Thus, we simply form the matrix product of C and A^{-1}, and find that this is a column matrix which is the same as X, so we read off our solution from this column matrix.

The only difficulty with this procedure is that it is probably more work to invert an n by n matrix than to solve the set of equations by other means. However, if you are going to use a computer and if you have a matrix inversion program available, you can solve the set of equations by using the matrix inversion program and then doing a single matrix product. See Section 8-4 for matrix inversion, and see Example 11-21 for a BASIC program that will solve three equations in three unknowns and can easily be modified to handle more equations.

Problem 9-16

Solve the set of simultaneous equations by matrix inversion.

$$
\begin{aligned}
2x_1 + x_2 \quad\;\; &= 1 \\
x_1 + 2x_2 + x_3 &= 2 \\
x_2 + 2x_3 &= 3
\end{aligned}
$$

Note that the inverse of the matrix has already been obtained in Example 8-10.   •

In order for a matrix to possess an inverse, it must be nonsingular. This means that its determinant cannot vanish. If the determinant vanishes, the equations are either linearly dependent or inconsistent.

Gauss–Jordan Elimination. This is a systematic procedure for carrying out the method of elimination. It is very similar to the Gauss-Jordan elimination method for finding the inverse of a matrix, described in Section 8-4. If the set of equations is written as in Eq. (9.37), we write the augmented matrix consisting of the A matrix and the C column vector. For a set of four equations, the augmented matrix is

$$\begin{bmatrix} a_{11} & a_{12} & a_{13} & a_{14} & c_1 \\ a_{21} & a_{22} & a_{23} & a_{24} & c_2 \\ a_{31} & a_{32} & a_{33} & a_{34} & c_3 \\ a_{41} & a_{42} & a_{43} & a_{44} & c_4 \end{bmatrix} \tag{9.39}$$

Row operations on this augmented matrix will not change the roots to the set of equations, since such operations are equivalent to multiplying one of the equations by a constant or to taking the difference of two equations. In Gauss–Jordan elimination, our aim is to transform the left part of the augmented matrix into the identity matrix, which will transform the right column into the four roots, since the set of equations will then be

$$\mathbf{EX} = \mathbf{C}' \tag{9.40}$$

The row operations are carried out exactly as in Section 8-4, except of course for having only one column in the right part of the augmented matrix.

Problem 9-17

Use Gauss–Jordan elimination to solve the set of simultaneous equations in Problem 9-16. Note that exactly those row operations will be required which were used in Example 8-10. •

There is a procedure known as *Gauss elimination*, in which row operations are carried out until the left part of the augmented matrix is in upper triangular form. The bottom row of the augmented matrix then provides the root for one variable. This is substituted into the equation represented by the next-to-bottom row, and it is solved, and the values substituted into the next equation, etc.

Linear Homogeneous Equations. In Section 9-4, we discussed pairs of linear homogeneous equations for two variables: We found that such a pair of equations needed to be linearly dependent in order to have a solution other than $x = 0$, $y = 0$. The same is true of sets with more than two variables.

A set of three equations in three unknowns which is homogeneous is written

$$a_{11}x_1 + a_{12}x_2 + a_{13}x_3 = 0$$
$$a_{21}x_1 + a_{22}x_2 + a_{23}x_3 = 0 \tag{9.41}$$
$$a_{31}x_2 + a_{32}x_2 + a_{33}x_3 = 0$$

If we attempt to apply Cramer's rule to this set of equations, without asking whether it is legitimate to do so, we find for example that

$$x_1 = \frac{\begin{vmatrix} 0 & a_{12} & a_{13} \\ 0 & a_{22} & a_{23} \\ 0 & a_{32} & a_{33} \end{vmatrix}}{\det(\mathbf{A})} \tag{9.42}$$

If $\det(\mathbf{A}) \neq 0$, this yields $x_1 = 0$, and similar equations will also give $x_2 = 0$ and $x_3 = 0$. This trivial solution is all that we can have if the determinant of $\mathbf{A}$ is nonzero (i.e., if the three equations are independent).

In order to find a possible nontrivial solution, we must first investigate the condition

$$\det(\mathbf{A}) = 0 \tag{9.43}$$

which in the 3 by 3 case is the same as

$$a_{11}a_{22}a_{33} - a_{11}a_{23}a_{32} - a_{12}a_{21}a_{33} - a_{13}a_{22}a_{31}$$
$$+ a_{12}a_{23}a_{31} + a_{13}a_{21}a_{32} = 0 \tag{9.44}$$

One case in which a set of linear homogeneous equations arises is the matrix eigenvalue problem. This is analogous to an eigenvalue equation for an operator, as in Eq. (8.3). The problem is to find a column vector, $\mathbf{X}$, and a scalar eigenvalue b, such that

$$\mathbf{BX} = b\mathbf{X} \tag{9.45}$$

where $\mathbf{B}$ is a given square matrix. Since the right-hand side of Eq. (9.45) is the same as $b\mathbf{EX}$, where $\mathbf{E}$ is the identity matrix, we can rewrite Eq. (9.45) as

$$(\mathbf{B} - b\mathbf{E})\mathbf{X} = \mathbf{0} \tag{9.46}$$

which is a set of linear homogeneous equations written in the notation of Eq. (9.37).

Example 9-12

Find the values of b and $\mathbf{X}$ that satisfy the eigenvalue equation

$$\begin{bmatrix} 1 & 1 & 0 \\ 1 & 1 & 1 \\ 0 & 1 & 1 \end{bmatrix} \begin{bmatrix} x_1 \\ x_2 \\ x_3 \end{bmatrix} = b \begin{bmatrix} x_1 \\ x_2 \\ x_3 \end{bmatrix} \tag{9.47}$$

Impose a normalization condition:

$$x_1^2 + x_2^2 + x_3^2 = 1 \tag{9.48}$$

Solution

In the form of Eq. (9.46),

$$\begin{bmatrix} 1-b & 1 & 0 \\ 1 & 1-b & 1 \\ 0 & 1 & 1-b \end{bmatrix} \begin{bmatrix} x_1 \\ x_2 \\ x_3 \end{bmatrix} = 0 \tag{9.49}$$

The condition that corresponds to Eq. (9.43) is

$$\begin{vmatrix} y & 1 & 0 \\ 1 & y & 1 \\ 0 & 1 & y \end{vmatrix} = y \begin{vmatrix} y & 1 \\ 1 & y \end{vmatrix} - 1 \begin{vmatrix} 1 & 1 \\ 0 & y \end{vmatrix} = y^3 - 2y = 0 \tag{9.50}$$

where we temporarily let $y = 1 - b$. Fortunately, this cubic equation can be solved by factorization. It has the three roots

$$y = 0, \qquad y = \sqrt{2}, \qquad y = -\sqrt{2}$$

or

$$b = 1, \qquad b = 1 - \sqrt{2}, \qquad b = 1 + \sqrt{2} \qquad \textbf{(9.51)}$$

The three roots in Eq. (9.51) are the eigenvalues we wanted to find. It is only when b is equal to one of these three values that Eq. (9.47) has a non-trivial solution. Since we have three values of b, we have three different eigenvectors to find. We do this by substituting in turn each value of b into Eq. (9.49) and solving the set of equations.

We begin with $b = 1$ and write

$$0 + x_2 + 0 = 0$$
$$x_1 + 0 + x_3 = 0 \qquad \textbf{(9.52)}$$
$$0 + x_2 + 0 = 0$$

It is now obvious that this set of equations is linearly dependent, since the first and third equations are the same. Remember that this is what we must have in order to have a nontrivial solution. Our solution is now

$$x_2 = 0$$
$$x_1 = -x_3 \qquad \textbf{(9.53)}$$

We have solved for two of the variables in terms of the third. Our normalization condition provides us with unique values. Imposing it, we find for our first eigenvector

$$\mathbf{X} = \begin{bmatrix} 1/\sqrt{2} \\ 0 \\ -1/\sqrt{2} \end{bmatrix} \qquad \textbf{(9.54)}$$

Note, however, that the negative of this could also have been taken.

We now seek our second eigenvector, for which $y = \sqrt{2}$, or $b = 1 - \sqrt{2}$. Equation (9.49) becomes

$$\sqrt{2}x_1 + x_2 + 0 = 0$$
$$x_1 + \sqrt{2}x^2 + x_3 = 0 \qquad \textbf{(9.55)}$$
$$0 + x_2 + \sqrt{2}x_3 = 0$$

With the normalization condition, the solution to this is

$$\mathbf{X} = \begin{bmatrix} 1/2 \\ -1/\sqrt{2} \\ 1/2 \end{bmatrix} \qquad \bullet \ \textbf{(9.56)}$$

Problem 9-18

a. Verify Eq. (9.56). Show that this is an eigenvector.
b. Find the third eigenvector for the problem of Example 9-12. •

There is a method in physical chemistry which leads to a set of simultaneous equations very similar to Eq. (9.46). This is the determination of molecular orbitals and their energies in the Hückel approximation. The condition analogous to Eq. (9.43) is called a *secular equation*, and the eigenvalue b in Eq. (9.46) is replaced by the orbital energy. You can read about this in some physical chemistry textbooks or in a book by Royer.[4]

ADDITIONAL PROBLEMS

9.19

Solve the quadratic equations.
a. $x^2 - 3x + 2 = 0$
b. $x^2 - 1 = 0$
c. $x^2 + 2x + 2 = 0$

9.20

Solve the following equations by factorization.
a. $4x^4 - 4x^2 - x - 1 = 0$
b. $x^3 + x^2 - x - 1 = 0$
c. $x^4 - 1 = 0$

9.21

Find the real roots of the following equations by graphing.
a. $x^3 - x^2 + x - 1 = 0$
b. $e^{-x} - 0.5x = 0$
c. $\sin(x)/x - 0.75 = 0$

9.22

Use Newton's method to calculate the pH of a 0.01 molar solution of lactic acid, $C_3H_6O_3$ at 25 °C. The acid dissociation constant, K_a, is equal to 1.38×10^{-4} mol liter^{-1} at this temperature. Use Eqs. (9.13) and (9.15) to calculate the pH and comment on the accuracy of these two approximations.

9.23

The difference in energy between the $n = 1$ states of a hydrogen atom and the $n = 2$ states is 10.20 eV. (1 eV = 1.6022×10^{-19} J.) Find the temperature at which the population of one of the $n = 2$ states is 10% of the population of one of the $n = 1$ states. Use trial and error, Newton's method, or the method of bisection.

9.24

Write a computer program in the BASIC language to carry out the method of bisection.

---

[4] Donald J. Royer, *Bonding Theory*, McGraw-Hill Book Company, New York, 1968, pp. 154–163.

9.25

Find the smallest positive root of the equation.

$$\sinh(x) - x^2 - x = 0$$

9.26

Solve the set of equations, using Cramer's rule.

$$3x_1 + x_2 + x_3 = 19$$
$$x_1 - 2x_2 + 3x_3 = 13$$
$$x_1 + 2x_2 + 2x_3 = 23$$

9.27

Solve the set of equations, using Gauss or Gauss–Jordan elimination.

$$x_1 + x_2 + x_3 = 9$$
$$2x_1 - x_2 - x_3 = 9$$
$$x_1 + 2x_2 - x_3 = 9$$

9.28

Determine which, if any, of the following sets of equations are inconsistent or linearly dependent. Draw a graph for each set of equations, showing both equations. Find the solution for any set that has a unique solution.

a. $x + 3y = 4$ b. $2x + 4y = 24$ c. $3x_1 + 4x_2 = 10$

$\quad\ \ 2x + 6y = 8$ $\qquad\quad x + 2y = 8$ $\qquad\quad 4x_1 - 2x_2 = 6$

9.29

Solve the set of equations

$$x^2 - 2xy + y^2 = 0$$
$$2x + 3y = 5$$

9.30

Solve the sets of equations.

a. $3x_1 + 4x_2 + 5x_3 = 25$

$\quad\ \ 4x_1 + 3x_2 - 6x_3 = -7$

$\quad\ \ x_1 + x_2 + x_3 = 6$

b.
$$\begin{bmatrix} 1 & 1 & 1 & 3 \\ 2 & 1 & 1 & 1 \\ 1 & 2 & 3 & 4 \\ 2 & 0 & 1 & 4 \end{bmatrix} \begin{bmatrix} x_1 \\ x_2 \\ x_3 \\ x_4 \end{bmatrix} = \begin{bmatrix} 6 \\ 5 \\ 10 \\ 7 \end{bmatrix}$$

9.31

Find the eigenvalues and eigenvectors for the matrix

$$\begin{bmatrix} 1 & 1 & 1 \\ 1 & 1 & 1 \\ 1 & 1 & 1 \end{bmatrix}$$

ADDITIONAL READING

J. V. Uspensky, *Theory of Equations*, McGraw-Hill Book Company, New York, 1948.
This book is an introduction to the theory and practice of solving equations. It includes discussion of single-variable and simultaneous equations, and also of numerical methods for approximations to roots.

Lois W. Griffiths, *Introduction to the Theory of Equations*, 2nd ed., John Wiley & Sons, New York, 1947.
This is another book in the same area as Uspensky.

S. D. Conte and Carl de Boor, *Elementary Numerical Analysis*, 2nd ed., McGraw-Hill Book Company, New York, 1972.
This is a standard beginning textbook in numerical analysis. It includes discussion of Newton's method and of the solution of simultaneous equations.

Robert W. Hornbeck, *Numerical Methods*, Quantum Publishers, New York, 1975.
This is a clearly written book which is available in an inexpensive paperbook version. It has a good discussion on numerical approximations to roots of equations, and another on simultaneous equations and matrix inversion. The discussions include warnings about difficulties that might arise.

Farrington Daniels and Robert A. Alberty, *Physical Chemistry*, 4th ed., John Wiley & Sons, New York, 1975.
This is a standard physical chemistry text, but it contains a section on the solution of chemical equilibrium expressions that has no counterpart in most modern physical chemistry texts.

CHAPTER TEN

The Treatment of Experimental Data

Section 10-1. INTRODUCTION

The usual aim of a physical chemistry experiment, whether carried out by a student or a professional chemist, is to determine a numerical value for one or more quantities. Sometimes the desired quantity can be measured directly. In other cases, the raw data must be analyzed mathematically, graphically, or numerically in order to calculate a value for some quantity other than the directly measured quantities. This process is called *data reduction*.

Once we have obtained a value for a quantity, we must determine how accurate that value is, since experimental error is always present. A number without any indication of its accuracy is not very useful. The difference between a measured value and the true value is the experimental error. Since we do not ordinarily know the true value, we must find a way to estimate the size of the experimental error. Statistical analysis is used for this purpose.

Statistics is the study of a large set of people, objects, or numbers, called a *population*. The entire population is not studied directly, because of its size or inaccessibility. A subset from the population, called a *sample*, is studied, and the probable properties of the population are inferred from the properties of the sample.

When statistical methods are applied to scientific experiments, the population is an imaginary set of many repetitions of the experiment. The actual set of several repetitions is considered to be a sample from this population. If the actual repetitions agree well with each other, the data are said to have good precision. If the data agree well with the correct value, the data are said to have good accuracy.

It is tempting to assume that a data set of high precision also has high accuracy, but this may be a poor assumption, and there are a lot of data in the scientific literature that do not now appear to be accurate as the experimenters thought at the time of publication.

The experimental errors that occur are classified as systematic errors and random errors. *Systematic errors* recur with the same direction on every repetition of the experiment, so they can affect the accuracy without affecting the precision. *Random errors* do not have the same direction every time, so they will affect the precision as well as the accuracy.

We discuss the estimation of random and systematic errors in this chapter, as well as the most common kinds of data reduction. After studying the chapter, you should be able to do all of the following:

1. Identify probable sources of error in a physical chemistry experiment and classify the errors as systematic or random.
2. Calculate the mean and standard deviation of a sample of numbers.
3. Calculate the probable error in the measured value of a directly measured quantity.
4. Carry out data reduction using mathematical formulas and do an error propagation calculation to determine the probable error in the final calculated quantity.
5. Carry out data reduction using graphical methods and determine the probable error in quantities obtained from the graphs.
6. Carry out data reduction using least squares methods and determine probable errors in quantities obtained by these methods.

Section 10-2. EXPERIMENTAL ERRORS IN DIRECTLY MEASURED QUANTITIES

Occasionally, a chemist requires a value for some property that can be measured directly. For such a measurement, no data reduction is necessary, and the error analysis is fairly simple. An example of such a measurement is the determination of the melting temperature of a pure substance. The simplest apparatus consists of a small bath containing a liquid in which the sample can be suspended in a small capillary tube held next to a thermometer. The bath is slowly heated and the reading of thermometer at the time of melting of the sample is recorded.

Example 10-1

List some of the possible experimental error sources in a simple melting-temperature determination. Classify each as systematic or random, and guess its magnitude.

Solution

1. Faulty thermometer calibration. This is systematic. With an inexpensive thermometer, this error might be as large as several tenths of a degree.
2. Lack of thermal equilibration between the liquid of the bath, the sample, and the thermometer. If the experimental procedure is the same for all repetitions, this will be systematic. The thermometer will likely be heated more slowly than the sample if the heating is done too rapidly. This error will probably be less than 1°C.
3. Failure to read the thermometer correctly. This is random. There are two kinds of error here. The first is more of a blunder than an experimental error, and amounts to counting the marks on the thermometer incorrectly and recording 87.5 °C instead of 88.5 °C, etc. This kind of error in a data point should easily be detected if the measurement is repeated several times. Once detected, such a data point would be discarded. The other kind of error is due to parallax, or looking at the thermometer at some angle other than a right angle. This might produce an error of about two-tenths of a degree Celsius.
4. Presence of impurities in the sample. This is systematic, since impurities that dissolve in the liquid always lower the melting point. If carefully handled samples of purified substances are used, this error should be negligible.
5. Failure to observe the onset of melting. This is systematic, although variable in magnitude. If the heating is done slowly, it should be possible to reduce this error to a few tenths of a degree.

Most of the described errors above are systematic, and most of them can be minimized by reducing the rate of heating. This suggests a possible procedure: an initial rough determination establishes the approximate value, and a final heating is done with slow and careful heating near the melting point. •

Problem 10-1

List as many sources of error as you can for some of the following measurements. Classify as systematic or random, and estimate the magnitude of each source of error.

a. The measurement of the diameter of a fine copper wire with a micrometer caliper.
b. The measurement of the length of a piece of glass tubing with a wooden meter stick.
c. The measurement of the mass of a porcelain crucible with a balance such as those found in an undergraduate laboratory.
d. The measurement of the mass of a silver chloride precipitate in a porcelain crucible.
e. The measurement of the resistance of an electrical heater using a Wheatstone bridge.

    f.   The measurement of the time required for an automobile to travel the distance between two highway markers 1 mi. apart, using a stopwatch. •

Although it is valuable to engage in the kind of educated guesswork involved in Example 10-1 and Problem 10-1, it is better to have some kind of objective way to estimate the magnitude of your experimental errors. This is more difficult to do with systematic errors than with random errors, because repeating the experiment with the same apparatus cannot give you information about systematic errors, which will simply recur with the same direction and probably the same magnitude. However, such repetitions will give information about random errors, since they will sometimes be in one direction and sometimes in the other.

There are two principal ways to gain information about systematic errors. One is to modify the apparatus and repeat the measurement, or to repeat the measurement with a different apparatus. The other is to use the same apparatus to measure a well-known quantity, observing the actual experimental error. If you do this, you should make several repetitions of the measurement to make sure that you are seeing systematic rather than random errors.

There are two principal kinds of apparatus modification that can be used to study systematic errors. One is to replace components of the apparatus. For example, in making a voltage measurement with a potentiometer, one compares the voltage with that of a standard cell. One could see if the same result is obtained with a different standard cell. If the result is different, you can assume that there probably is a systematic error about as large as the difference in the values.

The other kind of apparatus modification is to improve the apparatus, at least temporarily. For example, the apparatus may include some insulation that minimizes unwanted heat transfer. If the measurement gives a different value when the insulation is improved, there was probably a systematic error about as big as (or bigger than) the change in the result.

The methods of estimating systematic errors that we have just mentioned are often not available in physical chemistry laboratory courses. In this event, educated guesswork is nothing to be ashamed of. It is better to guess at your systematic errors than to ignore them, and by the time you gain some experience, you will be able to inspect an apparatus and do a good job of identifying sources of error and estimating the sizes of the errors.

Section 10-3. STATISTICAL TREATMENT OF RANDOM ERRORS

Since random errors recur with variable sign and magnitude, the principal tool for studying them is the statistical analysis of repetitions of the measurements. For the present, we will ignore systematic errors, and speak as though only random errors are present. From a set of data, we need to get an estimate of the correct value and the probable error in this estimate.

If we have a set of repetitions of a measurement, we use the *mean* of the set for our estimate of the correct value. The mean of a set of numbers is given by Eq. (4.40)

$$\bar{x} = \frac{1}{N} \sum_{i=1}^{N} x_i \qquad (10.1)$$

where N is the number of members of the set, and $x_1, x_2, \ldots$ are the members of the set.

The mean is one of three common averages. The *median* is a value such that half of the members of the set are greater than the median and half are smaller than the median. The *mode* is the value that occurs most frequently in the set.

In order to estimate the accuracy of our estimate, we need to study the dispersion, or spread of our set of data. The common quantity used to describe this is the *standard deviation*, which for a set of N numbers is given by

$$s = \left[\frac{1}{N-1} \sum_{i=1}^{N} (x_i - \bar{x})^2 \right]^{1/2} \qquad (10.2)$$

where $\bar{x}$ is the mean given by Eq. (10.1). The square of the standard deviation is called the *variance*. In Eq. (10.2), every term in the sum is positive or zero, since it is the square of a real quantity, so that the standard deviation can only vanish if every member of the set of numbers is equal to the mean. The more the members of the set differ from each other, the larger the standard deviation will be. In most cases, about two-thirds of the members of a set of numbers will have values between $\bar{x} - s$ and $\bar{x} + s$.

Example 10-2

Find the mean and the standard deviation of the set of numbers

$$32.41, \quad 33.76, \quad 32.91, \quad 33.04, \quad 32.75, \quad 33.23$$

Solution

$\bar{x} = \frac{1}{6}(32.41 + 33.76 + 32.91 + 33.04 + 32.75 + 33.23)$

$\quad = 33.02$

$s = \{\frac{1}{5}[(0.39)^2 + (0.74)^2 + (-0.11)^2 + (0.02)^2 + (-0.27)^2 + (0.21)^2]\}^{1/2}$

$\quad = 0.41$

Notice that one of the six numbers lies below 32.61, and that one lies above 33.43. ●

Problem 10-2

Find the mean, $\bar{x}$, and the standard deviation, s, for the following set of numbers.

2.876 m,   2.881 m,   2.864 m,   2.879 m,   2.872 m,   2.889 m,   2.869 m

Determine how many numbers lie below $\bar{x} - s$ and how many lie above $\bar{x} + s$.   ●

Consider now our population, the data from an infinite number of repetitions of our experiment. We assume that there are no systematic errors, so that the mean of the population is the correct value. Since there are infinitely many numbers in the population, the mean of the population, μ, is given by Eq. (4-49).

$$\mu = \int_a^b xf(x)\,dx \qquad (10.3)$$

where a is the smallest possible value for x and b is the largest possible value. In many cases, we will assume that $a = -\infty$ and that $b = +\infty$. The function $f(x)$ is the probability density, defined in Section 4-6:

probability of value between x and $x + dx = f(x)\,dx$ (10.4)

The *standard deviation of the population* is given by

$$\sigma = \left[\int_a^b (x - \mu)^2 f(x)\,dx \right]^{1/2} \qquad (10.5)$$

Notice that the sample standard deviation is not quite the square root of the mean value of the quantity $(x - \bar{x})^2$, while the population standard deviation is the square root of the mean of the quantity $(x - \mu)^2$. Mathematicians have shown that the $N - 1$ factor in the denominator of Eq. (10.2) is necessary in order for s to be an unbiased estimate of σ. This has to do with the fact that in a sample of N members, there are $N - 1$ independent pieces of information besides the mean. Since our population is inaccessible, we use the sample mean and the sample standard deviation as estimates of the population mean and population standard deviation.

The Standard Normal Distribution.   This probability distribution is commonly assumed to describe populations of various kinds, including the IQ scores of large numbers of people and velocities of molecules in a gas. In fact, many of the formulas used by statisticians are correct only if the population is described by the standard normal distribution, and hold only approximately for other distributions. The *Gaussian distribution* or *standard* *normal distribution* is given by

$$f(x) = \frac{1}{\sqrt{2\pi}\sigma} \exp\left[-\frac{(x-\mu)^2}{2\sigma^2}\right] \tag{10.6}$$

Example 10-3

Show that the distribution in Eq. (10.6) satisfies the normalization condition of Eq. (4.52) with the limits of integration equal to $-\infty$ and $+\infty$.

Solution

We have the following integral, which we treat by the method of substitution:

$$\int_{-\infty}^{\infty} \frac{1}{\sqrt{2\pi}\sigma} e^{-(x-\mu)^2/2\sigma^2}\, dx = \frac{1}{\sqrt{\pi}} \cdot \int_{-\infty}^{\infty} e^{-t^2}\, dt = 1 \qquad \bullet$$

Problem 10-3

Using Eqs. (10.3) and (10.5), calculate the mean and standard deviation of the standard normal distribution, showing that the symbols μ and σ in Eq. (10.6) have the same meaning as in Eqs. (10.3) and (10.5). $\qquad \bullet$

Figure 10-1 shows a graph of the standard normal distribution. Five values of x have been marked on the x axis: $x = \mu - 1.96\sigma$, $x = \mu - \sigma$, $x = \mu$, $x = \mu + \sigma$, and $x = \mu + 1.96\sigma$.

Example 10-4

Assuming the standard normal distribution, calculate the fraction of the population with x lying between $x = \mu - \sigma$ and $x = \mu + \sigma$.

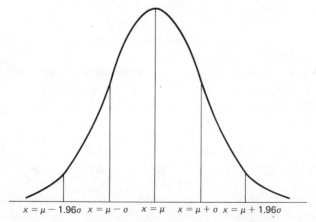

$x = \mu - 1.96\sigma$ $\quad x = \mu - \sigma$ $\quad x = \mu$ $\quad x = \mu + \sigma$ $\quad x = \mu + 1.96\sigma$

Figure 10-1. The standard normal, or Gaussian, probability distribution.

Solution

By integration of Eq. (10.6), we obtain

$$\text{fraction between } \mu - \sigma \text{ and } \mu + \sigma = \int_{\mu-\sigma}^{\mu+\sigma} \frac{1}{\sqrt{2\pi\sigma}} e^{-(x-\mu)^2/2\sigma^2} dx$$

$$= \int_{-\sigma}^{+\sigma} \frac{1}{\sqrt{2\pi\sigma}} e^{-y^2/2\sigma^2} dy$$

$$= \frac{2}{\sqrt{\pi}} \int_0^{1/\sqrt{2}} e^{-t^2} dt$$

The last integral is the error function with argument $1/\sqrt{2}$, which is described in Appendix 7. From the table in that appendix,

$$\text{fraction between } \mu - \sigma \text{ and } \mu + \sigma = \text{erf}\left(\frac{1}{\sqrt{2}}\right) = \text{erf}(0.707 \ldots) = 0.683 \ldots \bullet$$

Problem 10-4

Show that the fraction of a population lying between $\mu - 1.96\sigma$ and $\mu + 1.96\sigma$ is equal to 0.95 if the population is described by the standard normal distribution. $\bullet$

You should remember the facts obtained in Example 10-4 and Problem 10-4. With a standard normal distribution, 68% of the population lies within 1 standard deviation of the mean, 95% of the population lies within 1.96 standard deviations of the mean, and 99% of the population lies within 2.67 standard deviations of the mean. For other intervals, we can write

$$\text{fraction between } \mu - x_1 \text{ and } \mu + x_1 = \text{erf}\left(\frac{x_1}{\sqrt{2\sigma}}\right) \tag{10.7}$$

We assume that our population of imaginary repetitions of a measurement is governed by the standard normal distribution, which is also called the gaussian distribution, after Gauss, a famous German mathematician. There are a number of other probability distributions that are used in statistics, including the binomial distribution, the Poisson distribution, and the Lorentzian distribution.[1]

If a random experimental error arises as the sum of many contributions, the central limit theorem of statistics gives some justification for assuming that our final error will be governed by the standard normal distribution. The theorem states that if a number of random variables (independent variables) $x_1, x_2, \ldots x_n$ are governed by some probability distributions with

---

[1] See Philip R. Bevington, *Data Reduction and Error Analysis for the Physical Sciences*, McGraw-Hill Book Company, New York, 1969, Chap. 3, or Hugh D. Young, *Statistical Treatment of Experimental Data*, McGraw-Hill Book Company, New York, 1962, for discussions of various distributions.

finite means and finite standard deviations, then a weighted sum

$$y = \sum_{i=1}^{n} a_i x_i$$

is governed by a probability distribution that approaches a standard normal distribution as n becomes large.

Numerical Estimation of Random Errors. We now use the properties of the standard normal distribution in order to estimate random errors. A common practice among scientific workers is to make statements that have a 95% probability of being true. Such a statement is said to be at the 95% *confidence level*, or is sometimes said to be at the 5% *significance level*. Of course, every statement that we make at this confidence level has a 5% change of being wrong.

What we now want to do is to write a statement of the form

$$\mu = \text{correct value} = \bar{x} \pm \lambda \tag{10.8}$$

which has a 95% chance of being right. That is, we want to state an interval of size 2λ which has a 95% chance of containing the correct value. We call such an interval the 95% confidence interval, and we call the number λ the *probable error*.

We begin by thinking of our set of N measurements as only one possible set of N measurements. We could take infinitely many sets of N measurements from our population, and the means of these sets themselves form a new population. The mean of this new population is the same as the mean of the original population, and the standard deviation of the new population is given by

$$\sigma_m = \frac{\sigma}{\sqrt{N}} \tag{10.9}$$

where σ is the standard deviation of the original population. Notice that the new population has a smaller spread than the original population. The means of the sets of measurements cluster more closely about the population mean than do individual measurements. If the original population is governed by a standard normal distribution, the population of sample means will also be governed by a standard normal distribution.

If we knew the standard deviation of the original population, we would now be able to write

$$\lambda = \frac{1.96\sigma}{\sqrt{N}} \tag{10.10}$$

We do not know the population standard deviation. However, since we use the sample standard deviation as our estimate of the population standard

Table 10-1. Some Values of Student's t Factor*

Number of Observations, N	Number of Degrees of Freedom, $v = N - 1$	Maximum Value of Student's t Factor for the Probability Levels Indicated:			
		$t(v, 0.50)$	$t(v, 0.10)$	$t(v, 0.05)$	$t(v, 0.01)$
2	1	1.000	6.314	12.706	127.32
3	2	0.816	2.920	4.303	9.925
4	3	0.765	2.353	3.182	4.841
5	4	0.741	2.132	2.776	4.604
6	5	0.727	2.015	2.571	4.032
7	6	0.718	1.943	2.447	3.707
8	7	0.711	1.895	2.365	3.500
9	8	0.706	1.860	2.306	3.355
10	9	0.703	1.833	2.262	3.250
11	10	0.700	1.812	2.228	3.169
21	20	0.687	1.725	2.086	2.845
∞	∞	0.674	1.645	1.960	2.576

* R. A. Day and A. L. Underwood, Quantitative Analysis, 3rd ed., Prentice-Hall, Englewood Cliffs, N.J., 1974.

deviation, we could write as an approximation

$$\lambda \approx \frac{1.96s}{\sqrt{N}} \tag{10.11}$$

where s is the sample standard deviation. However, there is a statistically precise formula, derived by a mathematician named Gossett, which enables us to write an exact equivalent of Eq. (10.11).

Gossett defined the quantity

$$t = \frac{(\bar{x} - \mu)N^{1/2}}{s} \tag{10.12}$$

For some reason, Gossett wrote under the pseudonym "Student," so that the quantity t is called the *Student t factor*. In the Student t factor, μ is the population mean, x is the sample mean and s is the sample standard deviation. There is a different value of t for every sample, and although μ is not known, Student derived the probability distribution that t obeys.[2] From this distribution, which is called *Student's t distribution*, the maximum value of t corresponding to a given confidence level can be calculated for any value of N.

Table 10-1 gives these values for various values of N and for four different confidence levels. Notice that as N approaches infinity, the maximum Student t value for 95% confidence approaches 1.96, the factor in Eq. (10.10). This

[2] See Walter Clark Hamilton, *Statistics in Physical Science*, The Ronald Press Company, New York, 1964, pp. 78ff.

is because s approaches σ as N becomes large. Using a value from Table 10-1, we can write for the 95% confidence level

$$\lambda = \frac{t(v, 0.05)s}{\sqrt{N}} \tag{10.13}$$

where we use the notation that $t(v, 0.05)$ is the maximum value of t for the 95% confidence level (5% significance level) for a sample of $v + 1$ members. The quantity v is called the *number of degrees of freedom*, and equals $N - 1$.

Example 10-5

Assume that the melting temperature of calcium nitrate tetrahydrate, $Ca(NO_3)_2 \cdot 4H_2O$, has been measured 10 times, and that the results are:

42.70, 42.60, 42.78, 42.83, 42.58, 42.68, 42.65, 42.76, 42.73, 42.71

Ignoring systematic errors, determine the 95% confidence interval for the set of measurements.

Solution

Our estimate of the correct melting temperature is the sample mean:

$$\bar{T} = \tfrac{1}{10}(42.70 + 42.60 + 42.78 + 42.83 + 42.58 + 42.68$$
$$+ 42.65 + 42.76 + 42.73 + 42.71)$$
$$= 42.70$$

The sample standard deviation is

$$s = [\tfrac{1}{9}(0.00^2 + 0.10^2 + 0.08^2 + 0.13^2 + 0.12^2 + 0.02^2 + 0.05^2$$
$$+ 0.16^2 + 0.13^2 + 0.01^2)]^{1/2}$$
$$= 0.08$$

The value of $t(9, 0.05)$ is found from Table 10-1 to be equal to 2.26, so that

$$\lambda = \frac{(2.26)(0.08)}{\sqrt{10}} = 0.06$$

Therefore, we write

$$T = 42.70 \pm 0.06 \,^\circ C$$

This gives the 95% confidence interval, or the interval that has a 95% probability of containing the correct value. ●

Problem 10-5

Assume that the H—O—H bond angles in various crystalline hydrates have been measured to be 108°, 109°, 110°, 103°, 111°, and 107°. If these all come from the same population, give your estimate of the population mean and its 95% confidence interval. ●

Rejection of Discordant Data.   We now discuss the problem of what to do when a repetition of a measurement yields a value that differs greatly from the other members of the sample. For example, say that we repeated the measurement of the melting temperature of $Ca(NO_3)_2 \cdot 4H_2O$ in Example 10-5 one more time and obtained a value of 39.75 °C. If we include this eleventh data point, we get a sample mean of 42.43 °C and a sample standard deviation of 0.89 °C. Using the table of Student's t value in Table 10-1, we obtain a value for λ of 0.60 °C at the 95% confidence level. Some people think that the only honest thing to do is to report the melting temperature as 42.43 ± 0.60 °C.

However, if we assume that our sample standard deviation of 0.89 °C is a good estimate of the population standard deviation, our data point of 39.75 °C is 3.01 standard deviations away from the mean. From the table of the error function in Appendix 7, the probability of a randomly chosen member of a population differing from the mean by this much or more is 0.003, or 0.3%. There is considerable justification for asserting that such an improbable event was not really due to random experimental errors but to some kind of a mistake. If you decide to assume this, you discard the suspect data point and recompute the mean and standard deviation just as though the data point had not existed.

There is apparently no generally accepted rule for deciding when to discard a data point. Some people discard data points that are more than 2.7 standard deviations from the mean (remember that this means 2.7 standard deviations calculated with the suspect point left in). Pugh and Winslow[3] suggest that for a sample of N data points, a data point should be discarded if there is less than one chance in $2N$ that the point came from the same population. This would seem to discard too many points, since their rule would discard a point lying 1.96 standard deviations from the mean in a sample of 10 measurements. Since you would expect such a point to occur once in 20 times, the probability of occurring in a sample of 10 by random chance is fairly large. Perhaps a sufficiently conservative rule would be to discard a data point if there is one chance in $10N$ that it would occur by chance.

Section 10-4. DATA REDUCTION AND THE PROPAGATION OF ERRORS

The Combination of Errors.   Let us begin by assuming that we have measured two quantities, a and b, and have established a probable value and a 95% confidence interval for each:

$$a = \bar{a} \pm \lambda_a \tag{10.14a}$$

$$b = \bar{b} \pm \lambda_b \tag{10.14b}$$

[3] Emerson M. Pugh and George H. Winslow, *The Analysis of Physical Measurements*, Addison-Wesley Publishing Co., Reading, Mass. 1966.

We now want to obtain a probable value and a 95% confidence interval for c, the sum of a and b.

The probable value of c is simply the sum of $\bar{a}$ and $\bar{b}$:

$$\bar{c} = \bar{a} + \bar{b} \tag{10.15}$$

A very simple estimate of the probable error in c is simply the sum of λ_a and λ_b:

$$\lambda_c \approx \lambda_a + \lambda_b \tag{10.16}$$

However, Eq. (10.16) provides an overestimate, because there is some chance that errors in a and in b will cancel.

It has been shown by statisticians that if a and b are both governed by standard normal distributions, c is also governed by a standard normal distribution, and that the probable error in c is given by

$$\boxed{\lambda_c = (\lambda_a^2 + \lambda_b^2)^{1/2}} \tag{10.17}$$

This formula allows for the statistically correct amount of cancellation of errors.

Example 10-6

Two lengths have been measured as 24.8 ± 0.4 m and 13.6 ± 0.3 m. Find the probable value of their sum and its probable error.

Solution

The probable value is $24.8 + 13.6 = 38.4$ m, and the probable error is

$$\lambda = (0.4^2 + 0.3^2)^{1/2} = 0.5 \text{ m}$$

We thus write

$$\text{sum} = 38.4 \pm 0.5 \text{ m} \qquad \bullet$$

Problem 10-6

Two time intervals have been clocked by different observers as 56.57 ± 0.13 s and 75.12 ± 0.17 s. Find the probable value of their sum and its probable error. $\qquad \bullet$

Combination of Random and Systematic Errors. In Section 10-3, we discussed the use of statistics to determine the probable error due to random errors in the case that the measurements could be repeated a number of times. The probable error due to systematic errors can be estimated by apparatus modification or by guesswork. In the case of a directly measured quantity, these errors combine in the same way as the errors in Eq. (10.17), because they can add or cancel. If λ_r is the probable error due to random errors and λ_s is the probable error due to systematic errors, the total probable error is given by

$$\lambda_t = (\lambda_s^2 + \lambda_r^2)^{1/2} \tag{10.18}$$

In using this formula, you must try to make your estimate of the systematic error conform to the same level of confidence as your random error. If you use the 95% confidence level for the random errors, do not estimate the systematic errors at the 50% confidence level, which is what most people instinctively tend to do when asked what they think a probable error is. You might want to make a first guess at your systematic errors and then double it to avoid this underestimation.

Example 10-7

Assume that you estimate the total systematic error in the melting temperature measurement of Example 10-5 as 0.20 °C, at the 95% confidence level. Find the total expected error.

Solution

$$\lambda_t = (0.06^2 + 0.20^2)^{1/2} = 0.21 \,^{\circ}\text{C} \qquad \bullet$$

Data Reduction Using Mathematical Formulas. In most physical chemistry experiments, the quantity of interest is not measured directly. Other quantities are measured, and their values are used to calculate the value of the desired quantity. The simplest case is the use of a single mathematical formula. For example, in the Dumas method for determining the mass per mole of a volatile liquid,[4] one uses the formula

$$M = \frac{gRT}{PV} \tag{10.19}$$

where M is the mass per mole, g the measured mass of a sample of vapor of the substance contained in volume V at pressure P and temperature T, and R the gas constant.

Let us think of Eq. (10.19) as being an example of a general formula,

$$y = y(x_1, x_2, x_3, \ldots, x_n) \tag{10.20}$$

Let us assume that we have a 95% confidence interval for each of the independent variables, such that

$$x_i = \bar{x}_i \pm \lambda_i \qquad \text{(one equation for each value of } i\text{)}$$

Our problem is to take the uncertainties in x_1, x_2, etc., and calculate the uncertainty in y, the dependent variable. This is called the *propagation* of *errors*.

[4] Lawrence J. Sacks, *Experimental Chemistry*, Macmillan Publishing Co., New York, 1971, pp. 26–29.

If the errors are not too large, we can take an approach based on the differential calculus of several independent variables, which is discussed in Chapter 6. The fundamental equation of differential calculus is Eq. (6.11):

$$dy = \left(\frac{\partial y}{\partial x_1}\right)dx_1 + \left(\frac{\partial y}{\partial x_2}\right)dx_2 + \left(\frac{\partial y}{\partial x_3}\right)dx_3 + \cdots + \left(\frac{\partial y}{\partial x_n}\right)dx_n \quad \textbf{(10.21)}$$

This equation gives an infinitesimal change in y due to infinitesimal changes in the independent variables.

If finite changes are made in the independent variables, we can write as an approximation

$$\Delta y = \left(\frac{\partial y}{\partial x_1}\right)\Delta x_1 + \left(\frac{\partial y}{\partial x_2}\right)\Delta x_2 + \cdots + \left(\frac{\partial y}{\partial x_n}\right)\Delta x_n \quad \textbf{(10.22)}$$

If we had some *known* errors in x_1, x_2, etc., we could use Eq. (10.22) to calculate a known error in y. Since all we have is probable errors in the independent variables and do not know whether these are positive or negative, one cautious way to proceed would be to assume that the worst might happen and all the errors would add:

$$\lambda_y \approx \left|\left(\frac{\partial y}{\partial x_1}\right)\lambda_1\right| + \left|\left(\frac{\partial y}{\partial x_2}\right)\lambda_2\right| + \cdots + \left|\left(\frac{\partial y}{\partial x_n}\right)\lambda_n\right| \quad \textbf{(10.23)}$$

In this equation, λ_y is the probable error in y.

Equation (10.23) overestimates the errors because there is always some statistical probability that the random errors will cancel instead of adding. An equation that is analogous to Eq. (10.17) is

$$\lambda_y \approx \left[\left(\frac{\partial y}{\partial x_1}\right)^2 \lambda_1^2 + \left(\frac{\partial y}{\partial x_2}\right)^2 \lambda_2^2 + \cdots + \left(\frac{\partial y}{\partial x_n}\right)^2 \lambda_n^2\right]^{1/2} \quad \textbf{(10.24)}$$

This will be our working equation for the propagation of errors through formulas. It is the proper equation to use if all the independent variables are governed by standard normal distributions. In that case, it becomes more and more nearly exact as the errors become small.

Example 10-8

Find the expression for the propagation of errors for the Dumas molecular weight determination. Apply this to the following set of data for *n*-hexane:

$$T = 373.15 \pm 0.25 \text{ K}$$
$$V = 206.34 \pm 0.15 \text{ ml}$$
$$P = 760 \pm 0.2 \text{ torr}$$
$$g = 0.585 \pm 0.005 \text{ g}$$

Solution

$$\frac{gRT}{PV}$$

$$M = \frac{(0.585 \text{ g})(0.082057 \text{ liter atm K}^{-1} \text{ mol}^{-1})(373.15 \text{ K})}{(1.000 \text{ atm})(0.20634 \text{ liters})}$$

$$= 86.81 \text{ g mol}^{-1}$$

We now obtain the analog of Eq. (10.24) for our equation.

$$\lambda_M = \left[\left(\frac{RT}{PV} \right)^2 \lambda_g^2 + \left(\frac{gR}{PV} \right)^2 \lambda_T^2 + \left(\frac{gRT}{P^2 V} \right)^2 \lambda_P^2 + \left(\frac{gRT}{PV^2} \right)^2 \lambda_V^2 \right]^{1/2} \qquad \textbf{(10.25)}$$

Substituting the numerical values into Eq. (10.25), we obtain

$$\lambda_M = 0.747 \text{ g mol}^{-1}$$

Our 95% confidence interval is thus given by

$$M = 86.81 \pm 0.75 \text{ g mol}^{-1}$$

The accepted value is 86.17 g mol^{-1}.

9.1

5.11

Problem 10-7

In the cryscopic determination of molecular weight,[5] the molecular weight (mass per mole) is given by

$$M = \frac{1000 \, gK_f}{G\Delta T_f}(1 - k_f \Delta T_f)$$

where G is the mass of the solvent, g the mass of the unknown solute, ΔT_f the amount by which the freezing point of the solution is less than that of the pure solvent, and K_f and k_f are constants characteristic of the solvent.

Assume that in a given experiment, a sample of an unknown substance was dissolved in benzene, for which $K_f = 5.12$ K molal^{-1} and $k_f = 0.011$ K^{-1}. For the following data, calculate M and its probable error:

0.1107

$$G = 13.185 \pm 0.003 \text{ g}$$
$$g = 0.423 \pm 0.002 \text{ g}$$
$$\Delta T_f = 1.263 \pm 0.020 \text{ K}$$

Section 10-5. GRAPHICAL PROCEDURES

We discuss a particular example in this section, the treatment of a set of measurements of the vapor pressure of a pure liquid as a function of the temperature. Since the temperature is controlled and the pressure measured, we write

$$P = P(T) \qquad \textbf{(10.26)}$$

This is assumed to be a continuous function, so the data points should lie somewhere near a smooth curve representing the correct values. Table

[5] David P. Shoemaker, Carl W. Garland, and Jeffrey I. Steinfeld, *Experiments in Physical Chemistry*, McGraw-Hill Book Company, New York, 1974, pp. 174–185.

Table 10-2. Experimental Values of the Vapor Pressure of Pure Ethanol at Various Temperature

t (°C)	T (K)	Vapor Pressure, P (torr)	Expected Error in P (torr)
25.00	298.15	55.9	3.0
30.00	303.15	70.0	3.0
35.00	308.15	93.8	4.2
40.00	313.15	117.5	5.5
45.00	318.15	154.1	6.0
50.00	323.15	190.7	7.6
55.00	328.15	241.9	8.0
60.00	333.15	304.15	8.8
65.00	338.15	377.9	9.5

10-2 contains a set of data for the vapor pressure of pure ethanol. Figure 10-2 is a graph on which the data have been plotted. A fairly smooth curve has been drawn near the points, passing through or nearly through the confidence intervals given in the table.

This curve, which we hope approximates the true curve, gives us two things: It gives us a means to interpolate between the data points, and it

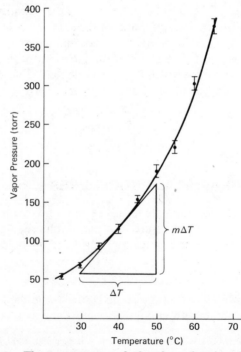

Figure 10-2.   The vapor pressure of ethanol as a function of temperature.

gives us "smoothed" values. When the curve misses the measured value, the value at the curve is probably a better estimate of the correct value than the measured value.

Furthermore, the curve can often be used for data reduction. In the present case, we have the Clapeyron equation of thermodynamics:

$$\frac{dP}{dT} = \frac{\Delta H}{T \, \Delta V} \qquad (10.27)$$

where ΔH is the enthalpy change of vaporization (at constant pressure, this is equal to the amount of heat required for vaporization), T the temperature on the Kelvin scale, and ΔV the volume change upon vaporization. If ΔV is known and the value of the derivative dP/dT can be determined, then the heat of vaporization can be calculated.

The derivative dP/dT is the slope of the tangent line at the point being considered (see Chapter 3). After the smooth curve has been drawn in by hand or with a French curve, a tangent line can be drawn in with a ruler. Two line segments can be drawn parallel to the coordinate axes, forming a right triangle. The slope is then the height of the triangle divided by its base.

Example 10-9

Find the heat of vaporization of ethanol at $40\,°C$ by graphically determining the slope of the tangent line at $40\,°C$ in the graph of Figure 10-2.

Solution

The tangent line and a suitable right triangle have already been drawn in Figure 10-2. In the units of the graph, the height of the triangle is 115 torr and the base is 20.0 K. In these units,

$$\frac{dP}{dT} = \frac{115 \text{ torr}}{20.0 \text{ K}} = 5.75 \text{ torr K}^{-1}$$

We assume that the vapor behaves as an ideal gas and that the volume of the liquid is negligible compared with the volume of the vapor.

$$\Delta H = \left(\frac{dP}{dT}\right) T \, \Delta V \approx \frac{RT^2}{P} \frac{dP}{dT} \qquad (10.28)$$

$$\approx \frac{(8.3143 \text{ JK}^{-1} \text{ mol}^{-1})(131.15 \text{ K})^2(5.75 \text{ torr K}^{-1})}{117.5 \text{ torr}}$$

$$\approx 40 \times 10^3 \text{ J mol}^{-1} = 40 \text{ kJ mol}^{-1} \qquad \bullet$$

Problem 10-8

The rate of a first-order chemical reaction obeys the equation

$$-\frac{dc}{dt} = kc \qquad \bullet \quad (10.29)$$

where c is the concentration of the reactant and k is a function of temperature called the rate constant. Following is a set of data for the reaction at $25\,^{\circ}C$:[6]

$$(CH_3)_3CBr + H_2O \rightarrow (CH_3)_3COH + HBr$$

Time (h)	Concentration of $(CH_3)_3CBr$ (mol/liter)
0	0.1039
3.15	0.0896
6.20	0.0776
10.0	0.0639
18.3	0.0353
30.8	0.0207
43.8	0.0101

Plot the data as given and draw a smooth curve through or nearly through the data points. Draw a tangent to the curve at some point and evaluate the rate constant.

Linearization. The procedure just used in Example 10-9 and Problem 10-8 suffers from two disadvantages. The first is that it is fairly difficult to draw a curve accurately, unless the curve is a straight line, and the second is that we had to construct a tangent line before determining the slope. Both of these disadvantages can be avoided if we linearize our graph. This means finding new variables such that the curve in the graph is expected to be a straight line. In the vapor pressure example, these variables are found as follows.

Equation (10.28) is a differential equation that can be solved by separation of variables if we assume that ΔH is a constant. The result is

$$\ln(P) = -\frac{\Delta H}{RT} + K \tag{10.30}$$

where K is a constant of integration.

Problem 10-9

Solve Eq. (10.28) to obtain Eq. (10.30) •

Equation (10.30) is the equation of a straight line if we use $1/T$ as the independent variable and $\ln(P)$ as the dependent variable. Figure 10-3 is a graph of the same data as Figure 10-2, using these variables. The expected errors in $\ln(P)$ were obtained by use of Eq. (10.24), with $y = \ln(P)$.

$$\lambda_y \approx \left(\frac{dy}{dP}\right)\lambda_P = \left(\frac{1}{P}\right)\lambda_P \tag{10.31}$$

---

[6] L. C. Bateman, E. D. Hughes, and C. K. Ingold, "Mechanism of Substitution at a Saturated Carbon Atom. Part XIX. A Kinetic Demonstration of the Unimolecular Solvolysis of Alkyl Halides," *J. Chem. Soc.* **1940**, 960.

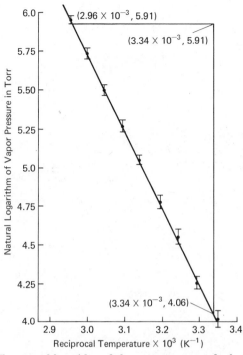

Figure 10-3. The natural logarithm of the vapor pressure of ethanol as a function of the reciprocal of the absolute temperature.

Problem 10-10

Calculate the expected error in $\ln(P)$ for a few data points in Table 10-2, using Eq. (10.31). •

A straight line passing nearly through the points has been drawn in the figure. In drawing such a line, it is a good idea to use a transparent ruler or a draftsman's triangle so that you can see the points on both sides of the line before drawing the line.

Example 10-10

Find the heat of vaporization of ethanol from the graph in Figure 10-3.

Solution

The necessary right triangle has already been drawn in Figure 10-3 and the coordinates of the vertices are given in the figure. If m is the slope, then

$$\Delta H = -mR = -(-4.87 \times 10^3 \text{ K})(8.314 \text{ J K}^{-1} \text{ mol}^{-1})$$
$$= 40.5 \times 10^3 \text{ J mol}^{-1}$$ •

Problem 10-11

Linearize the graph of the data in Problem 10-8. Do this by solving Eq. (10.29) to obtain

$$\ln(c) = -kt + K \qquad (10.32)$$

Find the value of k. •

Error Propagation. Consider the determination of the probable error in the heat of vaporization of ethanol from the vapor pressure data given. In Figure 10-3, the error bars give the expected errors in the values of the dependent variable. We assume that the errors in the independent variable are much smaller and can be neglected.

One way to proceed is to draw two additional lines on the graph, one that has the largest slope consistent with the data and one that has the smallest slope consistent with the data. Figure 10-4 is another graph of the values of $\ln(P)$ versus $1/T$, with these lines drawn in.

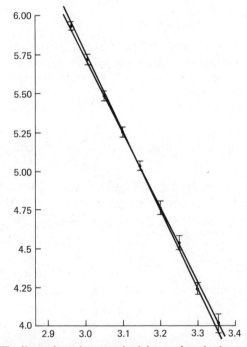

Figure 10-4. The lines of maximum and minimum slope in the graph of the natural logarithm of the vapor pressure of ethanol as a function of the reciprocal of the temperature.

Example 10-11

Find the slopes of the two extreme lines in Figure 10-4 and from them deduce the probable error in the heat of vaporization determined in Example 10-10.

Solution

Each slope is determined as in Example 10-10. The right triangles are not shown in Figure 10-4, but the results are

$$\Delta H(\text{minimum}) = 39.4 \times 10^3 \text{ J mol}^{-1}$$
$$\Delta H(\text{maximum}) = 42.2 \times 10^3 \text{ J mol}^{-1}$$

Notice that the difference between the minimum value of the heat of vaporization and the value of Example 10-10 is 1.1 kJ mol^{-1}, while the difference between the maximum value and the value of Example 10-10 is 1.7 kJ mol^{-1}. It is best to give a probable error of ± 1.7 kJ mol^{-1}. •

The method can be modified to include uncertainties in the independent variable. In this case, instead of drawing an error bar at each point, you draw an *error box*, whose width is equal to the size of the confidence interval for the independent variable and whose height is equal to the size of the confidence interval for the dependent variable. The lines of maximum and minimum slope are then drawn to pass through all or nearly all of the boxes.

In the graphical method, it is somewhat difficult to determine the confidence level of our error interval. If the error bars or error boxes are at the 95% confidence level, the error interval in the slope is probably somewhat larger than a 95% confidence interval.

Problem 10-12

Assume that each concentration of *tert*-butyl bromide in Problem 10-8 is uncertain by 1.5% of its value. Find the expected error in the rate constant, k, by graphical means. •

If you do not have information about the probable errors in the values of the dependent variable, you can still proceed to draw a line of maximum slope and one of minimum slope. One way to do this is to draw the line through the points and then to measure the vertical distances from the points to the line. This distances are called *residuals*. If the line were the correct function, these residuals would be the experimental errors.

Since we do not know the correct function, we assume that the residuals collectively have about the same sizes as the experimental errors. Pick a distance that is longer than about 95% of the residuals, and draw an error bar that extends this far above and below each point. Then proceed to draw the line of maximum slope and the line of minimum slope as before.

The determination of the uncertainty in the slope of a tangent line is more difficult. One way to proceed is to draw different curves which appear to be extremes that are compatible with the confidence intervals around the data points, and to draw tangents to the different curves and measure their slopes.

Other Graphical Data Reduction Procedures. We have discussed graphical procedures involving the plotting of a curve and the determination of the derivative of the function represented by the curve. It is also possible to use graphical methods of integration to reduce data. For example, the area under an absorption peak in an NMR spectrum is proportional to the number of nuclei producing the absorption, and the area under a peak in a gas or liquid chromatogram is proportional to the amount of the substance producing the peak.

There are several ways to carry out graphical integration. One tedious method is the counting of squares on the graph paper. Another method is to cut out the area and weigh it. The mass of the piece of paper will nearly be proportional to its area. It is also possible to use a mechanical device called a planimeter, which registers the area around which the stylus of the device is passed.

Error estimates can be obtained in graphical integration much as in graphical differentiation. If expected errors are available for individual data points, you can find the maximum error by incrementing each value by its expected error and determining the area again.

Section 10-6. NUMERICAL DATA REDUCTION PROCEDURES

Graphical techniques have largely lost favor, primarily because of the availability of rapid digital computers. Furthermore, numerical procedures are less subjective and are usually more accurate.

We concentrate on the numerical determination of functions and of derivatives of functions in this section. Numerical integration has already been discussed in Section 4-8.

Numerical Differentiation. Sometimes we need a value for the derivative of a function that is represented approximately by a set of data points. One procedure is based on choosing polynomial functions that provide "smoothed" values. For example, if we have the set of data points (x_1, y_1), (x_2, y_2), (x_3, y_3), etc., a smoothed value for the dependent variable to replace y_i is given by[7]

$$y_i = \tfrac{1}{35}[17y_i + 12(y_{i+1} + y_{i-1}) - 3(y_{i+2} + y_{i-2})] \qquad (10.33)$$

[7] A. G. Worthing and I. Geffner, *Treatment of Experimental Data*, John Wiley Sons, New York, 1943, p.7.

   This equation corresponds to the value of the parabolic function that most nearly fits the five data points included in the formula, and is valid only for equally spaced values of x. There are also similar formulas which involve a larger number of points.

   We now define a set of differences, which are used to calculate numerical approximations to derivatives. The first difference for the ith point is defined by

$$\Delta y_i = y_{i+1} - y_i \qquad (10.34a)$$

The second difference for the ith point is defined by

$$\Delta^2 y_i = \Delta y_{i+1} - \Delta y_i = y_{i+2} - 2y_{i+1} + y_i \qquad (10.34b)$$

The third difference for the ith point is defined by

$$\Delta^3 y_i = \Delta^2 y_{i+1} - \Delta^2 y_i \qquad (10.34c)$$

Higher-order differences are defined in a similar way. This set of differences is not the only one that can be defined. As you can see from Eqs. (10.34), this set involves only points with subscripts greater than or equal to i. Other schemes can be defined which use points on both sides of the ith data point. These definitions apply only to data sets for which the values of x are equally spaced. For data sets of ordinary accuracy, the values of y in Eqs. (10.34) should be the smoothed values given by Eq. (10.33).

   A value for the derivative dy/dx at the ith point is given by[8]

$$\left.\frac{dy}{dx}\right|_{x=x_i} = \frac{1}{w}(y_i - \tfrac{1}{2}\Delta^2 y_i + \tfrac{1}{3}\Delta^3 y_i - \tfrac{1}{4}\Delta^4 y_i + \cdots) \qquad (10.35)$$

where w is the spacing between values of x:

$$w = x_{i+1} - x_i \qquad (10.36)$$

Equation (10.35) is based on the Gregory–Newton interpolation formula,[9] which gives values of the smoothed function representing the data points.

   The second derivative is given by

$$\left.\frac{d^2 y}{dx^2}\right|_{x=x_i} = \frac{1}{w^2}(\Delta^2 y_i - \Delta^3 y_i + \tfrac{11}{12}\Delta^4 y_i - \tfrac{10}{12}\Delta^5 y_i + \cdots) \qquad (10.37)$$

We will not discuss applications of Eqs. (10.35) and (10.37). However, if the data of Table 10-2 are smoothed and the derivative at 40 °C is found by Eq. (10.35), a value for the heat of vaporization equal to 38 kJ mol^{-1} is found if the series is stopped with the third difference. This value is probably less reliable than the value obtained graphically.

Numerical Curve Fitting. The Method of Least Squares.   We now consider a numerical procedure for finding a continuous function to represent a set of data points such as the vapor pressure of a pure liquid at various values of

[8] Shoemaker, Garland, and Steinfeld, *Experiment in Physical Chemistry*, p. 42.
[9] Ibid.

the temperature. This was accomplished graphically in Section 10-5 by drawing a curve or a straight line through or nearly through the data points.

Say that our dependent variable is called y and our independent variable is called y. Our data points are $(x_1, y_1), (x_2, y_2), (x_3, y_3)$, etc., and we assume that there is some "correct" function

$$y = y(x) \qquad (10.38)$$

which we would like to approximate.

In the *method of least squares*, we do not consider all possible functions. We assume that we can take some family of functions, given by

$$y = f(x, a_1, a_2, \ldots, a_p) \qquad (10.39)$$

where the a's are parameters, and then find the values of the parameters which give that member of the assumed family which most nearly fits the data points.

For example, in the method of linear least squares, we make whatever change in variables we think is necessary to make our data points lie approximately on a straight line, and then find the straight line that best fits our points. In the vapor pressure example, this means choosing $\ln(P)$ as our dependent variable and $1/T$ as our independent variable, as in Section 10-5. A straight line is given by

$$y = a_1 x + a_2 \qquad (10.40)$$

and we seek that value of a_1 and that value of a_2 which give us the best fit.

We now discuss how the best values of the parameters are found. The residual for the ith data point is defined as the difference between the measured value and the value of the function at that point:

$$r_i = y_i - f(x_i, a_1, a_2, \ldots, a_p) \qquad (10.41)$$

When the function fits the points well, these residuals will collectively be small. The best fit is obtained when the sum of the squares of the residuals is minimized. We seek the minimum of

$$R = \sum_{i=1}^{N} [y_i - f(x_i, a_1, a_2, \ldots, a_p)]^2 \qquad (10.42)$$

where N is the number of data points.

This minimum occurs where all of the partial derivatives of R with respect to $a_1, a_2, \ldots, a_p$ vanish:

$$\frac{\partial R}{\partial a_i} = 0 \qquad (i = 1, 2, \ldots, p) \qquad (10.43)$$

This is a set of simultaneous equations, one for each parameter. The reason for naming the method the method of least squares is that we minimize the sum of the squares of the residuals. The method is also called the *regression method*.

If the simultaneous equations are nonlinear equations, they are solved by successive approximations.[10] For a family of linear functions, as in Eq. (10.40), or a family of polynomial functions, the equations are linear equations, and we can solve them by the methods of Chapter 9.

For the function of Eq. (10.40),

$$R = \sum_{i=1}^{N} (y_i - a_1 x_i - a_2)^2 \tag{10.44}$$

and the simultaneous equations are

$$\frac{\partial R}{\partial a_1} = 2 \sum_{i=1}^{N} (y_i - a_1 x_i - a_2)(-x_i) = 0 \tag{10.45a}$$

$$\frac{\partial R}{\partial a_2} = 2 \sum_{i=1}^{N} (y_i - a_1 x_i - a_2)(-x_i) = 0 \tag{10.45b}$$

This is a set of linear inhomogeneous simultaneous equations in a_1 and a_2. We write them in the usual format [see Eq. (9.22)].

$$S_{x2} a_1 + S_x a_2 = S_{xy} \tag{10.46a}$$

$$S_x a_1 + N a_2 = S_y \tag{10.46b}$$

where

$$S_x = \sum_{i=1}^{N} x_i \tag{10.47}$$

$$S_y = \sum_{i=1}^{N} y_i \tag{10.48}$$

$$S_{xy} = \sum_{i=1}^{N} x_i y_i \tag{10.49}$$

$$S_{x2} = \sum_{i=1}^{N} x_i^2 \tag{10.50}$$

From Cramer's rule, Eqs. (9.29) and (9.30):

$$a_1 = \frac{1}{D} (N S_{xy} - S_x S_y) \tag{10.51a}$$

$$a_2 = \frac{1}{D} (S_{x2} S_y - S_x S_{xy}) \tag{10.51b}$$

where

$$D = N S_{x2} - S_x^2 \tag{10.52}$$

These are our working equations.

[10] Shoemaker, Garland, and Steinfeld, *Experiments in Physical Chemistry*, pp. 44ff.

Example 10-12

Calculate the slope a_1 and the intercept a_2 for the least-squares line for the data in Table 10-2, using $\ln(P)$ as the dependent variable and $1/T$ as the independent variable. Calculate the heat of vaporization from the slope.

Solution

When the numerical work is done, the results are

$$a_1 = -4854 \text{ K}$$

$$a_2 = 20.28$$

$$\Delta H = -a_1 R = -(-4854 \text{ K})(8.3143 \text{ J K}^{-1} \text{ mol}^{-1})$$

$$= 40.36 \times 10^3 \text{ J mol}^{-1}$$

This compares well with the accepted value of $40.3 \times 10^3 \text{ J mol}^{-1}$. •

Problem 10-13

The following data give the vapor pressure of water at various temperatures.[11] Find the least-squares line for the data, using $\ln(P)$ for the dependent variable and $1/T$ for the independent variable. Find the heat of vaporization.

Temperature (°C)	Vapor Pressure (torr)
0	4.579
5	6.543
10	9.209
15	12.788
20	17.535
25	23.756

•

In some problems, it is not certain in advance what variables should be used for a linear least-squares fit. In the vapor pressure case, we had the thermodynamic relation, Eq. (10.30), which indicated that $\ln(P)$ and $1/T$ were the variables that should produce a linear relationship. In the analysis of chemical rate data, it may be necessary to try two or more hypotheses.

In a reaction involving one reactant, the concentration of the reactant is given by Eq. (10.32) if there is no back reaction and if the reaction is a first-order reaction. If there is no back reaction and the reaction is a second-order reaction, the concentration of the reactant is given by

$$\frac{1}{c} = kt + K \tag{10.53}$$

where k is the rate constant and K is a constant of integration. If there is no back reaction and the reaction is third order, the concentration of the

[11] R. Weast, Editor, *Handbook of Chemistry and Physics*, 51st ed., Chemical Rubber Publishing Co., Cleveland Ohio, 1971–72, p. D-143.

reactant is given by

$$\frac{1}{2c^2} = kt + K \tag{10.54}$$

If the order of a reaction is not known, it is possible to determine the order by trying different linear least-squares fits and finding which one most nearly fits the data. One way to see whether a given hypothesis produces a linear fit is to examine the residuals. Once the least-squares line has been found, the residuals can be calculated from Eq. (10.41), which for a linear function becomes

$$r_i = y_i - a_1 x_i - a_2 \tag{10.55}$$

If a given dependent variable and independent variable produce a linear fit, the points will deviate from the line only because of experimental error, and the residuals will be either positive or negative without any pattern.

However, if there is a general curvature to the data points in a graph, the residuals will have one sign at the ends of the graph, and the other sign in the middle. This is characteristic of a poor fit, and indicates that a different pair of variables should be tried or a nonlinear least-squares fit attempted.

Example 10-13

The following is a fictitious set of data for the concentration of the reactant in a chemical reaction with only one reactant. Determine whether the reaction is first, second, or third order. Find the rate constant and the initial concentration.

Point Number	Time (min)	Concentration (mol liter^{-1})
1	5	0.715
2	10	0.602
3	15	0.501
4	20	0.419
5	25	0.360
6	30	0.300
7	35	0.249
8	40	0.214
9	45	0.173

Solution

a. We first attempt a linear fit using $\ln(c)$ as the dependent variable and t as the independent variable. The result is

$$a_1 = -0.03504 \text{ min}^{-1} = -k$$
$$a_2 = -0.1592 = \ln[c(0)]$$
$$c(0) = 0.853 \text{ mol liter}^{-1}$$

The following set of residuals was obtained:

$$r_1 = -0.00109 \qquad r_6 = 0.00634$$
$$r_2 = 0.00207 \qquad r_7 = -0.00480$$
$$r_3 = -0.00639 \qquad r_8 = 0.01891$$
$$r_4 = -0.00994 \qquad r_9 = -0.01859$$
$$r_5 = 0.01348$$

This is a good fit, with no pattern of general curvature shown in the residuals.

b. We test the hypothesis that the reaction is second order by attempting a linear fit using $1/c$ as the dependent variable and t as the independent variable. The result is

$$a_1 = 0.1052 \text{ liter mol}^{-1} \text{ min}^{-1} = k$$

$$a_2 = 0.4846 \text{ liter mol}^{-1} = \frac{1}{c(0)}$$

$$c(0) = 2.064 \text{ mol liter}^{-1}$$

The following set of residuals was obtained:

$$r_1 = 0.3882 \qquad r_6 = -0.3062$$
$$r_2 = 0.1249 \qquad r_7 = -0.1492$$
$$r_3 = -0.0660 \qquad r_8 = -0.0182$$
$$r_4 = -0.2012 \qquad r_9 = 0.5634$$
$$r_5 = -0.3359$$

This is not such a satisfactory fit as in part a of the solution, since the residuals show a general curvature, beginning with positive values, becoming negative, and then becoming positive again.

c. We now test the hypothesis that the reaction is third order by attempting a linear fit using $1/(2c^2)$ as the dependent variable and t as the independent variable. The results are

$$a_1 = 0.3546 \text{ liter}^2 \text{ mol}^{-2} \text{ min}^{-1} = k$$
$$a_2 = -3.054 \text{ liter}^2 \text{ mol}^{-2}$$

This is obviously a bad fit, since the intercept, a_2, should not be negative. The residuals are

$$r_1 = 2.2589 \qquad r_6 = -2.0285$$
$$r_2 = 0.8876 \qquad r_7 = -1.2927$$
$$r_3 = -0.2631 \qquad r_8 = -0.2121$$
$$r_4 = -1.1901 \qquad r_9 = 3.8031$$
$$r_5 = -1.9531$$

Again, there is considerable curvature evident. The reaction is apparently first order, with the rate constant and initial concentration given in part a of the solution. ●

In addition to inspecting the residuals, we can calculate a quantity called the *correlation coefficient*, which gives information about the closeness of a least-squares fit. For linear least squares, the correlation coefficient is defined by

$$r = \frac{NS_{xy} - S_x S_y}{[(NS_{x2} - S_x^2)(NS_{y2} - S_y^2)]^{1/2}} \tag{10.56}$$

where S_x, S_y, S_{xy} and S_{x2} are defined in Eqs. (10.47) through (10.50) and

$$S_{y2} = \sum_{i=1}^{N} y_i^2 \tag{10.57}$$

If the data points lie exactly on the least-squares line, the correlation coefficient will be equal to 1 (if the slope is positive) or to -1 (if the slope is negative). If the slope happens to be exactly equal to zero, the correlation coefficient will not give you an indication of the closeness of the fit. If the data points are scattered randomly about the graph, or lie in a kind of circular pattern, so that no least-squares line can be found, the correlation coefficient will vanish. The magnitude of the correlation coefficient will be larger for a close fit than for a poor fit. In a fairly close fit, its magnitude might equal 0.99.

Example 10-14

Calculate the correlation coefficients for the three linear fits in Example 10-13.

Solution

Use of Eq. (10.56) gives the results:
a. For the first-order fit: $r = -0.9997$.
b. For the second-order fit: $r = 0.9779$.
c. For the third-order fit: $r = 0.9257$.
Again, the first-order fit is the best. ●

Problem 10-14

Do three linear least-squares fits on the data of Problem 10-8. Calculate the correlation coefficients for the three fits and show that the reaction is first order. ●

Error Propagation in Linear Least Squares.   We discuss two cases: (1) the expected error in each value of the dependent variable is known, and (2) the expected error in these values is not known. In both cases, we assume that the errors in the values of the independent variable are negligible.

Case 1. Let the expected error in the value of y_i be denoted by λ_i. Equations (10.51a) and (10.51b) are formulas for the slope and the intercept of the least-squares line in which the y's can be considered to be independent variables, so we can apply Eq. (10.24). The expected error in the slope is

$$\lambda_{a_1} = \left[\sum_{i=1}^{N} \left(\frac{\partial a_1}{\partial y_i} \right)^2 \lambda_i^2 \right]^{1/2} = \left[\frac{1}{D^2} \sum_{i=1}^{N} (Nx_i - S_x)^2 \lambda_i^2 \right]^{1/2} \quad \textbf{(10.58)}$$

where D and S_x are given by Eqs. (10.52) and (10.47). In the case that all of the expected errors in the y's are equal, Eq. (10.58) becomes

$$\lambda_{a_1} = \left(\frac{N}{D} \right)^{1/2} \lambda_y \quad \textbf{(10.59)}$$

where λ_y is the value of all the λ_i's.

The expected error in the intercept is

$$\lambda_{a_2} = \left[\sum_{i=1}^{N} \left(\frac{\partial a_2}{\partial y_i} \right)^2 \lambda_i^2 \right]^{1/2} = \left[\frac{1}{D^2} \sum_{i=1}^{N} (S_{x2} - S_x x_i)^2 \lambda_i^2 \right]^{1/2} \quad \textbf{(10.60)}$$

For the case that all of the λ_i's are equal,

$$\lambda_{a_2} = \left(\frac{S_{x2}}{D} \right)^{1/2} \lambda_y \quad \textbf{(10.61)}$$

where S_{x2} is given by Eq. (10.50).

Problem 10-15

Verify Eqs. (10.59) and (10.61).   •

Example 10-15

Assume, instead of the given errors, that the expected error in the logarithm of each vapor pressure in Table 10-2 is 0.040. Find the expected error in the least-squares slope and in the heat of vaporization.

Solution

From the data,

$$D = 1.327 \times 10^{-6} \, \text{K}^{-2}$$

so that

$$\lambda_{a_1} = \left(\frac{9}{1.327 \times 10^{-6} \ K^{-2}}\right)^{1/2} (0.040) = 104 \ K$$

$$\lambda_{\Delta H} = R\lambda_{a_1} = 8.7 \times 10^2 \ J \ mol^{-1} \qquad \bullet$$

Problem 10-16

Assume that the expected error in the logarithm of each concentration in Example 10-13 is 0.010. Find the expected error in the rate constant, assuming the reaction to be first order. $\qquad \bullet$

Case 2. Since we do not have information about the expected errors in the dependent variable, we assume that the residuals are a sample from the population of actual experimental errors. This is a reasonable assumption if systematic errors can be ignored.

The standard deviation of the N residuals is given by

$$s_r = \left(\frac{1}{N-2} \sum_{i=1}^{N} r_i^2\right)^{1/2} \qquad (10.62)$$

This differs from Eq. (10.2) in that a factor of $N - 2$ occurs in the denominator instead of $N - 1$. This is required because we have calculated two quantities, a least-squares slope and a least-squares intercept from the set of numbers. The number of remaining degrees of freedom is $N - 2$. Also, the mean of the residuals does not enter, because the mean of the residuals in a least-squares fit always vanishes. You might try to prove this if you have some spare time. It is what some people call "straightforward but tedious."

Equation (10.62) is now a consistent estimate of the standard deviation of the population of experimental errors, so we write at the 95% confidence level

$$\lambda_y = 1.96s_r \qquad (10.63)$$

using the fact that 95% of a population lies within 1.96 standard deviations of the mean if the standard normal distribution holds.

We can now write expressions similar to Eqs. (10.59) and (10.60):

$$\lambda_{a_1} = \left(\frac{N}{D}\right)^{1/2} (1.96s_r) \qquad (10.64)$$

and

$$\lambda_{a_2} = \left(\frac{1}{D} \sum_{i=1}^{N} x_i^2\right)^{1/2} (1.96s_r) \qquad (10.65)$$

Example 10-16

Calculate the residuals for the linear least-squares fit of Example 10-12. Find their standard deviation and the probable error in the slope and in the heat of vaporization, using the standard deviation of the residuals.

Solution

Numbering the data points from the 25 °C point (number 1) to the 65 °C point (number 9), we find the residuals:

$$r_1 = 0.0208 \qquad r_6 = -0.0116$$
$$r_2 = -0.0228 \qquad r_7 = -0.0027$$
$$r_3 = 0.0100 \qquad r_8 = 0.0054$$
$$r_4 = -0.0162 \qquad r_9 = 0.0059\ .$$
$$r_5 = 0.0113$$

The standard deviation of the residuals is found to be

$$s_r = 0.0154$$

```
LSQUAR   04:44 PM          04-Mar-80
100 DIM X(25), Y(25)
110 DIM R(25)
120 PRINT "THIS IS A SIMPLE LINEAR LEAST SQUARES PROGRAM."
122 PRINT "DO YOU WANT TO FIT LN(Y) TO 1/X (YES OR NO)?"
124 INPUT A$
130 PRINT "NUMBER OF DATA POINTS?"
140 INPUT N
142 S1 = 0
144 S1 = 0
146 S3 = 0
148 S4 = 0
149 S9 = 0
150 FOR I = 1 TO N
160 PRINT "TYPE IN X ,Y FOR POINT NUMBER"I
170 INPUT X(I), Y(I)
172 IF A$ = "YES" THEN 174 ELSE 180
174 Y(I) = LOG(Y(I))
176 X(I) = 1/X(I)
180 S1 = S1 + X(I)
190 S2 = S2 + Y(I)
200 S3 = S3 + X(I)*Y(I)
210 S4 = S4 + X(I)^2
215 S9 = S9 + Y(I)^2
220 NEXT I
230 D = N*S4 - S1^2
240 A1 = (N*S3 - S1*S2)/D
250 A2 = (S4*S2 - S1*S3)/D
260 PRINT "SLOPE IS";A1;"INTERCEPT IS";A2
270 PRINT "DO YOU WANT A LIST OF RESIDUALS (YES OR NO)?"
280 INPUT B$
295 S5 = 0
300 FOR I = 1 TO N
310 R(I) = Y(I)-A1*X(I)-A2
312 IF B$ = "YES" THEN 314 ELSE 315
314 PRINT "R("I") = "R(I)
315 S5 = S5 + R(I)^2
330 NEXT I
340 S = SQR(S5/(N-2))
350 PRINT "STANDARD DEVIATION OF RESIDUALS = "S
360 E1 = SQR(N/D)*1.96*S
370 PRINT "EXPECTED ERROR IN SLOPE (95% CONFIDENCE) = "E1
380 E2 = SQR(S4/D)*1.96*S
390 PRINT "EXPECTED ERROR IN INTERCEPT (95% CONFIDENCE) = "E2
400 R2 = (N*S3 - S1*S2)^2/((N*S4-S1^2)*(N*S9-S2^2))
410 PRINT "CORRELATION COEFFICIENT SQUARED = "R2
990 END

Ready
```

Figure 10-5.   A computer program to carry out a linear least square procedure.

```
RUN
LSQUAR  04:45 PM          04-Mar-80
THIS IS A SIMPLE LINEAR LEAST SQUARES PROGRAM.
DO YOU WANT TO FIT LN(Y) TO 1/X (YES OR NO)?
? YES
NUMBER OF DATA POINTS?
? 9
TYPE IN X ,Y FOR POINT NUMBER 1
? 298.15, 55.9
TYPE IN X ,Y FOR POINT NUMBER 2
? 303.15, 70.0
TYPE IN X ,Y FOR POINT NUMBER 3
? 308.15, 93.8
TYPE IN X ,Y FOR POINT NUMBER 4
? 313.15, 117.5
TYPE IN X ,Y FOR POINT NUMBER 5
? 318.15, 154.1
TYPE IN X ,Y FOR POINT NUMBER 6
? 323.15, 190.7
TYPE IN X ,Y FOR POINT NUMBER 7
? 328.15, 241.9
TYPE IN X ,Y FOR POINT NUMBER 8
? 333.15, 304.5
TYPE IN X ,Y FOR POINT NUMBER 9
? 338.15, 377.9
SLOPE IS-4854.26 INTERCEPT IS 20.2841
DO YOU WANT A LIST OF RESIDUALS (YES OR NO)?
? YES
R( 1 ) =  .207758E-1
R( 2 ) = -.228281E-1
R( 3 ) =  .100212E-1
R( 4 ) = -.162287E-1
R( 5 ) =  .113171E-1
R( 6 ) = -.116618E-1
R( 7 ) = -.272343E-2
R( 8 ) =  .540906E-2
R( 9 ) =  .591896E-2
STANDARD DEVIATION OF RESIDUALS =  .153639E-1
EXPECTED ERROR IN SLOPE (95% CONFIDENCE) =  78.4206
EXPECTED ERROR IN INTERCEPT (95% CONFIDENCE) =  .2471
CORRELATION COEFFICIENT SQUARED =  .999525

Ready
```

Figure 10-5 (*continued*)

Using the value of D from Example 10-15, the uncertainty in the slope is

$$\lambda_{a_1} = \left(\frac{9}{1.327 \times 10^{-6} \text{ K}^{-2}}\right)^{1/2} (1.96)(0.0154)$$

$$= 78.4 \text{ K}$$

The uncertainty in the heat of vaporization is

$$\lambda_{\Delta H} = R\lambda_{a_1} = (8.3143 \text{ J K}^{-1} \text{ mol}^{-1})(78.4 \text{ K}) = 652 \text{ J mol}^{-1} \qquad \bullet$$

Problem 10-17

Assuming the reaction in Problem 10-14 is first order, find the expected error in the rate constant, using the residuals. $\qquad \bullet$

If you have done any of the problems in this section by hand calculation, you have learned that the least-squares procedure is tedious. It is exactly the kind of thing that should be done with a computer or with a programmable calculator.

Figure 10-5 shows a computer program in BASIC that will carry out a linear least-squares fit. This program will calculate the expected error in the slope and the intercept from the residuals, as well as the square of the correlation coefficient. A similar program was used to work out the examples of this section.

Some Warnings About Linear Least-Squares Procedures. It is a poor idea to rely blindly on a numerical method. You should always determine whether your results are reasonable. It is possible to spoil your results by entering one number incorrectly or by failing to recognize a bad data point. Remember the first maxim of computing: "Garbage in, garbage out."

A straightforward way to make sure that a linear least-squares procedure has given you a good result is to make a plot of the data points and then to draw the least-squares line on the same graph. If some error has produced the wrong line, you will probably be able to tell it by looking at the graph. If the data points show a general curvature, you will probably be able to tell that as well from the graph. You should also be able to spot a bad data point

Another thing that you can do is to examine your residuals. It is a bad idea to use a linear least-squares program that does not provide you with the residuals. There are calculators that have a linear least-squares program built into them, but these do not generally provide you with residuals. As explained earlier, examination of the residuals can reveal a general curvature in your data points and can also reveal the presence of a bad data point.

A final warning is that in making a change in variables in order to fit a set of data to a straight line rather than to some other function, one is actually changing the relative importance, or weight, of the various data points.[12] Because of this, fitting $\ln(c)$ to a straight line

$$\ln(c) = -kt + K \qquad (10.66)$$

will not necessarily give the same value of k as will fitting c to the function

$$c = e^K e^{-kt} \qquad (10.67)$$

We now discuss a way to compensate for this and also to compensate for known errors of different sizes in our data points.

Weighting Factors in Linear Least Squares. Say that we have a set of data in which the probable errors in the values of the independent variable are negligible, and in which the probable errors in the values of the dependent variable are not all of the same size. In this case, instead of minimizing the sum of the squares of the residuals, one should minimize the sum of the squares of the residuals divided by a quantity proportional to the square of the probable error for each point. If σ_i is the standard deviation of the

---

[12] Donald E. Sands, "Weighting Factors in Least Squares," *J. Chem. Educ.* **51**, 473 (1974).

distribution from which r_i is drawn, we should minimize[13]

$$R' = \sum_{i=1}^{N} \frac{r_i^2}{\sigma_i^2} = \sum_{i=1}^{N} \frac{1}{\sigma_i^2}(y_i - a_1 x_i - a_2)^2 \tag{10.68}$$

The factors $1/\sigma_i^2$ in the sum are called *weighting factors*. The effect of this weighting is to give a greater importance, or greater weight, to those points which have a smaller expected error.

We can now proceed to minimize R', just as we minimized R. The equations are very similar to Eqs. (10.45) through (10.52), except that each sum includes the weighting factors. The results for the slope and intercept are

$$a_1 = \frac{1}{D'}(S_1' S_{xy}' - S_x' S_y') \tag{10.69a}$$

$$a_2 = \frac{1}{D'}(S_{x2}' S_y' - S_x' S_{xy}') \tag{10.69b}$$

where

$$D' = S_1' S_{x2}' - S_x'^2 \tag{10.70}$$

$$S_1' = \sum_{i=1}^{N} \frac{1}{\sigma_i^2} \tag{10.71a}$$

$$S_x' = \sum_{i=1}^{N} \frac{x_i}{\sigma_i^2} \tag{10.71b}$$

$$S_y' = \sum_{i=1}^{N} \frac{y_i}{\sigma_i^2} \tag{10.71c}$$

$$S_{xy}' = \sum_{i=1}^{N} \frac{x_i y_i}{\sigma_i^2} \tag{10.71d}$$

$$S_{x2}' = \sum_{i=1}^{N} \frac{x_i^2}{\sigma_i^2} \tag{10.71e}$$

Problem 10-18

Verify Eqs. (10.69) through (10.71).   •

There are two ways in which we will use Eq. (10.69). The first is in a case when the probable errors in the data are known and not equal, and the second is in a case when the probable errors are assumed to be equal and a change in variables is required for a linear fit.

Example 10-17

Find the least-squares line for the data of Table 10-2, assuming that the expected error is proportional to the standard deviation for each point.

[13] Philip R. Bevington. *Data Reduction and Error Analysis for the Physical Sciences*, McGraw-Hill Book Company, New York, 1969, pp. 106–108.

Solution

The expected errors in the logarithm of P were calculated by Eq. (10.31). These were substituted into Eqs. (10.69) through (10.71) in place of the σ_is. The results were

$$a_1 = -4872 \text{ K}$$
$$a_2 = 20.34$$

This gives a value of the heat of vaporization of 40.51 kJ mol^{-1}. These figures differ slightly from those of Example 10-12 and are presumably more nearly accurate. •

In the following example, we see what an inaccurate point can do if the unweighted least-squares procedure is used.

Example 10-18

Change the data set of Table 10-2 by adding a value of the vapor pressure at 70 °C of 421 ± 40 torr. Find the least-squares line using both the unweighted and weighted procedures.

Solution

After the point was added, the unweighted procedure was carried out as in Example 10-12, and the weighted procedure was carried out as in Example 10-17. The results were:
  Unweighted procedure:

$$a_1 = \text{slope} = -4752 \text{ K}$$
$$a_2 = \text{intercept} = 19.95$$

Weighted procedure:

$$a_1 = \text{slope} = -4855 \text{ K}$$
$$a_2 = \text{intercept} = 20.28$$

The data point of low accuracy has done more damage in the unweighted procedure than in the procedure with weighting factors. •

Let us now discuss the use of least squares with weighting factors when a change in variables is made. If the values of the original dependent variable are all uncertain by the same amount, an unweighted least-squares fit is appropriate if we use that variable in our procedure. However, if we take a function of the original variable, then the original expected errors, which are all equal, will not generally produce equal errors in the new variable, and the unweighted least squares is not appropriate.

Example 10-19

Assume that each concentration in the data set of Example 10-13 is uncertain by 0.0070 mol liter^{-1}. Calculate the probable error in each value of $\ln(c)$ and carry out a linear least-squares fit using $\ln(c)$ as the dependent variable and the square of the reciprocal of the probable error as the weighting factor.

Solution

For each point, we use the following, where $y = \ln(c)$:

$$\lambda_y = \frac{dy}{dc}\lambda_c = \frac{1}{c}\lambda_c$$

We have

Point Number	λ_y
1	0.010
2	0.012
3	0.014
4	0.017
5	0.019
6	0.023
7	0.028
8	0.033
9	0.040

Notice the differences in the accuracy of the logarithms. The more accurate points will now have greater weight.

The result of carrying out the procedure with weighting factors is:

$$a_1 = \text{slope} = -0.03494 \text{ min}^{-1} = -k$$
$$a_2 = \text{intercept} = -0.1612$$

If the errors are as stated, the present results are more reliable than those of Example 10-13. ●

This discussion suggests a possible procedure that you might want to use if you carry out a least-squares fit and a few points lie a long way from the line: Repeat the fit using weighting factors, utilizing the residuals from the first fit in place of the σ_i's of Eq. (10.69) through (10.71). This is not apparently a standard procedure, but it should give a better fit than the unweighted procedure.

Error Propagation in Weighted Least Squares. We can derive equations analogous to Eqs. (10.59) and (10.61) for the weighted least-squares procedure. We continue to assume that the errors in the values of the independent variable, x, are negligible, and use the symbol λ_i for the probable error in y_i. The standard deviations σ_1, σ_2, etc., in Eq. (10.71) are generally unknown, but will be proportional to λ_1, λ_2, etc., so we replace the σ_i's by the λ_i's. This will not affect the slope and the intercept. We now proceed as in the un-weighted case, writing as in Eq. (10.58)

$$\lambda_{a_1} = \left[\sum_{i=1}^{N} \left(\frac{\partial a_1}{\partial y_i} \right)^2 \lambda_i^2 \right]^{1/2} \tag{10.72}$$

$$= \left[\sum_{i=1}^{N} \frac{1}{D'^2} \left(\frac{S'_1 x_i}{\lambda_i^2} - \frac{S'_x}{\lambda_i^2} \right)^2 \lambda_i^2 \right]^{1/2}$$

$$= \frac{1}{D'} (S'^2_1 S'_{x2} - S'_1 S'^2_x)^{1/2}$$

where the sums are those of Eq. (10.71) except that the σ_i's have been replaced by the λ_i's.

Problem 10-19

Verify Eq. (10.72).

A similar procedure gives

$$\lambda_{a_2} = \frac{1}{D'} (S_1 S'^2_{x2} - S'_{x2} S'^2_x)^{1/2} \tag{10.73}$$

Problem 10-20

Verify Eq. (10.73).

Example 10-20

a. Find the expected error in the slope and intercept for the least-squares line found in Example 10-17.
b. Find the expected error in the slope and the intercept for both of the least-squares fits of Example 10-18.

Solution

a. The results are

$$\lambda_{a_1} = 97 \text{ K}$$
$$\lambda_{a_2} = 0.30$$

b. For the unweighted procedure, we find, from the residuals,

$$\lambda_{a_1} = 137 \text{ K}$$
$$\lambda_{a_2} = 0.43$$

For the weighted procedure, we find

$$\lambda_{a_1} = 96 \text{ K}$$
$$\lambda_{a_2} = 0.30$$

In the unweighted procedure, the introduction of the bad data point has raised the uncertainty in the slope from 78 K to 137 K, but in the weighted procedure, there was no effect. If a bad data point has a large known error, its effect in the weighted least-squares procedure is minimal. •

Problem 10-21

Modify the BASIC program in Figure 10-5 to carry out a weighted linear least-squares fit. Carry out the weighted procedure on the data of Problem 10-13, assuming that each vapor pressure is uncertain by 0.050 torr, so that the logarithms have different uncertainties. •

Linear Least Squares with Fixed Slope or Intercept. At times it is necessary to do a least-squares fit with the constraint that the slope or the intercept have a fixed value. For example, the Bouguer–Beer law states that the absorbance of a solution is proportional to the concentration of the colored substance. In fitting the absorbance of several solutions to their concentrations, one would specify that the intercept of the least-squares line had to be zero. In other cases, a particular slope might be required.

In the minimization of the sum of the squares of the residuals, one has only one equation if the intercept a_2 is fixed. This is the same as Eq. (10.45a):

$$\frac{dR}{da_1} = 2 \sum_{i=1}^{N} (y_i - a_1 x_i - a_2)(-x_i) = 0 \tag{10.74}$$

The solution to this is

$$a_1 = \frac{S_{xy} - a_2 S_x}{S_{x2}} \tag{10.75}$$

If the slope a_1 is required to have a fixed value, we have only one equation, which is the same as Eq. (10.45b).

$$\frac{dR}{da_2} = 2 \sum_{i=1}^{N} (y_i - a_1 x_i - a_2)(-1) = 0 \tag{10.76}$$

The solution to this is

$$a_2 = \frac{S_y - a_1 S_x}{N} \tag{10.77}$$

Notice that if the required slope is equal to zero,

$$a_2 = \frac{S_y}{N} = \bar{y} \tag{10.78}$$

where $\bar{y}$ is the mean of the y values.

ADDITIONAL PROBLEMS

10.22

A sample of 10 sheets of paper has been selected randomly from a ream (500 sheets) of paper. Regard the ream as a population, even though it has only a finite number of members. The width and length of each sheet of the sample were measured, with the following results:

Sheet Number	Width (in.)	Length (in.)
1	8.50	11.03
2	8.48	10.99
3	8.51	10.98
4	8.49	11.00
5	8.50	11.01
6	8.48	11.02
7	8.52	10.98
8	8.47	11.04
9	8.53	10.97
10	8.51	11.00

a. Calculate the mean length and its sample standard deviation, and the mean width and its sample standard deviation.
b. Give the expected ream mean length and width, and the expected error in each at the 95% confidence level.
c. Calculate the expected ream mean area from the width and length, and give the 95% confidence interval for the area.
d. Calculate the area of each sheet in the sample. Calculate from these areas the sample mean area and the standard deviation in the area.
e. Give the expected ream mean area and its 95% confidence interval from the results of part d.
f. Compare the results of parts c and e. Would you expect the two results to be identical? Why (or why not)?

10.23

The following is an actual set of student data on the vapor pressure of pure liquid ammonia, obtained in a physical chemistry laboratory course. Find the indicated heat of vaporization, using the graphical procedure once and the least-squares procedure once.

Temperature (°C)	Pressure (torr)
−76.0	51.15
−74.0	59.40
−72.0	60.00
−70.0	75.10
−68.0	91.70
−64.0	112.75
−62.0	134.80
−60.0	154.30
−58.0	176.45
−56.0	192.90

10.24

a. Ignoring the systematic errors, find the 95% confidence interval for the heat of vaporization, using the residuals, for the data in Problem 10.23.

b. Assuming that the apparatus used to obtain the data in Problem 10-23 was about like that found in most undergraduate physical chemistry laboratories, make a reasonable estimate of the systematic errors and find the 95% confidence interval for the heat of vaporization, including both systematic and random errors.

10.25

The vibrational contribution to the heat capacity of 1 mol of a gas is given in statistical mechanics by the formula

$$\bar{C}(\text{vib}) = R \sum_{i=1}^{3n-6} \frac{u_i^2 e^{-u_i}}{(1 - e^{-u_i})^2}$$

where $u_i = hv_i/kT$. Here v_i is the frequency of the ith normal mode of vibration, of which there are $3n - 6$ if n is the number of nuclei in the molecule (assumed nonlinear), h is Planck's constant, k is Boltzmann's constant, R is the gas constant, and T is the absolute temperature. The H_2O molecule has three normal modes. If their frequencies are given by

$$v_1 = 4.78 \times 10^{13} \pm 0.02 \times 10^{13} \text{ s}^{-1}$$
$$v_2 = 1.095 \times 10^{14} \pm 0.004 \times 10^{14} \text{ s}^{-1}$$
$$v_3 = 1.126 \times 10^{14} \pm 0.004 \times 10^{14} \text{ s}^{-1}$$

calculate the vibrational contribution to the heat capacity of H_2O vapor at 500 K and find the 95% confidence interval.

10.26

The nth moment of a probability distribution is defined by

$$M_n = \int (x - \mu)^n f(x)\, dx$$

The second moment is the variance, or square of the standard deviation. Show that for the standard normal distribution, $M_3 = 0$, and find the value of M_4. For this distribution, the limits of integration are $-\infty$ and $+\infty$.

10.27

Vaughan[14] obtained the following data for the dimerization of butadiene at 326 °C.

[14] W. E. Vaughan, "The Homogeneous Thermal Polymerization of 1,3-Butadiene," *J. Am. Chem. Soc.* **54**, 3863 (1932).

Time (min)	Partial Pressure of Butadiene (atm)
0	to be deduced
3.25	0.7961
8.02	0.7457
12.18	0.7057
17.30	0.6657
24.55	0.6073
33.00	0.5573
42.50	0.5087
55.08	0.4585
68.05	0.4173
90.05	0.3613
119.00	0.3073
259.50	0.1711
373.00	0.1081

Determine whether the reaction is first, second, or third order, using the least-squares method. Find the rate constant and its 95% confidence interval, ignoring systematic errors. Find the initial pressure of butadiene.

10.28

Make a graph of the partial pressure of butadiene as a function of time, using the data in Problem 10.27. Find the slope of the tangent line at 33.00 min and deduce the rate constant from it. Compare with the result from Problem 10.27.

10.29

Use Eq. (10.33) to "smooth" the data given in Example 10-13. Using the smoothed data and Eq. (10.35) find the derivative dc/dt at $t = 25$ min. Find the rate constant.

10.30

Assuming that the ideal gas law holds, find the number of moles of nitrogen gas in a container if

$$P = 0.856 \pm 0.003 \text{ atm}$$
$$V = 17.85 \pm 0.08 \text{ liter}$$
$$T = 297.3 \pm 0.08 \text{ K}$$

Find the expected error in the number of moles.

10.31

The Bouguer–Beer law (sometimes called the Lambert–Beer law) states that

$$A = abc$$

where A is the absorbance of a solution, defined as $\log_{10}(I_0/I)$ where I_0 is the incident intensity of light at the appropriate wavelength and I is the transmitted intensity; b is the length of the cell through which the light passes;

and c is the concentration of the absorbing substance. The coefficient a is called the *molar absorptivity*, ε, if the concentration is in moles per liter. The following is a set of data for the absorbance of a set of solutions of disodium fumarate at a wavelength of 250 nm. Using Eq. (10.75), find the least-squares value of ε if $b = 1.000$ cm.

A	0.1425	0.2865	0.4280	0.5725	0.7160	0.8575
(mol liter^{-1})	1.00×10^{-4}	2.00×10^{-4}	3.00×10^{-4}	4.00×10^{-4}	5.00×10^{-4}	6.00×10^{-4}

ADDITIONAL READING

Philip R. Bevington, *Data Reduction and Error Analysis for the Physical Sciences*, McGraw-Hill Book Company, New York, 1969.

This is a very nice book, which includes a lot of useful things, including a discussion of different probability distributions, including the standard normal distribution, and a discussion of weighted least-squares procedures.

R. A. Day, Jr., and A. L. Underwood, *Quantitative Analysis*, 3rd ed., Prentice-Hall, Englewood Cliffs, N.J., 1974.

This is a typical quantitative analysis textbook, and contains a discussion of experimental errors and some statistical techniques involving indirectly measured quantities.

Walter Clark Hamilton, *Statistics in Physical Science*, The Ronald Press Company, New York, 1964.

This is a useful small book which contains just about everything that a chemist might need to know about the use of statistics.

Eugene Jahnke and Fritz Emde, *Tables of Functions*, Dover Publications, New York, 1945.

This is a very useful paperback book which contains a lot of tables and formulas, including the error function and related quantities.

Emerson M. Pugh and George H. Winslow, *The Analysis of Physical Measurements*, Addison-Wesley Publishing Co., Reading, Mass., 1966.

This small book is available in paperback version, and contains quite a lot of information.

David P. Shoemaker, Carl W. Garland, and Jeffrey I. Steinfeld, *Experiments in Physical Chemistry*, 3rd ed., McGraw-Hill Book Company, New York, 1974.

This is a standard physical chemistry laboratory textbook, and contains a good section on the treatment of experimental errors as well as most of the experiments commonly done in physical chemistry courses.

Archie G. Worthing and Joseph Geffner, *Treatment of Experimental Data*, John Wiley & Sons, New York, 1943.

This is a useful source of formulas and facts about the application of statistics to experimental results.

Hugh D. Young, *Statistical Treatment of Experimental Data*, McGraw-Hill Book Company, New York, 1962.

This is a small paperback book, written at an elementary level, and explains on a physical basis what is behind some of the things discussed in this chapter.

CHAPTER ELEVEN

Computer Programming

Section 11-1. INTRODUCTION

This chapter can be studied at any time in your study of this book. Although it is placed last, there are some applications of computer programs in earlier chapters.

We discuss the BASIC computer language. This language was developed at Dartmouth College in the 1960s in an effort to make computer programming relatively easy without sacrificing versatility and power. The name is an acronym for **B**eginner's **A**ll-Purpose **S**ymbolic **I**nstruction **C**ode. Of the common computer languages, BASIC is probably the easiest to use. Furthermore, the operating systems of some minicomputers, such as the DEC 11/70 RSTS/E system which is used at the author's college, are constructed in such a way that BASIC programs are easier to compile and execute than programs in other languages. In the last section of this chapter, we comment on the FORTRAN language.

After studying this chapter, you should be able to write a computer program to carry out any calculation that is likely to arise in a physical chemistry course. Writing such programs consists of the following: (1) obtaining an algorithm, or method of solution, which provides mathematical formulas, etc., for the problem at hand; (2) mapping out in your mind, or on a flowchart, the sequence of operations necessary to implement the algorithm; (3) setting down the proper set of instructions to cause the computer to carry out the operations; and (4) entering and running the program on the computer.

Section 11-2. THE OPERATION OF A COMPUTER

The simplest computer consists of the following components:

1. **A central processing unit, where arithmetic and other operations are actually carried out.**

2. **One or more storage devices. Even the simplest computer has a core memory, in which numbers can be stored and from**

which they can be retrieved when needed. In addition, there may be disk, drum, and tape units.

3. **One or more input/output devices, which can be used to communicate information to the computer from the user (input) or from the computer to the user (output).**

Let us describe roughly what happens when the computer performs the addition of two numbers. The numbers can be stored in core memory in *binary*, not decimal, form. Each one occupies several binary devices that constitute a *word* or storage unit. The computer must locate the storage unit containing the first number and find the value by sensing the state of each binary device. It must then copy this number onto a similar set of binary devices in the *central processing unit* (CPU), which is called a *working register*. It then locates the second number and copies it onto a second working register.

The CPU, which is a complicated switching network, then adds the numbers together, ending with the result in one of the working registers. This result can then be stored in the core memory in some storage unit. The computer does this very rapidly, and can do many billions of such calculations without making any mistakes. You should however remember that it can only do what it is instructed to do. It has no means of telling what an instruction should have been if it is written incorrectly.

The thing that makes a computer so useful is that it is programmable. It is constructed to look through a certain section of core memory and to interpret the numbers stored there as instructions, telling it what operations to perform and where to find the numbers on which to perform the operations. These instructions are stored in a code called a *machine language*. The earliest computers had to have their sets of instructions, or programs, put into them in machine language.

Modern computers can also be programmed in machine language if the programmer wishes. However, they can also accept programs written in symbolic languages, such as BASIC, FORTRAN and ALGOL, which allow easily remembered symbols to be used to represent operations and locations in memory. This is made possible by a compiler program that makes a translation from symbolic language to machine language. The BASIC compilers for nearly all computers provide for nearly identical rules to be prescribed for the BASIC language, so that a program written for one computer can usually be run on another computer of a different brand with only minor modifications.

Section 11-3. VARIABLES AND OPERATIONS IN BASIC

The BASIC language is a simplified version of the FORTRAN language, whose name is an abbreviation for "formula translation." This language was designed to allow a programmer to write expressions closely resembling mathematical formulas written in the usual way.

Quantities that are represented by letter symbols in ordinary formulas are called *variables*. In BASIC, a variable can be represented by a single letter of the alphabet, or by a letter followed by a single digit, from 0 through 9. To the computer, such a variable name represents a particular storage unit in the core memory. If a variable called X1 is referred to in a program, the computer understands that whatever value is stored in the storage unit at the moment is to be used.

Example 11-1

Which of the following are permissible names for variables in BASIC?

YX,   D,   4Y,   H12,   ABD,   Y8

Solution

YX is not permissible because it has two letters.

D is permissible.

4Y is not permissible, because the first symbol must be a letter, not a number.

H12 is not permissible for most BASIC compilers, but would be accepted by some computers.

ABD is not permissible because it has three letters.

Y8 is permissible   •

Notice that only capital letters are used.

In BASIC, constants are written as integers or as decimal fractions, or in a modified version of scientific notation. The following are all valid expressions for constants:

> 2
> 3.14159
> 7.5E20,   which means 7.5×10^{20}
> 6.8796E-5,   which means 6.8796×10^{-5}

The modified version of scientific notation is necessary because computer terminals and key punch machines do not have a provision to write exponents as superscripts.

Do not write constants as fractions. If you wrote, for example, 1/3, the computer might or might not come up with the correct quantity, depending on what else was in the expression, because the computer would interpret the slash (/) as an instruction to perform a division.

The BASIC language uses the following symbols for the four elementary arithmetic operations:

addition	+	
subtraction	−	
multiplication	*	(asterisk)
division	/	(slash)

Only the symbol for multiplication is different from what we commonly use in mathematical formulas. The following are all valid BASIC expressions:

```
X + Y
X3 − 3.14159
A4/7.5 (means the variable A4 divided by 7.5, not A times 4 over 7.5)
4.5*A
```

If a variable or a constant is to be raised to a power, the symbol ↑ is used[1] for exponentiation.[2] Some systems allow a double asterisk to be used as an alternative (this is the exponentiation symbol in FORTRAN). The following are valid BASIC expressions

$$X\uparrow2 \quad \text{(means } X^2)$$
$$2\uparrow A \quad \text{(means } 2^A)$$
$$A\uparrow X \quad \text{(means } A^X)$$

In order to make a simple arithmetic expression into something the computer can execute, we must tell the computer what to do with the result of the expression. This is done by the *arithmetic assignment statement*, which has the form

$$\text{LET variable} = \text{expression} \tag{11.1}$$

An example of this would be

$$\text{LET X1} = \text{Y/Z2} \tag{11.2}$$

This would cause the computer to take whatever value is currently stored in the storage unit called Y, divide it by whatever value is currently stored in the storage unit called Z2, and copy the result into the storage unit called X1. Whatever value previously was assigned to the variable X1 is erased, so the arithmetic assignment statement is sometimes called a *replacement operator*.

The word LET is optional, so that Eq. (11.2) can be written

$$X1 = Y/Z2 \tag{11.3}$$

This looks just like a mathematical equality, but it is somewhat different from one. The old value of X1 may or may not have been equal to Y divided by Z2, but the new value of X1 is caused to be equal to Y divided by Z2. In fact, it is reasonable to write

$$X1 = X1 + Y/Z2 \tag{11.4}$$

This statement will cause the old value of X1 to be replaced by the old value of X1 plus Y divided by Z2. If the equals sign meant what it does in

---

[1] Some terminals have a caret (^) in place of this vertical arrow.

[2] A positive quantity can be raised to any power, but a negative quantity can only be raised to an integer power.

ordinary formulas, Eq. (11.4) would be nonsense unless $Y = 0$. As a BASIC statement, it makes good sense.

As in Eq. (11.4), one BASIC statement can contain more than one arithmetic operation. However, it can contain only one equal sign, or arithmetic assignment operator.

All BASIC compilers use the same method of deciding which arithmetic operation to do first, as follows:

1. Exponentiations are done first.
2. Multiplications and divisions are done next.
3. Additions, subtractions, and negations are done next. Negation is a unary operation consisting of changing the sign of a quantity. This is treated as a subtraction, not as a multiplication by -1.
4. Operations in an expression are performed from left to right.

Multiplication and division are equal in rank, so that a division will be performed before a multiplication if it is to its left in the expression, and after it if it is to its right. If there are some unperformed operations left in the expression when the expression has been completely scanned from left to right, the computer will then perform the operations in reverse order, from right to left.

Example 11-2

Determine the outcome of the following:

$$X = 2/3/5 + Y\uparrow3 - 7.5$$

Solution

First, Y is cubed, giving a result depending on what the current value of Y is. This result is temporarily saved. Then 2 is divided by 3, giving 0.66666 . . . , and this result is divided by 5. The result of cubing Y is added to this, and 7.5 is subtracted from this result.   ●

Problem 11-1

a. In Example 11-2, assume that $Y = 4.35$. Find the numerical value to be assigned to X.

b. Determine the outcome of

$$X = X*3 - 4 + B/C\uparrow3$$

if the old value of X is 3, the value of B is 7 and the value of C is 2.   ●

Problem 11-2

Write BASIC statements to accomplish the following.

a. Divide Y by the quantity A times B.

b. Multiply X squared by C divided by D.

c. Raise B to the power $X + Y$.   ●

It is possible to modify the order in which operations are performed by using parentheses, which are used in the same way as in ordinary mathematical formulas. The rule is that all operations inside a pair of parentheses will be performed before the resulting quantity is combined with anything else. It is possible to nest parentheses inside parentheses to just about any extent that might be necessary. However, you must be careful to put a mate with every parenthesis.

Example 11-3

Determine the outcome of the following:

$$X = 2/(3*5) + Y\uparrow(3 + W)*(A + B)$$

Solution

The computer first computes the value of each quantity enclosed in parentheses and saves the results. Then Y is raised to the power $3 + W$ and this result is saved. Then 2 is divided by the product of 3 and 5 which was already taken. The product of Y^{3+W} and $A + B$ is computed and added to the result of the first term.   ●

Problem 11-3

Determine the outcome of the following.
a. $X = (((A + B)\uparrow 2 + (A*B))\uparrow 3 - (C/D)\uparrow 0.5)$
b. $X = ((D/E) + (F*G)\uparrow 3)\uparrow 2$
Find the value that will result if $A = 1$, $B = 2$, $C = 3$, $D = 4$, $E = 5$, $F = 6$, and $G = 7$.   ●

Example 11-4

Write BASIC statements to evaluate the following.
a. $Y = [(A^2 + B^2)/2C]^N$
b. $Z = [AC + BC - DC(A + E)]/[(X + 1) - (2D)^2]$

Solution

a. $Y = ((A\uparrow 2 + B\uparrow 2)/(2*C))\uparrow N$
b. $Z = C*(A + B + D*(A + E))/(X + 1 - (2*D)\uparrow 2)$ ●

Problem 11-4

Write BASIC statements to evaluate the following formulas.

a. $R = \left[\dfrac{V}{\frac{4}{3}\pi}\right]^{1/3}$ (choose a variable name for π)

b. $R = (X^2 + Y^2)^{1/2}$

Example 11-5

Write BASIC statements to find the values of the roots of the quadratic equation

$$ax^2 + bx + c = 0 \tag{11.5}$$

Treat only the case that both roots are real.

Solution

The roots are given by the formula

$$x = \frac{-b \pm (b^2 - 4ac)^{1/2}}{2a} \tag{11.6}$$

There are two roots, which we name X1 and X2. We use A, B, and C, for the quantities called a, b, and c in the formula.

```
D  = (B↑2 − 4*A*C)↑.5
X1 = (−B + D)/(2*A)
X2 = (−B − D)/(2*A)                    •
```

Problem 11-5

a. Write BASIC statements to evaluate the mean of six numbers, called N1, N2, N3, N4, N5, N6.
b. Write BASIC statements to evaluate the square root of the mean of the squares of the six numbers.   •

The BASIC systems of some computers allow for a variable to be designated as an *integer variable*, which will take on only integer values. The advantage is that the computer uses less storage space to store an integer. An integer variable is denoted by a variable name ending in a percent symbol, as for example B%. The following is a legitimate BASIC statement using such variables:

$$B\% = A\%*C\% + 3$$

The computer will not round an integer variable to the nearest integer. It will discard the fractional part, which is called *truncating* the number. For example, the statement

$$B\% = 5/6$$

will assign a value of zero to the variable B%.

Section 11-4.   LIBRARY FUNCTIONS

If you need values for one of the common functions, such as the sine, the cosine, logarithms, etc., nearly all BASIC compilers have programs written into them to evaluate these functions for you. These functions are called *library functions*, or *predefined functions*, and are used by putting the name of the function into a BASIC statement in the correct way. For example, if you need the natural logarithm of the variable X3, you can write the statement

$$Y = LOG(X3) \tag{11.7}$$

Library functions can also be inserted into statements containing arithmetic operations, as in the statement

$$Z3 = (3*LOG(X) + (A*B)\uparrow 2) \tag{11.8}$$

Table 11-1 lists the library functions that are available in BASIC for a fairly small computer.

Table 11-1. Predefined Mathematical Functions in BASIC

Name	Function	Result	Example
ABS	Absolute value or magnitude	Positive constant	$Y = ABS(X)$
ATN	Arctangent (in radians)	Constant, $-\dfrac{\pi}{2} < c < \dfrac{\pi}{2}$	$Y = ATN(X)$
COS	Cosine of an angle in radians	Constant, $-1 < c < 1$	$Y = COS(X)$
EXP	Exponential function, e^x	Positive constant	$Y = EXP(X)$
INT	Truncation (drops decimal part of the argument)*	Integer constant	$Y = INT(X)$
LOG	Natural logarithm (base e) of positive argument	Constant	$Y = LOG(X)$
LGT or CLG or LOG10	Common logarithm (base 10) of positive argument	Constant	$Y = LGT(X)$
RND	Random number generation	Constant $0 < c < 1$	$Y = RND$ or $Y = RND(X)^{\dagger}$
SGN	Sign of argument (-1 if argument is negative, 0 if argument is 0, $+1$ if argument is positive)	$-1, 0, 1$	$Y = SGN(X)$
SIN	Sine of an angle in radians	Constant, $-1 < c < 1$	$Y = SIN(X)$
SQR	Square root of positive argument	Positive constant	$Y = SQR(X)$
TAN	Tangent of angle in radians	Constant	$Y = TAN(X)$

   * Some systems may have a FIX function that truncates either a positive or negative number If so, INT will do the same for a positive number. Check to see what it does with a negative number.
   † Argument optional, and makes no difference.

Example 11-6

Write statements to determine the angle A from its cosine, using the arctangent function. Assume that the cosine is stored under the name X.

Solution

We use the identities

$$\tan(A) = \frac{\sin(A)}{\cos(A)}$$
$$\sin^2(A) + \cos^2(A) = 1$$

We write the statements

```
S = SQR(1 − X↑2)
A = ATN(S/X)                        •
```

Notice that it is not necessary for the argument of the function to be a single variable. All of the BASIC functions will accept an expression as the argument, as long as it is written according to the BASIC rules. However, a function is not allowed to have the same function in its argument:

$$Y = SQR(SQR(A)) \qquad \textit{not permitted} \qquad \textbf{(11.9)}$$

It is permissible to put other functions in the argument of a function. For example, you can write the statement

$$Z = LOG(EXP(X) − A) \qquad\qquad \textbf{(11.10)}$$

Example 11-7

Write BASIC statements to evaluate the side of a triangle, $A1$, opposite the angle A, given the other two sides, $B1$ and $C1$, and the angle A, using

$$A_1^2 = B_1^2 + C_1^2 - 2B_1 C_1 \cos(A) \qquad\qquad \textbf{(11.11)}$$

Solution

$$A1 = SQR(B1↑2 + C1↑2 − 2*B1*C1*COS(A)) \qquad • \quad \textbf{(11.12)}$$

Problem 11-6

Write BASIC statements to evaluate the three angles of a triangle if the three sides are given. You will need Eq. (11.11), and if your computer does not have the arccosine function, you will need the statements in the solution to Example 11-6.   •

Section 11-5. TRANSFER STATEMENTS AND LOOPS

When a computer program has been translated into machine language and is stored in the appropriate part of core memory, the computer will execute the statements in the sequence in which they are stored, unless instructed to do otherwise. In this section, we discuss *transfer statements*, which cause the computer to "branch," or to go to a different point in the program, and *loops*, which cause the computer to go through a part of a program repeatedly.

There are three kinds of transfer statements in ordinary BASIC:

1. The unconditional GO TO statement.
2. The computed or conditional GO TO statement.
3. The IF–THEN statement.

In order to tell the computer in what order to store our statements, we must put a statement number in front of every BASIC statement. The computer will store them in numerical order, and unless otherwise instructed, will execute them in that order. As you write a program, it is a very good idea to skip some numbers between consecutive statements so that you can later insert more statements if you need to do so.

The Unconditional GO TO Statement. This statement does just what its name implies. The statement has the form

$$\text{GO TO s} \qquad\qquad (11.13)$$

where s stands for the number of the statement that you want the computer to execute next. For example, if you want the computer to go to statement number 170, the GO TO statement is, if its own number is 120,

$$\text{120 GO TO 170}$$

This will cause the computer to proceed to statement 170 instead of to whatever statement has the number next larger than 120.

If you want a program to be repeated an indefinite number of times, one way to do this is to put a GO TO statement at the end of the program. If the first executable statement is statement number 100, such a statement would be

$$\text{970 GO TO 100}$$

The Computed (Conditional) GO TO Statement. The unconditional GO TO statement does not allow for decision making. The computed GO TO statement allows us to have the computer branch to one statement if some condition is met, or to another if it is not, or to choose one of several statements. The general form of the computed GO TO statement is

$$\text{GO TO } (s_1, s_2, \ldots, s_m), \text{ exp} \qquad\qquad (11.14)$$

where $s_1, s_2, \ldots,$ represent statement numbers to which the computer can branch, and exp represents some arithmetic expression which the computer can evaluate. If the truncated value of the expression is 1, the computer branches to statement number s_1. If the truncated value of the expression is

2, the computer branches to statement number s_2, etc. By the *truncated value* of an expression, we mean the value obtained if the decimal part is discarded. For example, if an expression has the value 2.785, its truncated value is 2. The following statement will cause the computer to go to statement number 340 if the variable K has a truncated value of 1, to statement number 570 if K has a truncated value of 2, and to statement number 690 if K has a truncated value of 3:

$$510 \text{ GO TO } (340, 570, 690), \text{ K} \qquad (11.15)$$

The IF–THEN Statement. This statement is also used for conditional branching, and is easier to use than the computed GO TO statement if only two choices are needed. We will discuss only three types of IF–THEN statements.

The simplest IF–THEN *statement* has the form

$$\text{IF exp1 rel exp2 THEN s} \qquad (11.16)$$

where expl is an arithmetic expression, exp2 is a second arithmetic expression, and rel stands for a *relational operator*, or logical operator. The combination expl rel exp2 is called a *logical expression*, and can either be true or false. If the logical expression is true, the computer will branch to the statement whose number is s, and if it is false, the computer will continue to the next statement after the IF–THEN statement.

There are six or seven relational operators, which are listed in Table 11-2. The approximate equality does not exist in all BASIC compilers. Its purpose is to prevent round-off error from spoiling a comparison between two expressions that are supposed to be equal, but are not quite equal because of accumulation of small errors due to the fact that the computer only carries a fixed number of significant digits.

An example of an IF–THEN statement is

$$250 \text{ IF } X > = 0 \text{ THEN } 490 \qquad (11.17)$$

which will cause the computer to branch to statement number 490 if X is either zero or positive, or will let it go on to whatever statement follows statement number 250 if X is negative.

Table 11-2. Relational Operators in BASIC

Relation	Algebraic Symbol	BASIC Symbol
Is equal to	$=$	$=$
Is greater than	$>$	$>$
Is less than	$<$	$<$
Is greater than or equal to	$\geq$	$> =$
Is less than or equal to	$\leq$	$< =$
Is not equal to	$\neq$	$< >$
Is approximately equal to	$\approx$	$= =$

Another version of the IF–THEN statement includes an executable statement within the IF–THEN statement itself. It has the form

$$\text{IF exp1 rel exp2 THEN: S} \qquad \qquad \textbf{(11.18)}$$

where S represents an executable statement. For example, if you want to replace a negative value of X by zero, but want to leave X alone if it has a a positive or zero value, you can use the statement

$$390 \text{ IF X} < 0 \text{ THEN: X} = 0 \qquad \qquad \textbf{(11.19)}$$

This statement does not produce any branching, but executes the $X = 0$ statement only if the old value of X is negative.

A third version of the IF–THEN statement, which may not be included in all BASIC systems, has the form

$$\text{IF exp1 rel exp2 THEN } s_1 \text{ ELSE } s_2 \qquad \qquad \textbf{(11.20)}$$

If the logical expression expl rel exp2 is true, this will cause branching to statement number s_1, and if it is false, it will cause branching to statement number s_2. For example,

$$760 \text{ IF B} = = 0 \text{ THEN 590 ELSE 980}$$

will cause branching to statement 590 if B is approximately equal to zero, and to statement 980 otherwise.

Example 11-8

Write BASIC statements that will find the roots of the quadratic equation

$$ax^2 + bx + c = 0$$

if the roots are equal (case 1), if the roots are real and unequal (case 2), or if the roots are complex (case 3).

Solution

```
200 D = B↑2 − 4*A*C
210 IF D = 0 THEN 400
220 IF D > 0 THEN 500
230 D = −D
240 R = −B/(2*A)
250 I1 = SQR(D)/(2*A)
260 I2 = −SQR(D)/(2*A)
270 GO TO 600
400 X1 = −B/(2*A)
410 GO TO 600
500 X1 = (−B + SQR(D))/(2*A)
510 X2 = (−B − SQR(D))/(2*A)
520 GO TO 600
```

Problem 11-7

An angle measured in degrees is stored under the variable name A. Write BASIC statements that will set a variable called K equal to 1 if the angle is in the first quadrant, equal to 2 if the angle is in the second quadrant, etc. Assume that the angle is not larger than 360°. •

Example 11-9

Write BASIC statements that will calculate the sum of all the integers from 1 to 100.

Solution

```
100 S = 0
110 I = 1
120 S = S + I
130 I = I + 1
140 IF I < = 100 THEN 120
```
•

Problem 11-8

Modify the set of statements in Example 11-9 so that they will add up the squares of all the integers from 1 to 50. •

Looping. The BASIC language allows for the kind of repeated calculation done in Example 11-9 and Problem 11-8 to be done automatically. Use of this feature is called *looping* and the statements that cause the computer to repeat the procedure are called a FOR–NEXT loop. The general form of the statements is

$$\text{FOR i = exp1 TO exp2 STEP exp3}$$
$$\text{executable statements} \qquad\qquad (11.21)$$
$$\text{NEXT i}$$

The quantity i stands for a variable called the loop variable which has as its initial value the value of the arithmetic expression expl. Each time the computer reaches the statement NEXT i, the loop variable is incremented by the value of the arithmetic expression exp3 and the computer branches back to the first executable statement after the FOR statement. When i has a value greater than the value of exp2, the computer leaves the loop and goes on to the next statement after the NEXT i statement. For most purposes, exp1, exp2, and exp3 are chosen to be integers, but this is not necessary.

The loop that performs the same calculation as the statements in Example 11-9 is

```
100 S = 0
110 FOR I = 1 TO 100 STEP 1
120 S = S + I
130 NEXT I
```

Statement 100 is said to *initialize* the value of S. Such a statement is unnecessary with some computers if the initial value of S is to be zero. A variable cannot appear on the right-hand side of an equals sign in a statement unless there is a value assigned to that variable prior to that statement. Some BASIC systems will automatically assign a value of zero to a variable if the program has not yet evaluated it, but others refuse the instruction and give you an error message if you put an unevaluated variable on the right-hand side of an equals sign. It is best to initialize all variables yourself, and if you want an initial value other than zero, it is of course essential to do this.

Problem 11-9

Write a FOR–NEXT loop to sum up the squares of all the integers from 1 to 20, and to find the mean value of these squares. •

It is possible to branch out of a loop before all the repetitions have been executed, by inserting a computed GO TO statement or an IF–THEN statement, as in the following example of summation of a series.

Example 11-10

Write a FOR–NEXT loop to sum up the series
$$1 - \tfrac{1}{2} + \tfrac{1}{3} - \tfrac{1}{4} + \cdots$$
to four significant digits.

Solution

This is a convergent alternating series, so the difference between any partial sum and the whole series is less than the magnitude of the next term after the partial sum. Our error must be less than 5×10^{-5}, so we demand that the first neglected term be less than 2×10^{-5} in magnitude, to be cautious. Our loop is

```
100 S = 1
110 FOR N = 2 TO 100000 STEP 2
120 IF 1/N < 2E-5 THEN 150
130 S = S − 1/N + 1/(N + 1)
140 NEXT N
```

Notice that we have "stepped" by two units to accommodate the alternating signs. •

Problem 11-10

Write a FOR-NEXT loop to sum up the series of Eq. (3) of Appendix 6. Stop summing when a term divided by the partial sum is less than 10^{-4}. •

It is possible to "nest" one FOR–NEXT loop inside another, so that the computer will run through the inner loop a number of times each time that it goes through the outer loop.

Example 11-11

Write BASIC statements to sum up all the products *nm*, where *n* and *m* both range from 1 to 100.

Solution

```
 90 S = 0
100 FOR N = 1 TO 100 STEP 1
110 FOR M = 1 TO 100 STEP 1
120 S = S + N*M
130 NEXT M
140 NEXT N
```
•

Notice that the entire inner loop (the M loop) must lie completely inside the outer loop (the N loop). It is possible to branch from the inner loop to the outer loop, but not from the outer loop into the inner loop, unless the inner loop has been executed at least once. Unless you absolutely sure what you are doing, do not branch into a loop at all.

Problem 11-11

Modify the loops of Example 11-11 so that all terms with $m > n$ are omitted.
•
If a FOR–NEXT loop is imbedded inside another loop, it can have a FOR–NEXT loop imbedded inside it, and so forth, just about as deep as you might need to go.

Here are a few comments about the use of loops:

1. After the computer has left a loop, whether by branching or by completing the specified repetitions, the loop variable can be used and will have the value that it had upon leaving the loop. With some computers, if the loop has been executed the maximum number of times, the loop variable will have the same value as the maximum value, given by exp2 in Eq. (11.21). In others, it will have been incremented one time beyond this.
2. If the increment in the loop variable is 1, the STEP 1 part of the loop statement can be omitted.
3. A negative increment is permissible, but if exp3 in Eq. (11.21) is negative, exp2 must be smaller than expl.
4. The values of exp2 and exp3 can be changed during the execution of the loop if desired.

Section 11-6.   ARRAYS. VECTORS AND MATRICES

BASIC compilers provide a convenient way to handle lists of numbers, such as components of vectors, elements of matrices, etc. The input and output devices of computers do not have provision for reading and printing subscripts, so the indices are enclosed in parentheses. The components of the vector **A** are denoted by A(1), A(2), etc. Although enclosed in parentheses, the indices are still often called *subscripts*. The elements of a matrix are denoted by B(1, 2), B(7, 8), B(3, 4), etc. As usual, the first index denotes the row and the second index denotes the column of the matrix.

Such lists are called *arrays*. A vector is a one-dimensional array, and an ordinary matrix is a two-dimensional array. Most systems also permit three-dimensional arrays.

Example 11-12

Write BASIC statements to form the scalar product of the two vectors **A** and **B**, once without using a loop and once with a loop.

Solution

Without a loop

```
100 DIMENSION A(3), B(3)
110 C = A(1)*B(1) + A(2)*B(2) + A(3)*B(3)
```

With a loop

```
100 DIMENSION A(3), B(3)
105 C = 0
110 FOR I = 1 TO 3
120 C = C + A(I)*B(I)
130 NEXT I
```

The DIMENSION statement is necessary to reserve space in memory for the elements of the arrays. The number in parentheses is the number of memory units to be reserved. It can be larger than the number actually used, but cannot of course be any smaller. The word DIMENSION can be abbreviated to DIM. For a matrix with no more than 10 rows and 15 columns, the necessary DIMENSION statement would be

```
100 DIM A(10, 15)
```

Example 11-13

Write BASIC statements that will compute the product of two 3 by 3 matrices, **A** and **B**.

Solution

If **C** = **AB**, the element of **C** for the *i*th row and *j*th column is given by

$$C_{ij} = \sum_{k=1}^{3} A_{ik}B_{kj}$$

We assume that **A** and **B** are already in memory, but write a DIMENSION statement for them anyway.

```
100 DIM A(3, 3), B(3, 3), C(3, 3)
110 FOR I = 1 TO 3
120 FOR J = 1 TO 3
130 C(I, J) = 0
140 FOR K = 1 TO 3
150 C(I, J) = C(I, J) + A(I, K)*B(K, J)
160 NEXT K
170 NEXT J
180 NEXT I
```
●

Problem 11-12

Write BASIC statements to compute the product of two 3 by 3 matrices as in Example 11-13, but without using loops. This will require nine executable statements, one for each element of **C**. ●

Problem 11-13

Write BASIC statements to form the trace of the matrix **A**, with eight rows and eight columns. ●

Although it is possible to write loops to carry out matrix operations, BASIC compilers have the capability to carry these operations out automatically. For example, matrix multiplication can be done by the statement

$$\text{MAT LET C} = \text{A*B} \qquad (11.22)$$

or

$$\text{MAT C} = \text{A*B} \qquad (11.23)$$

The matrices **A** and **B** must be capable of being multiplied together. That is, **A** must have as many columns as **B** has rows.

For other matrix operations, we have the following statements (remember that the word LET is optional):

To take the sum of two matrices:

$$\text{MAT LET C} = \text{A} + \text{B} \qquad (11.24)$$

To take the difference of two matrices:

$$\text{MAT LET C} = \text{A} - \text{B} \qquad (11.25)$$

To multiply a matrix by a scalar expression:

$$\text{MAT LET C} = (\text{exp})*\text{A} \qquad (11.26)$$

where exp stands for a scalar expression. This scalar expression must be enclosed in parentheses. For example,

$$\text{MAT C} = (\text{SQR}(X/U))*A$$

To replace the matrix **B** by the matrix **A**:

$$\text{MAT LET B} = \text{A} \qquad (11.27)$$

Some systems may not recognize this assignment statement. In such a system, one must use the alternative statement

$$\text{MAT LET B} = (1)*\text{A} \qquad (11.28)$$

To form the inverse of a matrix **A**:

$$\text{MAT LET B} = \text{INV(A)} \qquad (11.29)$$

Of course, the matrix **A** must be square and nonsingular.

In addition to the matrix operation statements in Eqs. (11.22) through (11.29), there are some special matrix assignment statements to produce commonly needed matrixes:

To produce the identity matrix:

$$\text{MAT LET A} = \text{IDN} \qquad (11.30)$$

To produce a matrix that has every element equal to unity:

$$\text{MAT LET A} = \text{CON} \qquad (11.31)$$

To produce a matrix with every element equal to zero (the null matrix):

$$\text{MAT LET A} = \text{ZER} \qquad (11.32)$$

In order to use the statements in Eqs. (11.30) through (11.32), the appropriate DIMENSION statement must have been included earlier in the program. In Eq. (11.30), the matrix **A** must have been dimensioned as a square matrix, since the identity matrix must be square.

It is also possible to redimension matrices within a program, as long as the matrix is made smaller than its original size. It is also possible to use BASIC expressions to calculate a new dimension for the matrix. The statement to use is

$$\text{MAT LET A} = \text{RDM(exp1, exp2)} \qquad (11.33)$$

The computer will compute the two expressions exp1 and exp2, truncate them, and reduce the number of rows to the truncated value of exp1 and the number of columns to the truncated value of exp2.

The special assignment statements in Eqs. (11.30) through (11.33) can also be used to redimension a matrix at the same time as the special assignment is done. The modified statements are

$$\text{MAT LET A} = \text{CON(exp1, exp2)} \qquad (11.34)$$

$$\text{MAT LET A} = \text{IDN(exp1, exp1)} \qquad (11.35)$$

$$\text{MAT LET A} = \text{ZER(exp1, exp2)} \qquad (11.36)$$

In all these statements, the word LET is optional.

Example 11-14

Write BASIC statements that will solve the set of three linear inhomogeneous equations

$$a_{11}x_1 + a_{12}x_2 + a_{13}x_3 = c_1 \qquad \text{(11.37a)}$$

$$a_{21}x_1 + a_{22}x_2 + a_{23}x_3 = c_2 \qquad \text{(11.37b)}$$

$$a_{31}x_1 + a_{32}x_2 + a_{33}x_3 = c_3 \qquad \text{(11.37c)}$$

Solution

The necessary DIMENSION statement is

    100 DIM A(3, 3), C(3, 1) X(3, 1), B(3, 3)

The column vectors **C** and **X** must be dimensioned as 3 by 1 matrices in order to be included in matrix multiplications. The necessary BASIC statements are, after the matrices **A** and **C** have been evaluated,

    110 MAT B = INV(A)
    120 MAT X = B*C

The matrix inversion must be the only operation in a BASIC statement. The following is not permitted:

    MAT X = INV(A)*C   *invalid statement*   •

Problem 11-14

Write BASIC statements that will accomplish the following.
a. Take a given 5 by 5 matrix called **A** and find its inverse.
b. Take the resulting matrix and multiply it by another 5 by 5 matrix called **B**.
c. Set the last two rows of this matrix equal to rows of zeros.
d. Redimension the resulting matrix to a 3 by 5 matrix.
e. Multiply the resulting matrix by a given 5 by 3 matrix called **D**.   •

Section 11-7. STRING VARIABLES, CONSTANTS, AND FUNCTIONS

A *string variable* represents a memory unit or units in which letters and other symbols can be stored. That is, although they are stored as numbers according to some code, the computer will translate them into letters, etc., when reading them out.

A string variable is denoted in BASIC by a variable name that ends in a dollar sign. For example, A$ and X$ are legitimate names for string variables.

The literal information stored in a string variable is called a *string constant*. A string constant is always enclosed in quotation marks when written into a BASIC statement. For example, the string constant YES is written "YES".

The principal BASIC statement used with string variables is the assignment statement. It has the form, for example, of

$$100 \text{ LET B\$} = \text{``YES''} \tag{11.38}$$

The string constant in one variable can be assigned to another, as in

$$100 \text{ LET A\$} = \text{B\$} \tag{11.39}$$

This statement will cause whatever string constant is stored in B$ to be copied onto A$. The word LET is optional.

Example 11-15

Write BASIC statements that will assign "HELLO" to A$ and "GOODBYE" to B$ and then interchange the string constants.

Solution

```
100 A$ = "HELLO"
110 B$ = "GOODBYE"
120 C$ = A$
130 A$ = B$
140 B$ = C$
```

Notice the use of a third string variable, C$, which was used as a "scratch variable," to store the original content of A$ while the content of B$ was copied into A$.   •

Problem 11-15

Write BASIC statements that will assign string constants of your choice to the string variables A$, B$, C$, and D$, and then put the content of A$ into B$, the content of B$ into C$, the content of C$ into D$, and the content of D$ into A$.   •

Some BASIC systems have a string addition, or concatenation, operator, which puts two string constants or variables together.

Example 11-16

Assign the string constant "HELLO" to A$ and then change this to "HELLO, DOLLY".

Solution

```
100 A$ = "HELLO"
110 A$ = A$ + ", DOLLY"
```

An alternative is

```
100 A$ = "HELLO"
110 B$ = ", DOLLY"
120 A$ = A$ + B$
```                                                                    •

One common use of string variables is in computed GO TO statements. By putting a particular string constant in the variable, you control where the computer branches when the program is executed. This is done in the following way:

100 B$ = "YES"
110 IF B$ = "YES" THEN 500   (11.40)
120 executable statements
    ⋮
500 other executable statements

It is also possible to compare string variables, as in the statement

100 IF A$ = B$ THEN 500   (11.41)

Most BASIC systems also have *string functions*, which will do such things as discard part of the string constant in a string variable, create specific string variables, and map numbers into strings, and vice versa.

Section 11-8. INPUT AND OUTPUT

The process of putting information into the memory of the computer is called *input*. The process of communicating information from the computer to the user is called *output*.

The most convenient input/output device for small and medium-sized programs is a *remote terminal*. These devices are used with computers having *time-sharing systems*, which are arranged so that several users can operate the computer at the same time from different remote terminals. The computer automatically switches from one person's program to another's. Since each person has his own input/output device, and since the input/output processes are much slower than the computation processes for most programs, each user may appear to have exclusive use of the computer.

A typical remote terminal resembles a glorified typewriter. A typewriter keyboard is used for input, and a cathode-ray tube similar to a television screen may be used for output. Other terminals have character printers which type the output on paper. The terminal is used both to put the program into the computer and to cause execution of the program.

When the computer is not being used in the time-sharing mode, a card reader may be used for input, and a line printer, which prints an entire line of characters at one, may be used for output.

In this section, we discuss the input/output statements that must be written into the program. We first discuss the statements for terminal operation, and then the statements to be used with a card reader and printer.

Input Procedures.   It would be possible to write a simple program without any input statements. If this is done, the program itself must contain assignment statements that put the desired values into all variables. This is inconvenient, because these statements must be replaced if the program is run again with different values for the variables. It is better to include statements that will cause the computer to accept values for the variables at the time of execution.

The statement that does this is called the INPUT *statement*. It has the form

$$\text{INPUT variables} \tag{11.42}$$

where the word variables in this statement stands for a list of the names of the variables for which values are desired. For example, if values are needed for the variables X, Y, Z, A1 and A2, the input statement could be

100 INPUT X, Y, Z, A1, A2

Notice that the variable names in the list are separated by commas. When the program is being executed, this statement will cause the computer to pause and prompt the user by printing a question mark. The user then types in the value he wants X to have, then a comma, then the value he wants Y to have, then a comma, then the value for Z, another comma, and so forth. The user then presses the carriage return on the keyboard and execution continues. If the input statement requires five numbers and the user gives it less than this many, the computer will print another question mark and wait for the other numbers.

The INPUT statement can be used for arrays as well as for ordinary (scalar) arithmetic variables, as in the following statements:

```
90 DIM A(10), B(10, 10)
100 INPUT A(1), A(2), A(3)
110 INPUT B(1, 1), B(1, 2), B(2, 1), B(2, 2)
```

Note that some of the memory units reserved by the DIMENSION statement will remain unused. This is permissible.

Another way to read in arrays is by using loops, as in the following example.

Example 11-17

Write a loop that will provide for input of a 5 by 5 matrix called **C** and a vector of 7 components called **A**.

Solution

```
100 DIM C(5, 5), A(7)
110 FOR I = 1 TO 5
120 FOR J = 1 TO 5
130 INPUT C(I, J)
140 NEXT J
150 NEXT I
160 FOR I = 1 TO 7
170 INPUT A(I)
180 NEXT I
```

It is acceptable to use the symbol I for the loop variable for two different loops, just as long as one is not imbedded in the other.   ●

When the statements in Example 11-17 are executed, the computer will prompt the user by printing a question mark each time an element is to be typed in, if the carriage return is pressed after each number. If the different numbers are typed in on the same line, you must separate them by commas.

Problem 11-16

Write BASIC statements that will provide for the input of a 7 by 4 matrix called **A** and a 4 × 7 matrix called **B** and will then form the product of these matrices.   ●

BASIC systems provide for automatic input of matrices. The required statement is of the form

$$\text{MAT INPUT variables} \tag{11.43}$$

For example, the 5 by 5 matrix of Example 11-17 could be put into the computer by use of the statements

$$\begin{aligned} &100 \text{ DIM C(5, 5)} \\ &110 \text{ MAT INPUT C} \end{aligned} \tag{11.44}$$

The matrix elements are assigned values one row at a time. Thus, if the following set of numbers were typed in for the matrix **C**:

1, 2, 3, 4, 5, 6, 7, 8, 9, 10, 11, 12, 13, 14, 15, 16,

17, 18, 19, 20, 21, 22, 23, 24, 25

the resulting matrix would be

$$\begin{bmatrix} 1 & 2 & 3 & 4 & 5 \\ 6 & 7 & 8 & 9 & 10 \\ 11 & 12 & 13 & 14 & 15 \\ 16 & 17 & 18 & 19 & 20 \\ 21 & 22 & 23 & 24 & 25 \end{bmatrix}$$

String variables can be put into the computer during execution of the program in the same way as arithmetic variables. The statement is of the form of Eq. (11.42):

$$\text{100 INPUT B\$}$$

During execution of the program, the computer will stop at this statement, print a question mark, and wait for you to type in a string constant.

Example 11-18

Write BASIC statements that will provide for input of two 4 by 4 matrices, called **A** and **B**, and will provide for the user to type in the word "YES" if **A** is to be replaced by **B**.

Solution

```
100 DIM A(4, 4), B(4, 4)
110 MAT INPUT A
120 MAT INPUT B
130 INPUT Y$
140 IF Y$ = "YES" THEN 150 ELSE 200
150 MAT A = B
200 other executable statements
```
•

Problem 11-17

Write BASIC statements that will provide for input of 10 numbers as a one-dimensional array. Write statements that will take the arithmetic mean of the numbers if the user types in "ARITHMETIC MEAN" and take the geometric mean if the user types in "GEOMETRIC MEAN".　　　　•

In addition to the INPUT statement, the combination of a READ *statement* and a DATA *statement* can be used. The difference is that the INPUT statement is used to put values in during the execution of the program, while the READ–DATA statements allow the numbers to be typed in when the program is typed in. The statements have the form

$$\begin{array}{ll} \text{READ} & \text{variables} \\ \text{DATA} & \text{values} \end{array} \qquad \textbf{(11.45)}$$

For example, if the variables X and Y are to be given the values 10.35 and 17, the statements would read

```
100     READ X, Y
300     DATA 10.35, 17
```

The DATA statement is an example of a nonexecutable statement. These statements do not produce an action of the computer during execution of the program, but are included to provide information to the computer or to the user. DATA statements can be placed anywhere in the program, since they will not be used until the program is executed. The best place for them is at the end of the program, where they can easily be located and changed when necessary.

The computer will consider all the numbers in DATA statements as a block, and will begin reading at the first of the first DATA statement and will continue until the READ statements have all been executed. If all the data have been read in and there are still more READ statements, the computer will stop.

Example 11-19

Determine the outcome of the READ and DATA statements:

```
100 READ A, B, C, D
110 READ X
300 DATA 12, 14.1, 16
320 DATA 75, 83.57, 4, 12.001
```

Solution

The variable A will be given the value 12, the variable B will be given the value 14.1, the variable C will be given the value 16, and the variable D will will be given the value 75. When the second READ statement is executed, the variable X will be given the value 83.57. The values 4 and 12.001 will remain ready to be used for the next READ statement. If no more READ statements occur, they will remain unused.　　　　　　　　　　　　●

Problem 11-18

Determine the outcome of the following:

```
100 READ A1, A2, A3, X, Y, Z
300 DATA 1, 2, 3, 75
310 DATA 17, 17, 17
```
　　　　　　　　　　　　●

The READ and DATA statements can also be used with arrays, either by writing loops in your program or by using the MAT READ statement with a DATA statement.

Problem 11-19

Write a loop containing a READ statement that will read from a DATA statement and cause the 3 by 3 array **A** to take the values

$$\begin{bmatrix} 12.3 & 13.7 & 19.9 \\ 75.1 & 23.9 & 37.1 \\ 14.7 & 10.7 & 7.7 \end{bmatrix}$$
　　　　　　　　　　　　●

The MAT READ *statement* has the form

MAT READ array name　　　　　　　　　　　　(11.46)

The MAT READ statement will read in the elements of the array in the same order as the MAT INPUT statement, reading across the first row, then the second row, etc. The following statements will accomplish the same thing as the loop you were asked to write in Problem 11-19:

```
100 DIM A(3, 3)
110 MAT READ A
300 DATA 12.3, 13.7, 19.9, 75.1, 23.9, 37.1
310 DATA 14.7, 10.7, 7.7
```

All the numbers could have been put in a single DATA statement if there had been room on the line.

Only one array can be read in with a single MAT READ statement. The following is not a legitimate statement:

MAT READ A, B　　　*not permissible*

Problem 11-20

Write MAT READ and DATA statements to read in the matrices **A** and **B**. Write statements to take the product **AB** = **C**.   •

Output Procedures. Every program must include some output statements if there is some result to be communiciated to the user. The principal BASIC output statement in the PRINT statement, which has the form

PRINT list   (11.47)

where the word *list* stands for a list of carriage controls, variable names, arithmetic expressions, messages, and output controls.

In its simplest version, the PRINT statement contains only the names of variables, and will cause the values of these variables to be printed out on the screen of the terminal or to be typed out by the character printer of the terminal. For example, the following PRINT statement will cause the value of the variable X to be printed out:

400 PRINT X

Several variables can have their values printed out with a single PRINT statement. The following statement will cause the current values of five variables to be printed out on the same line:

400 PRINT X, Y, Z, A1, A2

Notice that the variable names in the list are separated by commas. This will produce standard spacing, with 15 spaces on the line used for each variable. Since many CRT terminals (cathode-ray-tube, or TV-screen terminals) have about 75 spaces per line, this means that only five numbers will appear on one line. If the output list contains more than five variables, the computer will put five on the first line and the rest on other lines.

String variables can also be printed out in the same way as arithmetic variables. For example, if we have

100 A$ = "YES"
110 PRINT A$

the terminal will print out the word YES when statement number 110 is executed.

In addition to the values of variables, messages can be printed with the PRINT statement. The required statement has the form

100 PRINT "message"   (11.48)

where the word *message* stands for whatever string of characters you want printed out. Such messages can be very helpful in a program that is to be run from a remote terminal, because messages can be printed out to tell the user what the program does and to remind him which variable the computer is ready to accept a value for.

Example 11-20

Modify the statements in Example 11-8 so that the computer will print out a statement of what the program does and prompt the user when input is required, and will print the results.

Solution

```
100 PRINT "THIS PROGRAM FINDS THE ROOTS OF THE"
110 PRINT "QUADRATIC EQUATION AX↑2 + BX + C = 0."
120 PRINT "WHAT IS THE VALUE OF THE COEFFICIENT A?"
130 INPUT A
140 PRINT "WHAT IS THE VALUE OF THE COEFFICIENT B?"
150 INPUT B
160 PRINT "WHAT IS THE VALUE OF THE COEFFICIENT C?"
170 INPUT C
200 D = B↑2 − 4*A*C
210 IF D = 0 THEN 400
220 IF D > 0 THEN 500
230 D = −D
240 PRINT "ROOTS ARE COMPLEX."
250 R = −B/(2*A)
260 I = SQR(D)/(2*A)
270 PRINT "REAL PART OF ROOTS IS:"
280 PRINT R
290 PRINT "IMAGINARY PARTS OF ROOTS ARE:"
300 PRINT I, −I
310 GO TO 600
400 PRINT "ROOTS ARE REAL AND EQUAL."
410 X1 = −B/(2*A)
420 PRINT "VALUE OF ROOTS IS:"
430 PRINT X1
440 GO TO 600
500 PRINT "ROOTS ARE REAL AND UNEQUAL."
510 X1 = (−B + SQR(D))/(2*A)
520 X2 = (−B − SQR(D))/(2*A)
525 PRINT "THE ROOTS ARE:"
530 PRINT X1, X2
600 END
```

Notice the END statement, which signals to the computer that the end of the program is reached. •

Problem 11-21

Modify the statements in Example 11-13 to provide for input of the matrices **A** and **B** and for output of the matrix **C**. Include necessary messages.

•

The PRINT statements in Example 11-20 are unnecessarily clumsy, because it is possible to include messages and output of variables in the same PRINT statement. For example, statements 525 and 530 in the example could be combined as

525 PRINT "X1 =", X1, "X2 =", X2

If the value of XI is 2.000 and the value of X2 is 4.000, the line of output would look like

X1 =   2.000   X2 =   4.000

Notice that there is a space between the equals sign and the number for each variable. This is because the computer puts each item in a separate "printing zone" of 15 spaces.

If you wish to pack your output more closely, you can do so by separating the items by semicolons (;) instead of commas. The statement would then look like

525 PRINT "X1 ="; X1, "X2 ="; X2

Some systems allow even closer packing by omission of any separating mark between the items. The statement would then look like

525 PRINT "X1 =" X1, "X2 =" X2

The following are legitimate PRINT statements that will give closely packed output:

690 PRINT A$; X; Y; Z
970 MAT PRINT C;

Notice the semicolon in the MAT PRINT statement. This will allow a matrix with more than five columns to be printed with all the elements of one row on the same line. Some systems require a comma to be typed in this position in order to print the matrix in the ordinary rectangular format. If no comma is put in, each element will appear on its own line.

There are other means to change the format of your output. For example, carriage control symbols can be placed at the first of the list of symbols in the PRINT statement. These symbols are enclosed in quotation marks, much like messages. The most commonly used symbols are:

" " (a blank enclosed in quotes)—produces single spacing
"0" produces double spacing
"-" produces triple spacing
"1" produces a skip to the top of the next page with a line printer, or skips six lines with a remote terminal

In addition, the PRINT statement without a list produces a bank line, and a list followed by a comma produces a line of output without a return to the next line, so that the next item of output will go on the same line if there is room for it.

Some BASIC systems may have additional input/output features, such as the use of statements called FORMAT statements, which determine where things are printed on the line. Some systems provide for printing a message by including it in an INPUT statement, and this is handy for reminding the user what variable is to be given a value.

Example 11-21

Add suitable input and output statements to the statements in Example 11-14.

Solution

```
100 PRINT "THIS PROGRAM SOLVES A SET OF THREE"
110 PRINT "SIMULTANEOUS LINEAR INHOMOGENEOUS"
120 PRINT "ALGEBRAIC EQUATIONS OF THE FORM"
125 PRINT "AX = C WHERE A IS A 3BY3 MATRIX"
130 PRINT "AND X AND C ARE COLUMN VECTORS."
135 PRINT "X CONTAINS THE UNKNOWNS."
140 DIM A(3, 3), C(3, 1), X(3, 1), B(3, 3)
150 PRINT "TYPE IN THE ELEMENTS OF A BY ROWS."
160 MAT INPUT A
170 PRINT "TYPE IN THE ELEMENTS OF C."
180 MAT INPUT C
190 MAT B = INV(A)
200 MAT X = B*C
210 PRINT "THE ROOTS ARE:"
220 MAT PRINT X
230 END
```
•

Input and Output for Batch Processing.   We have been discussing input and output statements used in programs to be run on a time-sharing system. Programs can also be put into the computer and run using a *card reader*, which is a device that senses the locations of holes in computer cards. This mode of operation is sometimes called *batch processing*. Instead of typing the program on the keyboard of a terminal, one prepares cards on a keypunch machine. The cards are stacked and fed into the card reader. Ordinarily, the output is printed on a *line printer*.

The INPUT statement cannot be used in batch processing, because the user will not be at a terminal to provide values for variables while the program is being executed. The READ and DATA statements are used in exactly the same way as on a terminal, and the PRINT statement is also used in the same way.

Example 11-22

Modify the statements of Example 11-20 so that they can be used for batch operation.

Solution

The modified statements are shown in Figure 11-1, which shows the cards on which the statements are typed. As you can see, each card has the holes, which the computer can read, and also ordinary characters printed along the top edge, which the user can read. If these characters were omitted, it would make no difference to the computer. The cards are stacked so that you must read from bottom to top. •

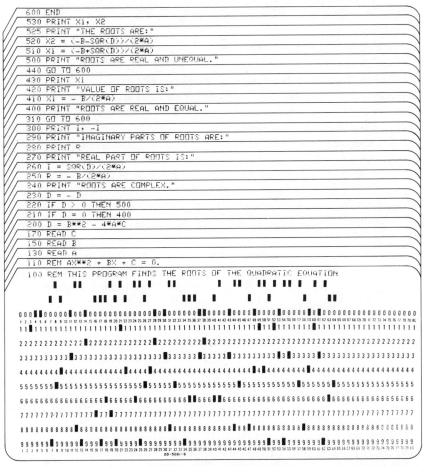

Figure 11-1. The program for Example 11-23, typed on cards. The cards are shown spread out so that the program reads from bottom to top in the figure, which is from top to bottom of the deck of cards.

Problem 11-22

Modify the statements of Example 11-21 so that they can be used for batch processing. •

Section 11-9. WRITING AND RUNNING A BASIC PROGRAM

We have discussed the construction of the body of a BASIC program. We must now discuss the construction of an entire program.

Before you begin writing any statements, you must get clearly in mind how the procedure, or algorithm, is to be applied. In a simple case, the algorithm might be a single formula. In more complicated cases, it might involve complicated mathematical procedures. You should collect all the formulas you will need, and you should analyze the different cases which might arise so that you can plan the needed branching in the program.

Many programmers find it convenient to construct a *flowchart prior* to writing the program. A flowchart schematically charts out the path that the computer will take through the problem. It contains variously shaped "boxes" connected by arrows. Rectangles are usually used to represent the calculation of values for variables, parallelograms are used for input and output, diamonds for decision making, etc. We will not discuss flowcharting, but Figure 11-2 shows a flowchart for Example 11-20. Whether you use a flowchart or not, plan your approach before you start writing statements.

The first phase is to construct all the BASIC statements that you will need in the program. You should write out the statements on paper just as you will type them, so that you will not tie up a terminal or a keypunch machine any longer than necessary.

The next phase is to prepare the program, either by typing it into the computer on a terminal or by typing it onto cards. The following procedure is used with the RSTS/E operating system on the Digital Equipment Corporation 11/70 computer. Other systems will be similar.

1. Turn on the terminal and wait a moment for it to warm up.
2. Type in the word HELLO or the word LOGIN and press the carriage return. The computer will print out a message and ask for an account number.
3. Type in the account number you have arranged for, and press the carriage return. The computer will ask for a password, which is supposed to be known only to authorized users of the particular account.
4. Type in the password and press the carriage return. The computer will print a message and wait for instructions.
5. Type SWITCH BASIC to call in the BASIC compiler.
6. Type in the word NEW and press the carriage return. The computer will print a message asking for the new program's name.

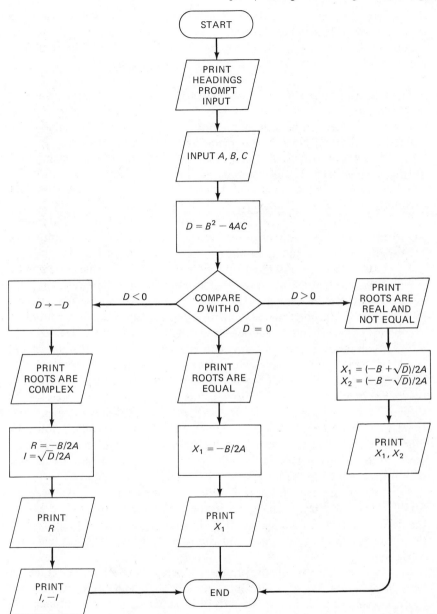

Figure 11-2.   Flowchart for the program of Example 11-20, which solves a quadratic equation.

7. Type in the program's name, which must have six or fewer characters in it, and press the carriage return. You cannot use a name that you have already used for a program currently stored in your account. The computer is now ready for you to type in the statements of your program.

The following rules should be remembered as you type in your program:

1. Every statement must have a number, including nonexecutable statements such as DIMENSION statements and DATA statements.

2. Only one statement may be put on a line, so press the carriage return after each statements. [Some systems, such as the RSTS/E system, allow for additional statements to be put in if they are separated by a backslash (\), but do not do this unless you are sure that your system allows it.]

3. The computer will rearrange the statements so that their numbers run from the smallest number to the largest. Leave gaps in your numbering so that additional statements can be inserted later if needed.

4. If you type in a statement that violates the BASIC rules, such as by having unpaired parentheses, two equals signs, etc., the computer will print an error message, informing you of the fact when you press the carriage return. Type in the corrected statement, using the same statement number. The computer will replace the defective statement with the corrected one.

5. If you need to delete a statement, type in the statement number and press the carriage return. This will replace the statement by a blank.

6. After typing all of the program, type whatever statements are required by your system to signal the computer that the program is complete. Some systems may require both a STOP statement and an END statement, and some may require only an END statement. These statements must be numbered, and of course must have higher numbers than any other statements in the program. It is a good idea to choose large numbers so that new parts can be added to the program later.

7. If you want to keep the program on file for later use, type the word SAVE. This will store the program in the BASIC language on the part of the disk file reserved for your account. If you do not have time to type in the entire program at one time, you can save part of a program and then type the rest of it later.

8. If you want to execute the program, type the word RUN. The compiler will translate the program into machine language, and the computer will begin execution. If there are any INPUT statements, the computer will stop at these statements and wait for you to type in values for the appropriate variables. If there are mistakes in the program that cause errors during execution, the computer will print out messages. If the mistakes prevent execution, the computer will stop, but in most cases, it will proceed to execute the program. However, the mistakes in the program will almost certainly mean that the results will be wrong. If there are no mistakes in the

program, the computer will complete execution of the program, print the output, and then tell the user that it is ready for something else.

9. If you make changes on a program that has already been saved, you can do this by typing the corrected version of each statement you want to change. You can then run the corrected version of the program by typing RUN, but in order to change the version of the program stored on the disk, you must type REPLACE.

10. If you want to erase a program from storage, you can do this by typing the word UNSAVE.

11. If you want to run a program that has previously been stored, type the word OLD and the name of the program and press the carriage return. This will copy the program into core memory from the disk. Then type the word RUN.

12. If you want to see a list of the statements making up a program, type the word LIST and the terminal will print out the entire program. If the program is not already in the core memory of the computer, you must first type OLD and the name of the program and press the carriage return.

13. If you want to save the machine-language version of the program, type the word COMPILE.

These 13 rules are correct for the RSTS/E operating system of the DEC 11/70 computer, and may be correct or nearly correct for other computers.

Example 11-23

Type the program in Example 11-21 into the computer under the name SIMEQ and use it to solve the simultaneous equations

$$\begin{aligned}
x_1 + x_2 + x_3 &= 9 \\
2x_1 + x_2 + 2x_3 &= 15 \\
3x_1 + 2x_2 + x_3 &= 16
\end{aligned}$$

Solution

The output sheet from the actual computer run is shown in Figure 11-3.

Problem 11-23

Type the program in Example 11-20 into the computer under the name QUADEQ and use it to solve the following two quadratic equations:
a. $3x^2 + 2x + 2 = 0$
b. $7x^2 - 5x + 1 = 0$

Example 11-24

Write and run a BASIC program to sort data points. That is, the program should take ordered pairs of numbers and rearrange them so that the value of the first variable, x, in any pair is smaller than the value of x in the next pair.

```
SIMEQ    04:49 PM        04-Mar-80
100 PRINT "THIS PROGRAM SOLVES A SET OF THREE SIMULTANEOUS"
110 PRINT "LINEAR INHOMOGENEOUS ALGEBRAIC EQUATIONS OF"
120 PRINT "THE FORM AX = C WHERE A IS A 3 BY 3 MATRIX AND"
130 PRINT "X AND C ARE COLUMN VECTORS.  X CONTAINS THE UNKNOWNS."
140 DIM A(3,3), C(3,1), X(3,1), B(3,3)
150 PRINT "TYPE IN THE ELEMENTS OF A BY ROWS."
160 MAT INPUT A
170 PRINT "TYPE IN THE ELEMENTS OF C."
180 MAT INPUT C
190 MAT B = INV(A)
200 MAT X = B*C
210 PRINT "THE ROOTS ARE:"
220 MAT PRINT X
230 END

Ready


RUN
SIMEQ    04:49 PM        04-Mar-80
THIS PROGRAM SOLVES A SET OF THREE SIMULTANEOUS
LINEAR INHOMOGENEOUS ALGEBRAIC EQUATIONS OF
THE FORM AX = C WHERE A IS A 3 BY 3 MATRIX AND
X AND C ARE COLUMN VECTORS.  X CONTAINS THE UNKNOWNS.
TYPE IN THE ELEMENTS OF A BY ROWS.
? 1,1,1,2,1,2,3,2,1
TYPE IN THE ELEMENTS OF C.
? 9,15,16
THE ROOTS ARE:
 2
 3
 4

Ready
```

Figure 11-3.   Program and Output for Example 11-23.

Solution

We use the "push-down" method, in which we begin by comparing the first value of x with the second. If the first is larger, we exchange the data points (exchanging both the x and y values to keep a given value of y with the same value of x). The second value of x, which is larger than the first, is compared with the third value, and exchanged with it if larger. This process is continued throughout the list, which puts the largest value of x in the last place in the list. The process is then repeated beginning at the first of the list but stopping one point before the end of the list, which will leave the second largest value of x in the next-to-last position. The process is then repeated a third time, stopping two points before the end of the list, putting the third largest value of x in the third place from the end, and so on.

We require two loops, one inside the other, to accomplish this. The program is shown in Figure 11-4. It is written to handle a maximum of 100 data points, but can easily be modified to handle more than this by changing the DIMENSION statement. The output from a run with 10 contrived data points is also shown.   •

Running a Program in the Batch Processing Mode.   If you are going to enter your program through a card reader rather than through a remote terminal,

```
SORT     04:37 PM        04-Mar-80
100 PRINT "THIS PROGRAM SORTS DATA POINTS CONSISTING OF "
110 PRINT "ORDERED PAIRS OF NUMBERS, SO THAT THE DATA"
120 PRINT "POINTS ARE PUT IN ORDER OF ASCENDING VALUE"
130 PRINT "OF THE FIRST VARIABLE."
140 DIM X(100), Y(100)
150 PRINT "TYPE IN THE NUMBER OF DATA POINTS."
160 INPUT N
170 FOR I = 1 TO N STEP 1
180 PRINT "TYPE IN X , Y FOR POINT NUMBER "I,"(SEPARATE BY COMMA)."
190 INPUT X(I), Y(I)
200 NEXT I
210 FOR I = 1 TO N-1 STEP 1
220 FOR J = 1 TO N-I STEP 1
230 IF X(J) <= X(J+1) THEN 300
240 T = X(J)
250 S = Y(J)
260 X(J) = X(J+1)
270 Y(J) = Y(J+1)
280 X(J+1) = T
290 Y(J+1) = S
300 NEXT J
310 NEXT I
340 FOR I = 1 TO N STEP 1
350 PRINT "X("I") = "X(I); "Y("I") = "Y(I)
360 NEXT I
400 END

Ready
```

```
RUN
SORT     04:38 PM        04-Mar-80
THIS PROGRAM SORTS DATA POINTS CONSISTING OF
ORDERED PAIRS OF NUMBERS, SO THAT THE DATA
POINTS ARE PUT IN ORDER OF ASCENDING VALUE
OF THE FIRST VARIABLE.
TYPE IN THE NUMBER OF DATA POINTS.
? 10
TYPE IN X , Y FOR POINT NUMBER  1        (SEPARATE BY COMMA).
? 10,2
TYPE IN X , Y FOR POINT NUMBER  2        (SEPARATE BY COMMA).
? 9,3
TYPE IN X , Y FOR POINT NUMBER  3        (SEPARATE BY COMMA).
? 6,1
TYPE IN X , Y FOR POINT NUMBER  4        (SEPARATE BY COMMA).
? 8,4
TYPE IN X , Y FOR POINT NUMBER  5        (SEPARATE BY COMMA).
? 7,5
TYPE IN X , Y FOR POINT NUMBER  6        (SEPARATE BY COMMA).
? 1,2
TYPE IN X , Y FOR POINT NUMBER  7        (SEPARATE BY COMMA).
? 2,1
TYPE IN X , Y FOR POINT NUMBER  8        (SEPARATE BY COMMA).
? 5,3
TYPE IN X , Y FOR POINT NUMBER  9        (SEPARATE BY COMMA).
? 4,7
TYPE IN X , Y FOR POINT NUMBER  10       (SEPARATE BY COMMA).
? 3,3
X( 1 ) =    1 Y( 1 ) =    2
X( 2 ) =    2 Y( 2 ) =    1
X( 3 ) =    3 Y( 3 ) =    3
X( 4 ) =    4 Y( 4 ) =    7
X( 5 ) =    5 Y( 5 ) =    3
X( 6 ) =    6 Y( 6 ) =    1
X( 7 ) =    7 Y( 7 ) =    5
X( 8 ) =    8 Y( 8 ) =    4
X( 9 ) =    9 Y( 9 ) =    3
X( 10 ) =   10 Y( 10 ) =   2

Ready
```

Figure 11-4.   Program and Output for Example 11-24.

the program itself is written in the same way, except that the INPUT statement cannot be used. The READ and DATA statements are used for the input of values for variables.

The program is typed on computer cards, using a key-punch machine. Each statement is typed on a separate card, and if you make a mistake, you replace the card with a corrected card. The key-punch machine has the capability of copying a card or a part of a card, so you can copy the correct part and retype the rest. As you type the cards, you will not receive any error messages, since the key-punch machine does not examine your statements. It is a good idea to proofread your cards before running the program.

You must prepare a few control cards. There will likely to be a card to specify the user's account number, the maximum amount of computer time to be used, etc. Next will come a card with a password, and some cards to tell the computer to prepare the BASIC compiler and whether to run the program after compilation. These control cards come at the first of the program, and there will also be some control cards at the end, to release the computer for the next user's program.

After the deck of cards is prepared, you can feed it into the card reader, if your computer center permits you to do this yourself. Some centers require you to submit your deck to be run by a computer operator, at least during the busy time of the day. When the program is fed in, the computer will print out error messages if any are required, or will execute the program if the program is correct. The output will probably come from the line printer.

Section 11-10. SOME OTHER FEATURES OF THE BASIC LANGUAGE

Remarks. Sometimes it is convenient to have messages in your program which the computer will ignore. The purpose of these might be to remind you of the function of different parts of the program, or to tell you what variables need to have values read in. The REMARK statement is used for this purpose. Its form is

REMARK message

The work REMARK can be abbreviated to REM. For example, you might write the statement

150 REM THIS PART OF THE PROGRAM READS IN THE DATA

When the program is executed, the computer will simply ignore this statement, but it will be included in a list of the program.

Some BASIC systems allow remarks to be appended to other statements by use of a symbol that separates the remark from the executable part of the statement. Some systems use a slash followed by an asterisk to separate the remark, and others use an exclamation mark. An example might be

$$520 \ X = -B/(2*A) \qquad /* \ X \ IS \ THE \ DOUBLE \ ROOT$$

or

$$520 \ X = -B/(2*A) \qquad ! \ X \ IS \ THE \ DOUBLE \ ROOT$$

Programmer-Defined Functions and Subroutines. In Section 11-4, we discussed predefined functions, which are provided as part of the compiler program. You can also define additional functions. The form of the defining statement is

$$110 \ DEF \ FNA(X) = expression \qquad (11.50)$$

In this statement, X is the independent variable for the function. It is possible to have a constant function of X, so that the expression does not contain X. The letter A in the function name can be replaced by any of the other letters of the alphabet, so that in a single program it is possible to have 26 different functions. The defining statement in Eq. (11.50) can be placed anywhere in the program, since it is not executed until it is called for elsewhere in the program.

The function is used by including the name of the function in a BASIC statement in much the same way as a LOG function, etc.

Example 11-25

Write a program that will produce a table of values of the function $y = x\ln(x) - x$ for integral values of x from 1 to 100.

Solution

```
100 PRINT "THIS PROGRAM GENERATES VALUES FOR THE"
110 PRINT "FUNCTION Y = XLN(X) − X"
120 DIM (Y100), X(100)
130 DEF FNA(Z) = Z*LOG(Z) − Z
140 FOR I = 1 TO 100 STEP 1
150 X(I) = I
160 Y(I) = FNA(X(I))
170 PRINT "IF X = "X(I); "Y(X) = "Y(I)
180 NEXT I
190 END
```

Problem 11-24

Write a program to make a table of values of the function $y = e^x(e^x - 1)^{-2}$.

If you have a function that cannot be expressed by one BASIC statement, you must use an internal subroutine, which is a self-contained program.

Using such a subroutine is referred to as "calling" the subroutine, and is done with a statement like

$$110 \text{ GOSUB } s \qquad\qquad (11.51)$$

where s represents the first statement number in the subroutine. During execution of the main program, when the computer reaches the statement in Eq. (11.51), it branches to the first statement of the subroutine, executes the statements of the subroutine, and then returns to the next statement following the GOSUB statement. The last statement in the subroutine must be the RETURN statement.

Example 11-26

Write a subroutine to evaluate $n! = n$ factorial $= n(n - 1)(n - 2) \cdots (2)(1)$.

Solution

```
700 REM THIS SUBROUTINE CALCULATES N FACTORIAL
710 F = 1
720 IF N < = 1, THEN 760
730 FOR J = 2 TO N STEP 1
740 F = F*J
750 NEXT J
760 RETURN
```
●

This subroutine could be called only in a part of the program where N has already been given a value. The following might be part of a program that uses this subroutine:

```
350 PRINT "WHAT IS THE VALUE OF N?"
360 INPUT N
370 GOSUB 710
380 Y = 2↑N/F
390 more statements
```

Notice that the variable F which has been given a value by the subroutine is used after the subroutine has done its work. Statement 380 will calculate $2^n/n!$.

Problem 11-25

Write a subroutine that will find the mean of the elements of a one-dimensional array with N elements. Write the statements that would call the subroutine and use the result as the value for a variable X. ●

The TAB Function. This function is used to place items of output at specified places on the line of output. The form of the statement is

$$110 \text{ PRINT TAB(exp); list} \qquad\qquad (11.52)$$

where exp stands for any BASIC expression. The computer will compute the value of the expression, truncate it (drop the fractional part to make an integer), and take the absolute value (magnitude) to obtain a positive integer. The output information in the list is then printed beginning in the column whose number is the positive integer just mentioned. For example, if you want to print the value of the variable X starting in column 30, you could use the statement

720 PRINT TAB(30); X

The TAB function can be used to make approximate graphs of functions. Choose a mark, such as a plus sign or an asterisk, and have the computer indicate a curve by printing a number of these.

Example 11-27

Use the TAB function to draw a plot of the cosine function from 0 radians to 2π radians.

Solution

We divide the interval into 20 subintervals, with a spacing equal to $\pi/10 = 0.314159\ldots$. Each such increment will correspond to one line on the terminal, and the dependent variable will be put into the TAB function

```
PLOTC    04:53 PM        04-Mar-80
100 PRINT "PLOT OF THE COSINE FUNCTION."
110 FOR X = 0 TO 6.28318 STEP 0.314159
120 PRINT TAB(30+25*COS(X));"+"
130 NEXT X
140 END

Ready

RUN
PLOTC    04:53 PM        04-Mar-80
PLOT OF THE COSINE FUNCTION.
                                                    +
                                                  +
                                              +
                                          +
                                      +
                                  +
                              +
                          +
                    +
                +
            +
        +
      +
        +
            +
                +
                      +
                          +
                              +
                                  +
                                      +
                                          +
                                              +
                                                +
                                                  +

Ready
```

Figure 11-5. The program and output for Example 11-28.

so that the graph will have its abscissa pointing downward and its ordinate to the right, 90 degrees from the usual position.

```
100 PRINT "PLOT OF THE COSINE FUNCTION."
110 FOR X = 0 TO 6.28318 STEP 0.314159
120 PRINT TAB(30 + 25*COS(X));"+"
130 NEXT X
140 END
```

The output from this program is shown in Figure 11-5. Notice that in order to spread the graph out, we have multiplied the cosine function by .25 so that the graph will occupy 50 spaces of the output lines, and that we have kept the graph on the page by adding 30 to the cosine. The graph thus ranges from the 5th column to the 55th.

●

Section 11-11.   THE FORTRAN LANGUAGE

At the present time (1980), the most commonly used computer language among practicing physical chemists is FORTRAN. BASIC is a simplified version of FORTRAN. Most arithmetic assignment statements are almost identical in the two languages. The same symbols are used for addition, subtraction, multiplication, and division. For exponentiation, the double asterisk is used in FORTRAN, but this is also used as an option in some BASIC systems. Thus, to assign the value of x^3 to y, the FORTRAN statement is

$$Y = X**3$$

In BASIC, the word LET is optional in arithmetic assignment statements. It is not used in FORTRAN.

Another difference is in the way that loops are written. Instead of the FOR and NEXT statements, FORTRAN uses a DO statement, and the loops are called DO *loops*. The loop has the form

```
DO 300      I = 1, 100, 1
executable statements
300 last executable statement in the loop
```

The three numbers following the equals sign in the DO statement are first the initial value of the variable I, next the final value of I, and then the increment in I for each passage through the loop. These numbers must be integers.

In FORTRAN, it is not necessary to number each statement, but it is necessary to number the last statement in a DO loop and every statement to which the computer is instructed to branch. The statements will be executed in the order entered, except for loops and branches.

Example 11-28

Convert the loop in Example 11-12 from BASIC to FORTRAN.

Solution

```
      DIMENSION A(3), B(3)
      C = 0.
      DO 130 I = 1, 3, 1
130   C = C + A(I)*B(I)
```

FORTRAN programs are usually written for batch processing and are thus typed on cards. We list a number of differences:

1. FORTRAN statements must begin in column 7 of the computer card or somewhere to the right of this column. Statement numbers can be placed in the first five columns, and the sixth column is reserved for a digit identifying a "continuation card," which contains part of an expression that was too long to fit on one card.

2. A comment card is identified by the letter C in the first column. The computer will ignore the information on such a card, which is analogous to the information in a REMARK statement in BASIC.

3. FORTRAN makes provision for variables that are allowed to take on integral values only. The names of such variables must begin with one of the letters of the alphabet from I through N, and the names of variables that can take on nonintegral values ("real" variables) must begin with one of the other letters of the alphabet. Some FORTRAN systems will not permit real and integer variables to be mixed in arithmetic expressions.

4. The predefined functions are much the same as in BASIC, but some of the names are different. For example, the name for the square root function is SQRT, and the name for the natural logarithm function is ALOG.

5. Variable names are not restricted to one letter or one letter and one or two digits. Most systems permit up to five or six characters, the first of which must be a letter of the alphabet.

6. The GO TO statement is the same, but the IF statement of FORTRAN is slightly different from the IF–THEN statement of BASIC. Instead of the form of Eq. (11-16), the IF statement of FORTRAN has the form

$$\text{IF (exp1 rel exp2) GO TO s} \qquad (11.53)$$

where the quantity in parentheses is a logical expression just as in Eq. (11.16), and s is the statement number to which the computer should branch if the logic expression is true. The relational operators are written differently from those of BASIC, and are shown in Table 11-3.

An example of such a statement might be

$$\text{IF (X.LE.2.E-10) GO TO 320}$$

Table 11-3. Relational Operators in FORTRAN

Relation	Algebraic Symbol	FORTRAN Symbol
Is equal to	$=$	.EQ.
Is greater than	$>$	.GT.
Is less than	$<$	.LT.
Is greater than or equal to	$\geq$	.GE.
Is less than or equal to	$\leq$	.LE.
Is not equal to	$\neq$	.NE.

Notice that there is a decimal point following the 2. In FORTRAN, any number representing the value of a real variable must have a decimal point in it, and any number representing the value of an integer variable is not permitted to have a decimal point in it.

7. Most FORTRAN compilers do not have the matrix operations built into them, which means that you have to write your own subroutines to multiply, add and invert matrices.

There are also differences in input and output procedures. FORTRAN programs are usually run in the batch-processing mode, so a statement analogous to the INPUT statement is usually not used. The READ statement of FORTRAN is similar to that of BASIC except that a FORMAT declaration must be used (nonexecutable statements are called *declarations* in FORTRAN). A FORMAT declaration tells the computer exactly where on a data card to look for a given number. A set of columns on a data card is called a *field* and it is necessary to specify whether the field contains an integer (an I field), a real number without a power of 10 multiplying it (an F field), or a real number with a power of 10 multiplying it (an E field). A field of 10 columns with four columns after the decimal point would be an F10.4 field or an E10.4 field. FORMAT declarations must also be used on output, and are used to contain any messages to be printed out.

The required use of FORMAT declarations make FORTRAN more difficult for the beginner to learn but also makes it more versatile. However, some BASIC systems provide for the use of FORMAT statements which are similar to those of FORTRAN.

If you are interested in learning to write programs in FORTRAN, you can look at any one of the manuals designed to teach you the language. The book by Organick, listed at the end of the chapter, has been one of the standard texts for a number of years.

ADDITIONAL PROBLEMS

11.26

Each of the parts of this problem consists of a formula and a BASIC statement that is supposed to calculate a value for the dependent variable in that formula. Find and correct any mistakes in the BASIC statements.

Formula	BASIC Statement
a. $z = \dfrac{xy}{ac}$	Z = X*Y/A*C
b. $y = mx + b$	Y = MX + B
c. $y = ax^2 + bx + c$	Y = A*X↑2 + B*X + C
d. $d = b^2 - 4ac$	D = B↑2 − 4A*C
e. $z = (ax^2 + b)^n/an$	Z = (A*X↑2 + B)*N/A/N
f. $y = e^{e^x}$	Y = EXP(EXP(X))
g. $y = \log_{10}(\cos(x))$	Y = LOG(COS(X))
h. $y = (a + x)^{-1}$	Y = 1/(A + X)

11.27

Write BASIC statements that will calculate a value for each of the dependent variables in the formulas.

a. $y = (A + B)/(C + D)$
b. $z = (17y^2 + 3x^2)^{1/2}/14ab$
c. $f = (a + b)^{3/2}/(c + d)$
d. $s = (a_1 + a_2 + a_3 + a_4 + \cdots + a_n)/n!$
e. $g = (b\sin^2(ax) + a\sin^2(bx))^{1/2}$
f. $h = (ax^2 + bx + c)^{-1}$
g. $s = \dfrac{1}{n-1}\left[\sum_{i=1}^{n}(x_i - m)^2\right]^{1/2}$

11.28

Write a BASIC program that will convert a list of distance measurements in meters to miles, feet, and inches. Include input and output statements, and label your output.

11.29

Write a complete BASIC program that will make a properly labeled table of values of the function $y(x) = \ln(x) + \cos(x)$ for values of x from 0 to 2π, at intervals of $\pi/100$.

11.30

Modify the program from Problem 11.29 to produce an approximate plot of the function.

11.31

Write a complete BASIC program to calculate the pH of an aqueous solution of a weak monobasic acid. Neglect the hydrogen ions from the water so that you can use the formula

$$\frac{(H^+)(A^-)}{(HA)} = K_a$$

where (H^+) is the hydrogen-ion concentration, (A^-) the acid-anion concentration, and (HA) the concentration of undissociated acid. This formula assumes that all activity coefficients are equal to unity. Allow for the value of K_a, the acid dissociation constant, and for the value of $(HA) + (A^-)$, the gross concentration of acid, to be read in. Since you will have a quadratic equation to solve, write the program to test the discriminant $b^2 - 4ac$ (where the quadratic equation is written $ax^2 + bx + c = 0$) to make sure the roots are real, and to reject any negative roots. Label your output, which should include (H^+) as well as pH, which is equal to $-\log_{10}(H^+)$ with our assumptions.

11.32

Modify the program that you wrote for Problem 11.31 so that it will produce a table of values of pH for various values of gross acid concentration.

11.33

Write a program that will carry out for you all the calculations for a gravimetric determination of sodium chloride in an unknown mixture. Write the program so that it will carry out the calculation for any number of repetitions of the determination on the same mixture. For each sample, arrange for the mass of the dried sample and its container and the mass of the empty container to be read in, as well as the mass of the precipitated silver chloride plus its container and the mass of the empty container. Have the program calculate the percentage of each sample that was chloride ion and the percentage that was sodium chloride, and have it calculate the mean and the standard deviation. Label all output.

11.34

Write a program that will prepare an approximate graph of the pressure of a nonideal gas described by the van der Waals equation of state

$$\left(P + \frac{a}{V^2}\right)(V - b) = RT$$

where R is the ideal gas constant and a and b are constants whose values are to be read into the computer for the particular gas. Arrange the program so that the user can choose between a graph of the pressure as a function of the volume of one mole, V, at fixed temperature, T, or a graph of the pressure as a function of T at fixed V. Give the user the option of receiving a table of values of the pressure as well as a graph.

ADDITIONAL READING

Leonard Soltzberg, Arvind A. Shah, John C. Saber, and Edgar T. Canty, *BASIC and Chemistry*, Houghton Mifflin Company, Boston, 1975.

> *This book is an introduction to the BASIC language and a manual for its use in chemistry problems. It contains a number of ready-to-use programs. It is available as a fairly inexpensive paperback book.*

C. Joseph Sass, *BASIC Programming and Applications*, Allyn and Bacon, Boston, 1976.

> *This is also an introduction to the BASIC language, but the applications contained are business-oriented. It is available as a paperback.*

Nesa L'abbe Wu, *BASIC; The Time-Sharing Language*, William C. Brown Company, Dubuque, Iowa, 1975.

This is a fairly complete introduction to BASIC, with business-oriented applications. It is available as a paperback.

T. R. Dickson, *The Computer and Chemistry*, W. H. Freeman and Company, San Francisco, 1968.

This is a fairly complete introduction to programming in the FORTRAN language, with applications to chemistry problems.

DeLos F. DeTar, *Computer Programs for Chemistry*, Vols. 1 and 2, W. A. Benjamin, New York, 1968.

These two volumes contain a large number of ready-to-use programs for problems in chemistry, including analysis of NMR spectra, analysis of chemical reaction rate data, etc. The programs are written in FORTRAN.

E. I. Organick, *A FORTRAN IV Primer*, Addison-Wesley Publishing Co., Reading, Mass., 1966.

This book is a systematic and thorough survey of programming in FORTRAN, and probably contains everything you might need to know about this language.

Values of Physical Constants[1]

Avogadro's number:
$$N_0 = 6.0220 \times 10^{23} \text{ mol}^{-1}$$

The ideal gas constant:
$$R = 8.3144 \text{ J K}^{-1} \text{ mol}^{-1} = 0.082056 \text{ liter atm K}^{-1} \text{ mol}^{-1}$$
$$= 1.9872 \text{ cal K}^{-1} \text{ mol}^{-1}$$

The magnitude of an electron's charge:
$$e = 1.60219 \times 10^{-19} \text{ C} = 4.8032 \times 10^{-10} \text{ electrostatic unit}$$

Planck's constant:
$$h = 6.6262 \times 10^{-34} \text{ J s} = 6.6262 \times 10^{-27} \text{ erg s}$$

Boltzmann's constant:
$$k = 1.3807 \times 10^{-23} \text{ J K}^{-1} = 1.3807 \times 10^{-16} \text{ erg K}^{-1}$$

The mass of an electron:
$$m_e = 9.1095 \times 10^{-31} \text{ kg} = 9.1095 \times 10^{-28} \text{ g}$$

The mass of a proton:
$$m_p = 1.67265 \times 10^{-27} \text{ kg} = 1.67265 \times 10^{-24} \text{ g}$$

The mass of a neutron:
$$m_n = 1.674954 \times 10^{-27} \text{ kg} = 1.67495 \times 10^{-24} \text{ g}$$

The speed of light:
$$c = 2.99792 \times 10^8 \text{ m s}^{-1} = 2.99792 \times 10^{10} \text{ cm s}^{-1}$$

The acceleration due to gravity near the earth's surface:
$$g = 9.80 \text{ m s}^{-2} = 980 \text{ cm s}^{-2}$$

[1] Vojtech Fried, Hendrik F. Hameka, and Uldis Blukis, *Physical Chemistry*, Macmillan Publishing Co., New York, 1977.

The gravitational constant:

$$G = 6.67 \times 10^{-11} \text{ m}^3 \text{ s}^{-2} \text{ kg}^{-1}$$

The permittivity of a vacuum:

$$\varepsilon_0 = 8.85419 \times 10^{-12} \text{ C}^2 \text{ N}^{-1} \text{ m}^{-2}$$

The permeability of a vacuum:

$$\mu_0 = 4\pi \times 10^{-7} \text{ NC}^{-2}\text{s}^2$$

Some Conversion Factors

1 pound = 0.4535924 kg

1 inch = 2.54 cm (exactly)

1 calorie = 4.184 J (exactly)

1 electron volt = 1.60219×10^{-19} J

1 erg = 10^{-7} J

1 atm = 760 torr

= 101,325 N m^{-2} = 101,325 pascal (Pa)

1 atomic mass unit (u) = 1.66057×10^{-27} kg

1 horsepower = 745.700 watt = 745.700 J s^{-1}

$$F = \frac{I}{mol} v$$

A Short Table of Derivatives[1]

In the following list, a, b, and c are constants, and e is the base of natural logarithms.

1. $\dfrac{d}{dx}(au) = a\dfrac{du}{dx}$

2. $\dfrac{d}{dx}(uv) = u\dfrac{dv}{dx} + v\dfrac{du}{dx}$

3. $\dfrac{d}{dx}(uvw) = uv\dfrac{dw}{dx} + uw\dfrac{dv}{dx} + vw\dfrac{du}{dx}$

4. $\dfrac{d(x^n)}{dx} = nx^{n-1}$

5. $\dfrac{d}{dx}\left(\dfrac{u}{v}\right) = \dfrac{1}{v}\dfrac{du}{dx} - \dfrac{u}{v^2}\dfrac{dv}{dx} = \dfrac{1}{v^2}\left(v\dfrac{du}{dx} - u\dfrac{dv}{dx}\right)$

6. $\dfrac{d}{dx}f(u) = \dfrac{df}{du}\dfrac{du}{dx}$ where f is some differentiable function of u and u is some differentiable function of x (the chain rule)

7. $\dfrac{d^2}{dx^2}f(u) = \dfrac{df}{du}\dfrac{d^2u}{dx^2} + \dfrac{d^2f}{du^2}\left(\dfrac{du}{dx}\right)^2$

8. $\dfrac{d}{dx}\sin(ax) = a\cos(ax)$

9. $\dfrac{d}{dx}\cos(ax) = -a\sin(ax)$

10. $\dfrac{d}{dx}\tan(ax) = a\sec^2(ax)$

---

[1] These formulas, and the other material in Appendices 3 through 7, are from H. B. Dwight, *Tables of Integrals and other Mathematical Data*, 4th ed., Macmillan Publishing Co., New York, 1961.

APPENDIX

11. $\dfrac{d}{dx}\operatorname{ctn}(ax) = -a\csc^2(ax)$

12. $\dfrac{d}{dx}\sec(ax) = a\sec(ax)\tan(ax)$

13. $\dfrac{d}{dx}\csc(ax) = -a\csc(ax)\operatorname{ctn}(ax)$

14. $\dfrac{d}{dx}\sin^{-1}\left(\dfrac{x}{a}\right) = \dfrac{1}{\sqrt{a^2 - x^2}}$ if x/a is in the first or fourth quadrant

$\qquad\qquad\qquad = \dfrac{-1}{\sqrt{a^2 - x^2}}$ if x/a is in the second or third quadrant

15. $\dfrac{d}{dx}\cos^{-1}\left(\dfrac{x}{a}\right) = \dfrac{-1}{\sqrt{a^2 - x^2}}$ if x/a is in the first or second quadrant

$\qquad\qquad\qquad = \dfrac{1}{\sqrt{a^2 - x^2}}$ if x/a is in the third or fourth quadrant

16. $\dfrac{d}{dx}\tan^{-1}\left(\dfrac{x}{a}\right) = \dfrac{a}{a^2 + x^2}$

17. $\dfrac{d}{dx}\operatorname{ctn}^{-1}\left(\dfrac{x}{a}\right) = \dfrac{-a}{a^2 + x^2}$

18. $\dfrac{d}{dx}e^{ax} = ae^{ax}$

19. $\dfrac{d}{dx}a^x = a^x\ln(a)$

20. $\dfrac{d}{dx}a^{cx} = ca^{cx}\ln(a)$

21. $\dfrac{d}{dx}u^y = yu^{y-1}\dfrac{du}{dx} + u^y\ln(u)\dfrac{dy}{dx}$

22. $\dfrac{d}{dx}x^x = x^x[1 + \ln(x)]$

23. $\dfrac{d}{dx}\ln(ax) = \dfrac{1}{x}$

24. $\dfrac{d}{dx}\log_a(x) = \dfrac{\log_a(e)}{x}$

25. $\dfrac{d}{dq}\displaystyle\int_p^q f(x)\,dx = f(q)$ if p is independent of q

26. $\dfrac{d}{dp}\displaystyle\int_p^q f(x)\,dx = -f(p)$ if q is independent of p

APPENDIX THREE

A Short Table of Indefinite Integrals

In the following, an arbitrary constant of integration is to be added to each equation. a, b, c, g, and n are constants.

1. $\displaystyle\int dx = x$

2. $\displaystyle\int x\,dx = \frac{x^2}{2}$

3. $\displaystyle\int \frac{1}{x}\,dx = \ln(|x|)$ do not carry this integration from negative to positive values of x

4. $\displaystyle\int x^n\,dx = \frac{x^{n+1}}{n+1}$ where $n \neq -1$

5. $\displaystyle\int (a + bx)^n dx = \frac{(a + bx)^{n+1}}{b(n+1)}$

6. $\displaystyle\int \frac{1}{(a + bx)}\,dx = \frac{1}{b}\ln(|a + bx|)$

7. $\displaystyle\int \frac{1}{(a + bx)^n}\,dx = \frac{-1}{(n-1)b(a + bx)^{n-1}}$

8. $\displaystyle\int \frac{x}{a + bx}\,dx = \frac{1}{b^2}\left[(a + bx) - a\ln(|a + bx|)\right]$

9. $\displaystyle\int \frac{a + bx}{c + gx}\,dx = \frac{bx}{g} + \frac{ag - bc}{g^2}\ln(|c + gx|)$

10. $\displaystyle\int \frac{1}{(a + bx)(c + gx)}\,dx = \frac{1}{ag - bc}\ln\left(\left|\frac{c + gx}{a + bx}\right|\right)$

11. $\displaystyle\int \frac{1}{a^2 + x^2}\,dx = \frac{1}{a}\tan^{-1}\left(\frac{x}{a}\right)$

12. $\int \dfrac{x}{(a^2 + x^2)^2}\,dx = \dfrac{-1}{2(a^2 + x^2)}$

13. $\int \dfrac{x}{(a^2 + x^2)}\,dx = \dfrac{1}{2}\ln(a^2 + x^2)$

14. $\int \dfrac{1}{a^2 - b^2 x^2}\,dx = \dfrac{1}{2ab}\ln\left(\left|\dfrac{a + bx}{a - bx}\right|\right)$

15. $\int \dfrac{x}{a^2 - x^2}\,dx = -\dfrac{1}{2}\ln(|a^2 - x^2|)$

16. $\int \dfrac{x^{1/2}}{a^2 + b^2 x}\,dx = \dfrac{2x^{1/2}}{b^2} - \dfrac{2a}{b^3}\tan^{-1}\left(\dfrac{bx^{1/2}}{a}\right)$

17. $\int \dfrac{1}{(a + bx)^{p/2}}\,dx = \dfrac{-2}{(p - 2)b(a + bx)^{(p - 2)/2}}$

18. $\int \dfrac{1}{(x^2 + a^2)^{1/2}}\,dx = \ln(x + (x^2 + a^2)^{1/2})$

19. $\int \dfrac{x}{(x^2 + a^2)^{1/2}}\,dx = (x^2 + a^2)^{1/2}$

20. $\int \dfrac{1}{(x^2 - a^2)^{1/2}}\,dx = \ln(x + (x^2 - a^2)^{1/2})$

21. $\int \dfrac{x}{(x^2 - a^2)^{1/2}}\,dx = (x^2 - a^2)^{1/2}$

22. $\int \sin(ax)\,dx = -\dfrac{1}{a}\cos(ax)$

23. $\int \sin(a + bx)\,dx = -\dfrac{1}{b}\cos(a + bx)$

24. $\int x\sin(x)\,dx = \sin(x) - x\cos(x)$

25. $\int x^2 \sin(x)\,dx = 2x\sin(x) - (x^2 - 2)\cos(x)$

26. $\int \sin^2(x)\,dx = \dfrac{x}{2} - \dfrac{\sin(2x)}{4} = \dfrac{x}{2} - \dfrac{\sin(x)\cos(x)}{2}$

27. $\int x\sin^2(x)\,dx = \dfrac{x^2}{4} - \dfrac{x\sin(2x)}{4} - \dfrac{\cos(2x)}{8}$

28. $\int \dfrac{1}{1 + \sin(x)}\,dx = -\tan\left(\dfrac{\pi}{4} - \dfrac{x}{2}\right)$

29. $\int \cos(ax)\,dx = \dfrac{1}{a}\sin(ax)$

30. $\int \cos(a + bx)\,dx = \dfrac{1}{b}\sin(a + bx)$

31. $\int x\cos(x)\,dx = \cos(x) + x\sin(x)$

32. $\int x^2 \cos(x)\,dx = 2x\cos(x) + (x^2 - 2)\sin(x)$

33. $\int \cos^2(x)\,dx = \dfrac{x}{2} + \dfrac{\sin(2x)}{4} = \dfrac{x}{2} + \dfrac{\sin(x)\cos(x)}{2}$

34. $\int x\cos^2(x)\,dx = \dfrac{x^2}{4} + \dfrac{x\sin(2x)}{4} + \dfrac{\cos(2x)}{8}$

35. $\int \dfrac{1}{1 + \cos(x)}\,dx = \tan\left(\dfrac{x}{2}\right)$

36. $\int \sin(x)\cos(x)\,dx = \dfrac{\sin^2(x)}{2}$

37. $\int \sin^2(x)\cos^2(x)\,dx = \dfrac{1}{8}\left[x - \dfrac{\sin(4x)}{4}\right]$

38. $\int \sin^{-1}\left(\dfrac{x}{a}\right)dx = x\sin^{-1}\left(\dfrac{x}{a}\right) + (a^2 - x^2)^{1/2}$

39. $\int \left[\sin^{-1}\left(\dfrac{x}{a}\right)\right]^2 dx = x\left[\sin^{-1}\left(\dfrac{x}{a}\right)\right]^2 - 2x + 2(a^2 - x^2)^{1/2}\sin^{-1}\left(\dfrac{x}{a}\right)$

40. $\int \cos^{-1}\left(\dfrac{x}{a}\right)dx = x\cos^{-1}\left(\dfrac{x}{a}\right) - (a^2 - x^2)^{1/2}$

41. $\int \left[\cos^{-1}\left(\dfrac{x}{a}\right)\right]^2 dx = x\left[\cos^{-1}\left(\dfrac{x}{a}\right)\right]^2 - 2x - 2(a^2 - x^2)^{1/2}\cos^{-1}\left(\dfrac{x}{a}\right)$

42. $\int \tan^{-1}\left(\dfrac{x}{a}\right)dx = x\tan^{-1}\left(\dfrac{x}{a}\right) - \dfrac{a}{2}\ln(a^2 + x^2)$

43. $\int x\tan^{-1}\left(\dfrac{x}{a}\right)dx = \dfrac{1}{2}(x^2 + a^2)\tan^{-1}\left(\dfrac{x}{a}\right) - \dfrac{ax}{2}$

44. $\int e^{ax}\,dx = \dfrac{1}{a}e^{ax}$

45. $\int a^x\,dx = \dfrac{a^x}{\ln(a)}$

46. $\int xe^{ax}\,dx = e^{ax}\left(\dfrac{x}{a} - \dfrac{1}{a^2}\right)$

47. $\int x^2 e^{ax}\,dx = e^{ax}\left[\dfrac{x^2}{a} - \dfrac{2x}{a^2} + \dfrac{2}{a^3}\right]$

48. $\int e^{ax}\sin(x)\,dx = \dfrac{e^{ax}}{a^2 + 1}\left[a\sin(x) - \cos(x)\right]$

49. $\int e^{ax}\cos(x)\,dx = \dfrac{e^{ax}}{a^2 + 1}\left[a\cos(x) + \sin(x)\right]$

50. $\int e^{ax} \sin^2(x)\, dx = \dfrac{e^{ax}}{a^2+4}\left[a\sin^2(x) - 2\sin(x)\cos(x) + \dfrac{2}{a}\right]$

51. $\int \ln(ax)\, dx = x\ln(ax) - x$

52. $\int x\ln(x)\, dx = \dfrac{x^2}{2}\ln(x) - \dfrac{x^2}{4}$

53. $\int \dfrac{\ln(ax)}{x}\, dx = \dfrac{1}{2}\left[\ln(ax)\right]^2$

54. $\int \dfrac{1}{x\ln(x)}\, dx = \ln(|\ln(x)|)$

55. $\int \tan(ax)\, dx = \dfrac{1}{a}\ln(|\sec(ax)|)$

$$= -\dfrac{1}{a}\ln(|\cos(ax)|)$$

56. $\int \cot(ax)\, dx = \dfrac{1}{a}\ln(|\sin(ax)|)$

A Short Table of Definite Integrals

In the following list, a, b, m, n, p, and r, are constants.

1. $\displaystyle\int_0^\infty x^{n-1} e^{-x}\, dx = \int_0^1 \left[\ln\left(\frac{1}{x}\right)\right]^{-1} dx = \Gamma(n) \qquad (n > 0)$

The function $\Gamma(n)$ is called the gamma function. It has the following properties:

$$\Gamma(n+1) = n\Gamma(n) \qquad \text{for any } n > 0$$
$$\Gamma(n) = (n-1)! \qquad \text{for any integral value of } n > 0$$
$$\Gamma(n)\Gamma(1-n) = \frac{\pi}{\sin(n\pi)} \qquad \text{for } n \text{ not an integer}$$
$$\Gamma(\tfrac{1}{2}) = \sqrt{\pi}$$

2. $\displaystyle\int_0^\infty \frac{1}{1+x+x^2}\, dx = \frac{\pi}{3\sqrt{3}}$

3. $\displaystyle\int_0^\infty \frac{x^{p-1}}{(1-x)^p}\, dx = \frac{\pi}{\sin(p\pi)} \qquad (0 < p < 1)$

4. $\displaystyle\int_0^\infty \frac{x^{p-1}}{a+x}\, dx = \frac{\pi a^{p-1}}{\sin(p\pi)} \qquad (0 < p < 1)$

5. $\displaystyle\int_0^\infty \frac{x^p}{(1+ax)^2}\, dx = \frac{p\pi}{a^{p+1}\sin(p\pi)}$

6. $\displaystyle\int_0^\infty \frac{1}{1+x^p}\, dx = \frac{\pi}{p\sin(\pi/p)}$

7. $\displaystyle\int_0^{\pi/2} \sin^2(mx)\, dx = \int_0^{\pi/2} \cos^2(mx)\, dx = \frac{\pi}{4} \qquad (m = 1, 2, \ldots)$

8. $\displaystyle\int_0^\pi \sin^2(mx)\, dx = \int_0^\pi \cos^2(mx)\, dx = \frac{\pi}{2} \qquad (m = 1, 2, \ldots)$

9. $\int_0^{\pi/2} \tan^p(x)\, dx = \int_0^{\pi/2} \text{ctn}^p(x)\, dx = \dfrac{\pi}{2\cos(p\pi/2)}$ $\qquad (p^2 < 1)$

10. $\int_0^{\pi/2} \dfrac{x}{\tan(x)}\, dx = \dfrac{\pi}{2}\ln(2)$

11. $\int_0^{\pi/2} \sin^p(x)\cos^p(x)\, dx = \dfrac{\Gamma\left(\dfrac{p+1}{2}\right)\Gamma\left(\dfrac{q+1}{2}\right)}{2\Gamma\left(\dfrac{p+q}{2}+1\right)}$ $\qquad (p+1>0, q+1>0)$

12. $\int_0^{\pi} \sin(mx)\sin(nx)\, dx = \begin{cases} 0 & \text{if } m \neq n \\ \dfrac{\pi}{2} & \text{if } m = n \end{cases}$ $\qquad$ (*m, n* integers)

13. $\int_0^{\pi} \cos(mx)\sin(nx)\, dx = \begin{cases} 0 & \text{if } m \neq n \\ \dfrac{\pi}{2} & \text{if } m = n \end{cases}$ $\qquad$ (*m, n* integers)

14. $\int_0^{\pi} \sin(mx)\cos(nx)\, dx = \begin{cases} 0 & \text{if } m = n \\ 0 & \text{if } m \neq n \text{ and } m+n \text{ is even} \\ \dfrac{2m}{m^2 - n^2} & \text{if } m \neq n \text{ and } m+ \text{ is odd} \\ & (m, n \text{ integers}) \end{cases}$

15. $\int_0^{\infty} \sin\left(\dfrac{\pi x^2}{2}\right) dx = \int_0^{\infty} \cos\left(\dfrac{\pi x^2}{2}\right) dx = 1/2$

16. $\int_0^{\infty} \sin(x^p)\, dx = \Gamma\left(1 + \dfrac{1}{p}\right)\sin\left(\dfrac{\pi}{2p}\right)$ $\qquad (p > 1)$

17. $\int_0^{\infty} \cos(x^p)\, dx = \Gamma\left(1 + \dfrac{1}{p}\right)\cos\left(\dfrac{\pi}{2p}\right)$ $\qquad (p > 1)$

18. $\int_0^{\infty} \dfrac{\sin(mx)}{x}\, dx = \begin{cases} \dfrac{\pi}{2} & \text{if } m > 0 \\ 0 & \text{if } m = 0 \\ -\dfrac{\pi}{2} & \text{if } m < 0 \end{cases}$

19. $\int_0^{\infty} \dfrac{\sin(mx)}{x^p}\, dx = \dfrac{\pi m^{p-1}}{2\sin(p\pi/2)\Gamma(p)}$ $\qquad (0 < p < 2, m > 0)$

20. $\int_0^{\infty} e^{-ax}\, dx = \dfrac{1}{a}$ $\qquad (a > 0)$

21. $\int_0^{\infty} xe^{-ax}\, dx = \dfrac{1}{a^2}$ $\qquad (a > 0)$

22. $\displaystyle\int_0^\infty x^2 e^{-ax}\,dx = \frac{2}{a^3}$ $(a > 0)$

23. $\displaystyle\int_0^\infty x^{1/2} e^{-ax}\,dx = \frac{\sqrt{\pi}}{2a^{3/2}}$ $(a > 0)$

24. $\displaystyle\int_0^\infty e^{-r^2 x^2}\,dx = \frac{\sqrt{\pi}}{2r}$ $(r > 0)$

25. $\displaystyle\int_0^\infty x e^{-r^2 x^2}\,dx = \frac{1}{2r^2}$ $(r > 0)$

26. $\displaystyle\int_0^\infty x^2 e^{-r^2 x^2}\,dx = \frac{\sqrt{\pi}}{4r^3}$ $(r > 0)$

27. $\displaystyle\int_0^\infty x^{2n+1} e^{-r^2 x^2}\,dx = \frac{n!}{2r^{2n+2}}$ $(r > 0, n = 1, 2, \ldots)$

28. $\displaystyle\int_0^\infty x^{2n} e^{-r^2 x^2}\,dx = \frac{(1)(3)(5)\ldots(2n-1)}{2^{n+1} r^{2n+1}}\sqrt{\pi}$ $(r > 0, n = 1, 2, \ldots)$

29. $\displaystyle\int_0^\infty x^a e^{-(rx)^b}\,dx = \frac{1}{br^{a+1}}\Gamma\left(\frac{a+1}{b}\right)$ $(a + 1 > 0, r > 0, b > 0)$

30. $\displaystyle\int_0^\infty \frac{e^{-ax} - e^{-bx}}{x}\,dx = \ln\left(\frac{b}{a}\right)$

31. $\displaystyle\int_0^\infty e^{-ax}\sin(mx)\,dx = \frac{m}{a^2 + m^2}$ $(a > 0)$

32. $\displaystyle\int_0^\infty x e^{-ax}\sin(mx)\,dx = \frac{2am}{(a^2 + m^2)^2}$ $(a > 0)$

33. $\displaystyle\int_0^\infty x^{p-1} e^{-ax}\sin(mx)\,dx = \frac{\Gamma(p)\sin(p\theta)}{(a^2 + m^2)^{p/2}}$ $(a > 0, p > 0, m > 0)$

    where $\sin(\theta) = m/r$, $\cos(\theta) = a/r$, $r = (a^2 + m^2)^{1/2}$

34. $\displaystyle\int_0^\infty e^{-ax}\cos(mx)\,dx = \frac{a}{a^2 + m^2}$ $(a > 0)$

35. $\displaystyle\int_0^\infty x e^{-ax}\cos(mx)\,dx = \frac{a^2 - m^2}{(a^2 + m^2)^2}$ $(a > 0)$

36. $\displaystyle\int_0^\infty x^{p-1} e^{-ax}\cos(mx)\,dx = \frac{\Gamma(p)\cos(p\theta)}{(a^2 + m^2)^{p/2}}$ $(a > 0, p > 0)$

    where θ is the same as given in Eq. 33

37. $\displaystyle\int_0^\infty \frac{e^{-ax}}{x}\sin(mx)\,dx = \tan^{-1}\left(\frac{m}{a}\right)$ $(a > 0)$

38. $\displaystyle\int_0^\infty \frac{e^{-ax}}{x}[\cos(mx) - \cos(nx)]\,dx = \frac{1}{2}\ln\left(\frac{a^2 + n^2}{a^2 + m^2}\right)$ $(a > 0)$

39. $\displaystyle\int_0^\infty e^{-ax}\cos^2 mx)\,dx = \frac{a^2 + 2m^2}{a(a^2 + 4m^2)}$ $(a > 0)$

40. $\displaystyle\int_0^\infty e^{-ax}\sin^2(mx)\,dx = \frac{2m^2}{a(a^2 + 4m^2)}$ $(a > 0)$

41. $\displaystyle\int_0^1 \left[\ln\left(\frac{1}{x}\right)\right]^q dx = \Gamma(q + 1)$ $(q + 1 > 0)$

42. $\displaystyle\int_0^1 x^p \ln\left(\frac{1}{x}\right) dx = \frac{1}{(p + 1)^2}$ $(p + 1 > 0)$

43. $\displaystyle\int_0^1 x^p \left[\ln\left(\frac{1}{x}\right)\right]^q dx = \frac{\Gamma(q + 1)}{(p + 1)^{q+1}}$ $(p + 1 > 0, q + 1 > 0)$

44. $\displaystyle\int_0^1 \ln(1 - x)\,dx = -1$

45. $\displaystyle\int_0^1 x\ln(1 - x)\,dx = \frac{-3}{4}$

46. $\displaystyle\int_0^1 \ln(1 + x)\,dx = 2\ln(2) - 1$

47. $\displaystyle\int_0^\infty e^{-ax^2}\cos(kx)\,dx = \frac{\sqrt{\pi}}{2\sqrt{a}}\, e^{-k^2/(4a)}$

Some Mathematical Formulas and Identities

1. The arithmetic progression of the first order to n terms:

    $$a + (a + d) + (a + 2d) + \cdots + [a + (n - 1)d] = na + \tfrac{1}{2}n(n - 1)d$$

    $$= \frac{n}{2}(\text{1st term} + n\text{th term})$$

2. The geometric progression to n terms:

    $$a + ar + ar^2 + \cdots + ar^{n-1} = \frac{a(1 - r^n)}{1 - r}$$

3. The definition of the arithmetic mean of $a_1, a_2, \ldots, a_n$:

    $$\frac{1}{n}(a_1 + a_2 + \cdots + a_n)$$

4. The definition of the geometric mean of $a_1, a_2, \ldots, a_n$:

    $$(a_1 a_2 \cdots a_n)^{1/n}$$

5. The definition of the harmonic mean of $a_1, a_2, \ldots, a_n$: If $\bar{a}_H$ is the harmonic mean, then

    $$\frac{1}{\bar{a}_H} = \frac{1}{n}\left(\frac{1}{a_1} + \frac{1}{a_2} + \frac{1}{a_3} + \cdots + \frac{1}{a_n}\right)$$

6. If

    $$a_0 + a_1 x + a_2 x^2 + a_3 x^3 + \cdots + a_n x^n$$
    $$= b_0 + b_1 x + b_2 x^2 + b_3 x^3 + \cdots + b_n x^n$$

    for all values of x, then

    $$a_0 = b_0, a_1 = b_1, a_2 = b_2, \ldots, a_n = b_n$$

Trigonometric Identities

7. $$\sin^2(x) + \cos^2(x) = 1$$

8. $\tan(x) = \dfrac{\sin(x)}{\cos(x)}$

9. $\operatorname{ctn}(x) = \dfrac{1}{\tan(x)}$

10. $\sec(x) = \dfrac{1}{\cos(x)}$

11. $\csc(x) = \dfrac{1}{\sin(x)}$

12. $$\sec^2(x) - \tan^2(x) = 1$$

13. $$\csc^2(x) - \operatorname{ctn}^2(x) = 1$$

14. $$\sin(x + y) = \sin(x)\cos(y) + \cos(x)\sin(y)$$

15. $$\cos(x + y) = \cos(x)\cos(y) - \sin(x)\sin(y)$$

16. $\sin(2x) = 2\sin(x)\cos(x)$

17. $\cos(2x) = \cos^2(x) - \sin^2(x) = 1 - 2\sin^2(x)$

18. $\tan(x + y) = \dfrac{\tan(x) + \tan(y)}{1 - \tan(x)\tan(y)}$

19. $\tan(2x) = \dfrac{2\tan(x)}{1 - \tan^2(x)}$

20. $\sin(x) = \dfrac{1}{2i}(e^{ix} - e^{-ix})$

21. $\cos(x) = \frac{1}{2}(e^{ix} + e^{-ix})$

22. $\sin(x) = -\sin(-x)$

23. $\cos(x) = \cos(-x)$
24. $\tan(x) = -\tan(-x)$
25. $\sin(ix) = i\sinh(x)$
26. $\cos(ix) = \cosh(x)$
27. $\tan(ix) = i\tanh(x)$
28. $\sin(x \pm iy) = \sin(x)\cosh(y) \pm i\cos(x)\sinh(y)$
29. $\cos(x \pm iy) = \cos(x)\cosh(y) \mp i\sin(x)\sinh(y)$
30. $\cosh(x) = \frac{1}{2}(e^x + e^{-x})$
31. $\sinh(x) = \frac{1}{2}(e^x - e^{-x})$
32. $\tanh(x) = \dfrac{\sinh(x)}{\cosh(x)}$
33. $\mathrm{sech}(x) = \dfrac{1}{\cosh(x)}$
34. $\mathrm{csch}(x) = \dfrac{1}{\sinh(x)}$
35. $\mathrm{ctnh}(x) = \dfrac{1}{\tanh(x)}$
36. $\cosh^2(x) - \sinh^2(x) = 1$
37. $\tanh^2(x) + \mathrm{sech}^2(x) = 1$
38. $\mathrm{ctnh}^2(x) - \mathrm{scsh}^2(x) = 1$
39. $\sinh(x) = -\sinh(-x)$
40. $\cosh(x) = \cosh(-x)$
41. $\tanh(-x) = -\tanh(-x)$
42. Relations obeyed by any triangle with angle A opposite side a, angle B opposite side b, and angle C opposite side c:

> a. $A + B + C = 180° = \pi$ radians
>
> b. $c^2 = a^2 + b^2 - 2ab\cos(C)$
>
> c. $\dfrac{a}{\sin(A)} = \dfrac{b}{\sin(B)} = \dfrac{c}{\sin(C)}$

Infinite Series

Series with Constant Terms

1. $1 + \frac{1}{2} + \frac{1}{3} + \frac{1}{4} + \cdots = \infty$

2. $1 + \dfrac{1}{2^2} + \dfrac{1}{3^2} + \dfrac{1}{4^2} + \cdots = \dfrac{\pi^2}{6}$

3. $1 + \dfrac{1}{2^4} + \dfrac{1}{3^4} + \dfrac{1}{4^4} + \cdots = \dfrac{\pi^4}{90}$

4. $1 + \dfrac{1}{2^p} + \dfrac{1}{3^p} + \dfrac{1}{4^p} + \cdots = \zeta(p)$

The function $\zeta(p)$ is called the *Riemann zeta function*. See H. B. Dwight, *Tables of Elementary and Some Higher Mathematical Functions*, 2nd ed., Dover Publications, New York, 1958, for tables of values of this function.

5. $1 - \frac{1}{2} + \frac{1}{3} - \frac{1}{4} + \cdots = \ln(2)$

6. $1 - \dfrac{1}{2^p} + \dfrac{1}{3^p} - \dfrac{1}{4^p} + \cdots = \left(1 - \dfrac{2}{2^p}\right)\zeta(p)$

Power Series

7. Maclaurin's series: If there is a power series in x for $f(x)$, it is

$$f(x) = f(0) + \left.\frac{df}{dx}\right|_{x=0} x + \frac{1}{2!}\left.\frac{d^2f}{dx^2}\right|_{x=0} x^2 + \frac{1}{3!}\left.\frac{d^3f}{dx^3}\right|_{x=0} x^3 + \cdots$$

8. Taylor's series: If there is a power series in $x - a$ for $f(x)$, it is

$$f(x) = f(a) + \frac{df}{dx}\bigg|_{x=a} (x - a) + \frac{1}{2!}\frac{d^2f}{dx^2}\bigg|_{x=a} (x - a)^2 + \cdots$$

In eqs. (7) and (8), $\dfrac{df}{dx}\bigg|_{x=a}$ means the value of the derivative at $x = a$.

9. If, for all values of x,

$$a_0 + a_1 x + a_2 x^2 + a_3 x^3 + \cdots = b_0 + b_1 x + b_2 x^2 + b_3 x^3 + \cdots$$

then

$$a_0 = b_0, a_1 = b_1, a_2 = b_2, \text{ etc.}$$

10. The reversion of a series: if

$$y = ax + bx^2 + cx^3 + \cdots$$

and

$$x = Ay + By^2 + Cy^3 + \cdots$$

then

$$A = \frac{1}{a}, \qquad B = -\frac{b}{a^3}, \qquad C = \frac{1}{a^5}(2b^2 - ac),$$

$$D = \frac{1}{a^7}(5abc - a^2 d - 5b^3), \qquad \text{etc.}$$

See Dwight, *Table of Integrals and Other Mathematical Data* (cited above), for more coefficients.

11. Powers of a series: If

$$S = a + bx + cx^2 + dx^3 + \cdots$$

then

$$S^2 = a^2 + 2abx + (b^2 + 2ac)x^2 + 2(ad + bc)x^3$$
$$+ (c^2 + 2ae + 2bd)x^4 + 2(af + be + cd)x^5 + \cdots$$

$$S^{1/2} = a^{1/2}\left[1 + \frac{b}{2a}x + \left(\frac{c}{2a} - \frac{b^2}{8a^2}\right)x^2 + \cdots\right]$$

$$S^{-1} = a^{-1}\left[1 - \frac{b}{a}x + \left(\frac{b^2}{a^2} - \frac{c}{a}\right)x^2 + \left(\frac{2bc}{a^2} - \frac{d}{a} - \frac{b^3}{a^3}\right)x^3 + \cdots\right]$$

12.
$$\sin(x) = x - \frac{x^3}{3!} + \frac{x^5}{5!} - \frac{x^7}{7!} + \cdots$$

13.
$$\cos(x) = 1 - \frac{x^2}{2!} + \frac{x^4}{4!} - \frac{4^6}{6!} + \cdots$$

14. $\sin(\theta + x) = \sin(\theta) + x\cos(\theta) - \frac{x^2}{2!}\sin(\theta) - \frac{x^3}{3!}\cos(\theta) + \cdots$

15. $\cos(\theta + x) = \cos(\theta) - x\sin(\theta) - \frac{x^2}{2!}\cos(\theta) + \frac{x^3}{3!}\sin(\theta) + \cdots$

16. $\sin^{-1}(x) = x + \frac{x^3}{2 \cdot 3} + \frac{1 \cdot 3x^5}{2 \cdot 4 \cdot 5} + \frac{1 \cdot 3 \cdot 5x^7}{2 \cdot 4 \cdot 6 \cdot 7} + \cdots$

where $x^2 < 1$. The series gives the principal value, $-\pi/2 < \sin^{-1}(x) < \pi/2$.

17. $\cos^{-1}(x) = \frac{\pi}{2} - \left(x + \frac{x^3}{2 \cdot 3} + \frac{1 \cdot 3x^5}{2 \cdot 4 \cdot 5} + \cdots \right)$

where $x^2 < 1$. The series gives the principal values, $0 < \cos^{-1}(x) < \pi$.

18.
$$e^x = 1 + x + \frac{x^2}{2!} + \frac{x^3}{3!} + \frac{x^4}{4!} + \cdots \qquad (x^2 < \infty)$$

19. $a^x = e^{x\ln(a)} = 1 + x\ln(a) + \frac{(x\ln(a))^2}{2!} + \cdots$

20.
$$\ln(1 + x) = x - \frac{x^2}{2} + \frac{x^3}{3} - \frac{x^4}{4} \cdots \qquad (x^2 < 1 \text{ and } x = 1)$$

21. $\ln(1 - x) = -\left(x + \frac{x^2}{2} + \frac{x^3}{3} + \frac{x^4}{4} + \cdots \right) \qquad (x^2 < 1 \text{ and } x = -1)$

22. $\sinh(x) = x + \frac{x^3}{3!} + \frac{x^5}{5!} + \frac{x^7}{7!} + \cdots \qquad (x^2 < \infty)$

23. $\cosh(x) = 1 + \frac{x^2}{2!} + \frac{x^4}{4!} + \frac{x^6}{6!} + \cdots \qquad (x^2 < \infty)$

Some Integrals with Exponentials in the Integrands

We begin with the integral

$$\int_0^\infty e^{-x^2}\,dx = I$$

We compute the value of this integral by a trick:

$$I^2 = \left[\int_0^\infty e^{-x^2}\,dx\right]^2 = \int_0^\infty e^{-x^2}\,dx \int_0^\infty e^{-y^2}\,dy = \int_0^\infty \int_0^\infty e^{-(x^2+y^2)}\,dx\,dy$$

We now change to polar coordinates:

$$I^2 = \int_0^{\pi/2} \int_0^\infty e^{-\rho^2}\rho\,d\rho\,d\phi = \frac{\pi}{2}\int_0^\infty e^{-\rho^2}\rho\,d\rho$$

$$= \frac{\pi}{2}\int_0^\infty \frac{1}{2}e^{-z}\,dz = \frac{\pi}{4}$$

Therefore,

$$I = \int_0^\infty e^{-x^2}\,dx = \frac{\sqrt{\pi}}{2} \tag{1}$$

and

$$\int_0^\infty e^{-ax^2}\,dx = \frac{1}{2}\sqrt{\frac{\pi}{a}} \tag{2}$$

Another trick can be used to obtain the integral:

$$\int_0^\infty x^{2n}e^{-ax^2\,dx}$$

APPENDIX

where n is an integer. For $n = 1$,

$$\int_0^\infty x^2 e^{-ax^2} dx = -\int_0^\infty \frac{d}{da}\left[e^{-ax^2} \right] dx = -\frac{d}{da}\int_0^\infty e^{-ax^2} dx$$

$$= -\frac{d}{da}\left[\frac{1}{2}\sqrt{\frac{\pi}{a}} \right] = \frac{1}{4a}\sqrt{\frac{\pi}{a}} = \frac{\pi^{1/2}}{4a^{3/2}} \tag{3}$$

For n an integer greater than unity:

$$\int_0^\infty x^{2n} e^{-ax^2} dx = (-1)^n \frac{d^n}{da^n}\left[\frac{1}{2}\sqrt{\frac{\pi}{a}} \right] \tag{4}$$

Equations (3) and (4) depend on the interchange of the order of differentiation and integration. This can be done only if an improper integral is uniformly convergent. The integral in Eq. (2) is uniformly convergent for all real values of a greater than zero.

Similar integrals with odd powers of x are easier. By the method of substitution,

$$\int_0^\infty x e^{-ax^2} dx = \frac{1}{2a}\int_0^\infty e^{-y} dy = \frac{1}{2a} \tag{5}$$

We can apply the trick of differentiating under the integral sign just as in Eq. (4) to obtain

$$\int_0^\infty x^{2n+1} e^{-ax^2} dx = (-1)^n \frac{d^n}{da^n}\left(\frac{1}{2a} \right) \tag{6}$$

All the integrals of this appendix are related to the gamma function, defined in Eq. (1) of Appendix 4. For example,

$$\int_0^\infty x^{2n+1} e^{-x^2} dx = \tfrac{1}{2}\int_0^\infty y^n e^{-y} dy = \tfrac{1}{2}\Gamma(n+1) \tag{7}$$

The Error Function. The indefinite integral

$$\int e^{-x^2} dx$$

has never been expressed as a closed form (a formula not involving an infinite series or something equivalent). The definite integral for limits other than 0 and ∞ is therefore not obtainable in closed form. Because of the frequent occurence of such definite integrals, tables of numerical approximations have been generated.[1] One form in which the tabulation is done is as the

---

[1] Two commonly available sources are Eugene Jahnke and Fritz Emde, *Tables of Functions*, Dover Publications, New York, 1945, and Milton Abramowitz and Irene A. Stegun, eds., *Handbook of Mathematical Functions with Formulas, Graphs and Mathematical Tables*, U.S. Government Printing Office, Washington, D.C., 1964.

error function, denoted by erf(x) and defined by

$$\operatorname{erf}(x) = \frac{2}{\sqrt{\pi}} \int_0^x e^{-t^2} dt \qquad (8)$$

As you can see from Eq. (1),

$$\lim_{x \to \infty} \operatorname{erf}(x) = 1 \qquad (9)$$

The name "error function" is chosen because of its frequent use in probability calculations involving the standard normal probability distribution (see Chapter 10).

Another form giving the same information is the *normal probability integral*[2]

$$\frac{1}{\sqrt{2\pi}} \int_{-x}^x e^{-t^2/2} dt$$

Values of the Error Function

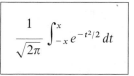

$$\operatorname{erf}(x) = \frac{2}{\sqrt{\pi}} \int_0^x e^{-t^2} dt*$$

x		0	1	2	3	4	5	6	7	8	9
0.0	0.0	000	113	226	338	451	564	676	789	901	*013
1	0.1	125	236	348	459	569	680	790	900	*009	*118
2	0.2	227	335	443	550	657	763	869	974	*079	*183
3	0.3	286	389	491	593	694	794	893	992	*090	*187
4	0.4	284	380	475	569	662	755	847	937	*027	*117
5	0.5	205	292	379	465	549	633	716	798	879	959
6	0.6	039	117	194	270	346	420	494	566	638	708
7		778	847	914	981	*047	*112	*175	*238	*300	*361
8	0.7	421	480	538	595	651	707	761	814	867	918
9		969	*019	*068	*116	*163	*209	*254	*299	*342	*385

* From Eugene Jahnke and Fritz Emde, *Tables of Functions*, Dover Publications, New York, p. 24.

[2] See, for example, Herbert B. Dwight, *Tables of Integrals and Other Mathematical Data*, 4th ed., Macmillan Publishing Co., New York, 1961.

Values of the Error Function (*continued*)

x		0	1	2	3	4	5	6	7	8	9
1.0	0.8	427	468	508	548	586	624	661	698	733	768
1		802	835	868	900	931	961	991	*020	*048	*076
2	0.9	103	130	155	181	205	229	252	275	297	319
3		340	361	381	400	419	438	456	473	490	507
4	0.95	23	39	54	69	83	97	*11	*24	*37	*49
5	0.96	61	73	84	95	*06	*16	*26	*36	*45	*55
6	0.97	63	72	80	88	96	*04	*11	*18	*25	*32
7	0.98	38	44	50	56	61	67	72	77	82	86
8		91	95	99	*03	*07	*11	*15	*18	*22	*25
9	0.99	28	31	34	37	39	42	44	47	49	51
2.0	0.995	32	52	72	91	*09	*26	*42	*58	*73	*88
1	0.997	02	15	28	41	53	64	75	85	95	*05
2	0.998	14	22	31	39	46	54	61	67	74	80
3		86	91	97	*02	*06	*11	*15	*20	*24	*28
4	0.999	31	35	38	41	44	47	50	52	55	57
5		59	61	63	65	67	69	71	72	74	75
6		76	78	79	80	81	82	83	84	85	86
7		87	87	88	89	89	90	91	91	92	92
8	0.9999	25	29	33	37	41	44	48	51	54	56
9		59	61	64	66	68	70	72	73	75	77

Some Useful Computer Programs

A Program for Numerical Integration. NUMINT. This program implements Simpson's rule or its equivalent for equally spaced data points, for unequally spaced data points, or for an integrand function, which the user can specify when running the program.

For the equally spaced data points or the integrand function, Simpson's rule in its usual form is used. For unequally spaced data points, the program fits parabolas to data points three at a time but uses the parabola to represent the integrand only in the first panel. Therefore, the number of panels can be even or odd for this option. Since the usual form of Simpson's rule requires the number of panels to be even, the unequally spaced point option can be used for equally spaced points if you have an odd number of panels.

```
100 PRINT "THIS PROGRAM CALCULATES AN INTEGRAL BY SIMPSON'S ONE-THIRD RULE."
110 PRINT "IT GIVES YOU A CHOICE OF ENTERING A LIST OF VALUES FOR THE"
120 PRINT "INTEGRAND, OR ENTERING AN INTEGRAND FUNCTION."
130 DIM X(101), Y(101), A(101), B(101), C(101)
140 INPUT "DO YOU WANT TO USE A TABLE OF EQUALLY SPACED DATA POINTS?";AS
150 IF AS = 'YES' THEN 500 ELSE 160
160 PRINT 'DO YOU WANT TO USE A TABLE OF DATA POINTS WHICH ARE NOT'
170 INPUT 'NECESSARILY EQUALLY SPACED'; CS
180 IF CS = 'YES' THEN 800 ELSE 200
200 INPUT "DO YOU HAVE AN INTEGRAND FUNCTION?";BS
210 IF BS = 'YES' THEN 220 ELSE 1500
220 INPUT 'NUMBER OF POINTS (MUST BE ODD, LESS THAN OR EQUAL TO 101)';N
230 INPUT 'LOWER LIMIT OF INTEGRATION'; X1
240 INPUT 'UPPER LIMIT OF INTEGRATION'; X9
250 W = (X9 - X1)/(N-1)
260 X(1) = X1
270 FOR I = 2 TO N STEP 1
280 X(I) = X(I-1) + W
290 NEXT I
293 PRINT "TYPE IN BASIC STATEMENTS WHICH WILL EVALUATE YOUR INTEGRAND"
294 PRINT "FUNCTION. NUMBER THE STATEMENTS WITH NUMBERS BETWEEN 300"
295 PRINT "AND 400. CALL YOUR INTEGRAND Y(I) AND YOUR INDEPENDENT"
296 PRINT "VARIABLE X(I). THEN TYPE CONT."
298 STOP
299 FOR I = 1 TO N STEP 1
410 NEXT I
420 S = Y(1)
430 FOR I = 2 TO N-1 STEP 2
440 S = S + 4*Y(I) + 2*Y(I+1)
```

```
450 NEXT I
460 S = (S - Y(N))*W/3
470 PRINT 'INTEGRAL FROM';X1;' TO ';X9;' IS EQUAL TO';S
480 GO TO 32767
500 INPUT 'NUMBER OF POINTS (MUST BE ODD)';N
510 INPUT "INTERVAL BETWEEN THE VALUES OF THE INDEPENDENT VARIABLE";W
520 FOR I = 1 TO N STEP 1
530 PRINT 'VALUE OF INTEGRAND FOR POINT NUMBER';I
540 INPUT Y(I)
550 NEXT I
560 S = Y(1)
570 FOR I = 2 TO N-1 STEP 2
580 S = S + 4*Y(I) + 2*Y(I+1)
590 NEXT I
600 FOR I = 1 TO N-1 STEP 1
610 B = B + Y(I)
620 T = T + (Y(I) + Y(I+1))/2
630 NEXT I
640 S = (S - Y(N))*W/3
650 B = B*W
660 T = T*W
670 PRINT 'INTEGRAL BY BAR GRAPH AREA =';B
680 PRINT 'INTEGRAL BY TRAPEZOIDAL RULE =';T
690 PRINT "INTEGRAL BY SIMPSON'S RULE =";S
700 GO TO 32767
800 PRINT 'YOU HAVE CHOSEN TO USE A LIST OF VALUES FOR THE INTEGRAND'
810 PRINT 'AND THE INDEPENDENT VARIABLE WHICH ARE NOT NECESSARILY'
820 PRINT 'EQUALLY SPACED.  THIS PROGRAM FITS PARABOLAS TO THE POINTS'
824 PRINT "THREE AT A TIME, MUCH LIKE SIMPSON'S RULE.  THE PARABOLA IS"
826 PRINT 'USED TO REPRESENT THE INTEGRAND BETWEEN THE FIRST AND SECOND'
828 PRINT 'POINTS, EXCEPT FOR THE LAST THREE POINTS, FOR WHICH THE '
829 PRINT 'PARABOLA IS USED FOR BOTH PANELS.'
830 INPUT 'NUMBER OF DATA POINTS (CAN BE EVEN OR ODD)';N
840 INPUT 'DO YOU WANT TO DO ANY OPERATIONS ON THE VARIABLES'; DS
850 IF DS = 'YES' THEN 860 ELSE 880
860 PRINT 'TYPE IN BASIC STATEMENTS TO OPERATE ON YOUR VARIABLES.'
870 PRINT 'USE STATEMENT NUMBERS BETWEEN 900 AND 940.  CALL YOUR'
875 PRINT 'INDEPENDENT VARIABLE X(I) AND YOUR INTEGRAND Y(I).'
876 PRINT 'THEN TYPE CONT.'
877 STOP
880 FOR I = 1 TO N
885 PRINT 'X(1) MUST BE LOWER LIMIT OF INTEGRATION.' IF I = 1
890 PRINT 'X(';N;')  MUST BE UPPER LIMIT OF INTEGRATION.' IF I = N
892 PRINT 'VALUE OF X(';I;'), Y(';I;')'
894 INPUT X(I), Y(I)
945 NEXT I
950 FOR I = 1 TO N-2
960 E = Y(I)-Y(I+2)-(Y(I+1)-Y(I+2))*(X(I)-X(I+2))/(X(I+1)-X(I+2))
970 G = X(I)*X(I)-X(I+2)*X(I+2)
980 H = (X(I+1)*X(I+1)-X(I+2)*X(I+2))*(X(I)-X(I+2))
990 G = G - H/(X(I+1)-X(I+2))
1000 F = Y(I+1)-Y(I+2)-E*(X(I+1)*X(I+1)-X(I+2)*X(I+2))/G
1010 A(I) = E/G
1020 B(I) = F/(X(I+1)-X(I+2))
1030 C(I) = Y(I+2)-E*X(I+2)*X(I+2)/G-F*X(I+2)/(X(I+1)-X(I+2))
1040 NEXT I
1050 Z = 0
1060 FOR I = 1 TO N-3
1070 Z = Z + A(I)*(X(I+1)^3-X(I)^3)/3+B(I)*(X(I+1)^2-X(I)^2)/2
1080 Z = Z + C(I)*(X(I+1)-X(I))
1090 NEXT I
1100 Z = Z + A(N-2)*(X(N)^3-X(N-2)^3)/3
1110 Z = Z + B(N-2)*(X(N)^2-X(N-2)^2)/2+C(N-2)*(X(N)-X(N-2))
1120 PRINT 'INTEGRAL FROM ';X(1);' TO ';X(N);' = ';Z
1130 GO TO 32767
1500 PRINT 'SORRY, THERE ARE NO OTHER OPTIONS IN THIS PROGRAM.'
32767 END
```

A Linear Least Squares Program.   SQUARE. This program performs a linear regression (least squares) fit to a set of data points. The program offers a number of options so that various functions of the numbers in the data points can be fit to a straight line. A Hammett Rho-Sigma fit is one of the options, as are fits to the integrated rate equations for first, second, and third order reaction kinetics for reactions with a single reactant. The user can also specify functions of his choice for the dependent and independent variables while running the program.

```
10 REM THIS PROGRAM IS A VERSATILE LEAST SQUARES PROGRAM DESIGNED TO
20 REM ACCOMODATE A VARIETY OF CHEMICAL CURVE-FITTING PROBLEMS.
50 DIM X(40)
55 DIM X1(40), Y1(40)
60 DIM Y(40)
70 DIM R(40)
75 DEF FNM = (N*S0 - X1*Y1)/(N*X2 - X1*X1)
85 DEF FNB = (X2*Y1 - X1*S0)/(N*X2 - X1*X1)
95 DEF FNR2 = ((N*S0 - X1*Y1)^2)/((N*X2 - X1*X1)*(N*Y2 - Y1*Y1))
100 PRINT 'WELCOME TO "SQUARE," A PROGRAM WHICH WILL GIVE YOU THE
110 PRINT 'PARAMETERS SPECIFYING THE CURVE WHICH BEST FITS THE DATA'
120 PRINT 'POINTS WHICH YOU TYPE IN.'
140 INPUT 'DO YOU WANT TO DO A HAMMETT RHO-SIGMA PLOT', AS
150 IF AS = 'YES' GO TO 1010
160 INPUT 'DO YOU WANT TO FIT REACTION KINETICS DATA', AS
170 IF AS = 'YES' GO TO 2010
180 INPUT 'DO YOU WANT TO DO A LINEAR FIT, Y = MX + B, TO DATA AS GIVEN',AS
190 IF AS = 'YES' GO TO 3010
200 PRINT 'DO YOU WANT TO DO A LINEAR FIT TO THE NATURAL LOGARITHM OF '
210 INPUT 'YOUR DEPENDENT VARIABLE, LN(Y) = MX + B',AS
220 IF AS = 'YES' GO TO 4010
230 PRINT 'DO YOU WANT TO DO A LINEAR FIT OF YOUR DEPENDENT VARIABLE TO THE'
240 INPUT 'NATURAL LOGARITHM OF YOUR INDEPENDENT VARIABLE, Y = M LN(X) + B', AS
250 IF AS = 'YES' GO TO 5010
260 INPUT 'DO YOU WANT A FIT TO Y = B X^A', AS
270 IF AS = 'YES' GO TO 6010
280 PRINT 'DO YOU WANT TO FIT THE NATURAL LOGARITHM OF YOUR DEPENDENT'
290 PRINT 'VARIABLE TO THE RECIPROCAL OF YOUR INDEPENDENT VARIABLE,'
300 INPUT 'LN(Y) = M/X + B'; AS
310 IF AS = 'YES' GO TO 7010
320 INPUT 'DO YOU WANT AN EXPONENTIAL FIT, Y = B EXP(AX)';AS
330 IF AS = 'YES' GO TO 8010
360 PRINT 'DO YOU WANT TO SPECIFY SOME FUNCTION OF YOUR VARIABLES TO BE'
370 INPUT 'FIT TO SOME OTHER FUNCTION OF YOUR VARIABLES';AS
380 IF AS = 'YES' THEN 10010 ELSE 32000
1010 PRINT 'YOU HAVE CHOSEN TO DO A LINEAR FIT IN A HAMMETT RHO-SIGMA PLUS'
1020 PRINT 'PLOT.  TYPE IN THE NUMBER OF DATA POINTS, INCLUDING ONE POINT'
1030 INPUT 'FOR BENZENE AT THE ORIGIN OF THE PLOT.',N
1040 PRINT 'TYPE IN YOUR DATA POINTS: FOR EACH SUBSTITUENT, TYPE IN FIRST'
1050 PRINT 'THE SIGMA PLUS VALUE, THEN A COMMA, AND THEN THE PARTIAL RATE'
1060 PRINT 'FACTOR (RATIO OF REACTION RATE OF SUBSTITUTED COMPOUND TO THE'
1070 PRINT "RATE FOR 1 POSITION ON BENZENE). REMEMBER 0,1 FOR BENZENE."
1075 PRINT 'PRESS THE RETURN KEY AFTER EACH DATA POINT.'
1080 FOR I = 1 TO N
1090 PRINT 'SIGMA PLUS, PARTIAL RATE FACTOR FOR SUBSTITUENT NUMBER';I
1100 INPUT X(I), Y(I)
1105 Y(I) = LOG10(Y(I))
1110 NEXT I
1115 Q% = 2
1120 GO TO 2195
1190 PRINT 'SLOPE (RHO) = ';M;' INTERCEPT (DELTA) = ';B
1200 IF B < 0 THEN 1205 ELSE 1209
1205 PRINT 'LOG10(F) = ';M; 'TIMES SIGMA PLUS   ';B
1206 GO TO 1230
1209 PRINT 'LOG10(F) = ';M; 'TIMES SIGMA PLUS  + ';B
1230 BS = 'YES'
1240 GO TO 2250
2010 PRINT 'YOU HAVE CHOSEN TO FIT TIME-CONCENTRATION DATA TO INTEGRATED'
2020 PRINT 'RATE LAWS.  TYPE IN THE REACTION ORDER (1, 2, OR 3) WHICH
2030 INPUT 'YOU WANT TO TRY TO FIT', O
2040 INPUT 'TYPE IN THE NUMBER OF DATA POINTS', N
2050 INPUT 'DO YOU WANT A LIST OF RESIDUALS?' , BS
2052 IF O = 1 GO TO 2060
2054 IF O = 2 GO TO 2105
2056 IF O = 3 GO TO 2150
2060 FOR I = 1 TO N
2070 PRINT 'TYPE IN TIME, CONCENTRATION FOR POINT NUMBER'; I
2080 INPUT X(I), Y(I)
2090 Y(I) = LOG(Y(I))
2100 NEXT I
2102 GO TO 2195
2105 FOR I = 1 TO N
```

```
2110 PRINT 'TYPE IN TIME, CONCENTRATION FOR POINT NUMBER'; I
2120 INPUT X(I), Y(I)
2130 Y(I) = 1/Y(I)
2140 NEXT I
2145 GO TO 2195
2150 FOR I = 1 TO N
2160 PRINT 'TYPE IN TIME, CONCENTRATION FOR POINT NUMBER'; I
2170 INPUT X(I), Y(I)
2180 Y(I) = 1/(2*Y(I)^2)
2190 NEXT I
2195 FOR I = 1 TO N
2200 X1=X1+X(I)\X2=X2+X(I)^2\Y1=Y1+Y(I)\Y2=Y2+Y(I)^2\S0=S0+X(I)*Y(I)
2210 NEXT I
2220 M = FNM\B = FNB\R2 = FNR2
2225 IF Q% = 2 GO TO 1190
2230 PRINT 'SLOPE =';M, 'INTERCEPT =';B
2231 IF Q% = 1 GO TO 3080
2233 IF Q% = 3 GO TO 4095
2234 IF Q% = 4 GO TO 5110
2236 IF Q% = 5 GO TO 6110
2237 IF Q% = 6 GO TO 7110
2239 IF Q% = 7 GO TO 8110
2243 IF Q% = 9 GO TO 10240
2250 IF BS = 'YES' THEN 2252 ELSE 2256
2252 PRINT 'RESIDUALS - DEVIATIONS FROM THE LINE:'
2256 FOR I = 1 TO N
2260 R(I) = Y(I) - M*X(I) - B
2270 IF BS = 'YES' THEN 2280 ELSE 2290
2280 PRINT 'RESIDUAL R(';I;') = ';R(I)
2290 R1 = R1 + R(I)*R(I)
2300 NEXT I
2310 S = SQR(R1/(N-2))
2320 PRINT 'STANDARD DEVIATION OF THE RESIDUALS = '; S
2330 D = N*X2 - X1*X1
2340 E1 = SQR(N/D)*1.96*S
2350 PRINT 'UNCERTAINTY IN THE SLOPE (95% CONFIDENCE) = '; E1
2360 E2 = SQR(X2/D)*1.96*S
2370 PRINT 'UNCERTAINTY IN THE INTERCEPT (95% CONFIDENCE) = ';E2
2380 R2 = FNR2
2390 PRINT 'CORRELATION COEFFICIENT SQUARED = '; R2
2395 IF Q% >= 1 GO TO 32767
2400 IF O = 1 GO TO 2450
2410 IF O = 2 GO TO 2500
2420 IF O = 3 GO TO 2550
2450 K1 = -M
2455 PRINT 'FIRST ORDER RATE CONSTANT K = '; K1
2460 CO = EXP(B)
2470 PRINT 'INITIAL CONCENTRATION CO = '; CO
2480 GO TO 32767
2500 PRINT 'SECOND ORDER RATE CONSTANT K = '; M
2510 CO = 1/B
2520 PRINT 'INITIAL CONCENTRATION CO = '; CO
2530 GO TO 32767
2550 PRINT 'THIRD ORDER RATE CONSTANT K = '; M
2560 CO = SQR(1/(2*B))
2570 PRINT 'INITIAL CONCENTRATION CO = '; CO
2580 GO TO 32767
3010 PRINT 'YOU HAVE CHOSEN TO FIT A STRAIGHT LINE TO YOUR DATA AS GIVEN.'
3015 INPUT 'TYPE IN THE NUMBER OF DATA POINTS.',N
3020 PRINT 'DATA PAIRS MUST BE TYPED IN WITH THE INDEPENDENT VARIABLE FIRST.'
3025 Q% = 1
3030 FOR I = 1 TO N
3040 PRINT 'TYPE IN X(';I;'), Y(';I;')'
3050 INPUT X(I), Y(I)
3060 NEXT I
3070 GO TO 2195
3080 INPUT 'DO YOU WANT A LIST OF RESIDUALS';BS
3081 IF B >= 0 THEN 3082 ELSE 3086
3082 PRINT 'Y = ';M; 'TIMES X + ';B
3084 GO TO 2250
3086 PRINT 'Y = ';M; 'TIMES X   ';B
3090 GO TO 2250
4010 PRINT 'YOU HAVE CHOSEN TO DO A LINEAR FIT TO LN(Y).'
4020 INPUT 'TYPE IN THE NUMBER OF DATA POINTS.',N
4030 PRINT 'TYPE IN DATA POINTS WITH X(I) FIRST, THEN Y(I) (NOT LN(Y)).'
4040 Q% = 3
4050 FOR I = 1 TO N
4060 PRINT 'TYPE IN X(';I;'), Y(';I;')'
4065 INPUT X(I), Y(I)
```

```
4070 Y(I) = LOG(Y(I))
4080 NEXT I
4090 GO TO 2195
4095 INPUT 'DO YOU WANT A LIST OF RESIDUALS';BS
4097 IF B >= 0 THEN 4100 ELSE 4110
4100 PRINT 'LN(Y) = ';M;' TIMES X   + ';B
4101 GO TO 2250
4110 PRINT 'LN(Y) = ';M;' TIMES X   ';B
4120 GO TO 2250
5010 PRINT 'YOU HAVE CHOSEN TO FIT Y = M LN(X) + B.'
5020 INPUT 'TYPE IN THE NUMBER OF DATA POINTS.';N
5030 Q% = 4
5040 PRINT 'TYPE IN DATA POINTS WITH X(I) (NOT LN(X(I)), THEN Y(I).'
5050 FOR I = 1 TO N
5060 PRINT 'TYPE IN X(';I;'),   Y(';I;')'
5070 INPUT X(I), Y(I)
5080 X(I) = LOG(X(I))
5090 NEXT I
5100 GO TO 2195
5110 INPUT 'DO YOU WANT A LIST OF RESIDUALS';BS
5120 IF B >= 0 THEN 5130 ELSE 5150
5130 PRINT 'Y = ';M;' TIMES LNX  + ';B
5140 GO TO 2250
5150 PRINT 'Y = ';M;' TIMES LNX    ';B
5160 GO TO 2250
6010 PRINT 'YOU HAVE CHOSEN TO FIT Y = BX^A.'
6020 INPUT 'TYPE IN THE NUMBER OF DATA POINTS.';N
6030 PRINT 'TYPE IN THE DATA POINTS WITH X(I) FIRST, THEN Y(I).'
6040 Q% = 5
6050 FOR I = 1 TO N STEP 1
6060 PRINT 'TYPE IN X(';I;'),   Y(';I;')'
6070 INPUT X(I), Y(I)
6080 Y(I) = LOG(Y(I)) \ X(I) = LOG(X(I))
6090 NEXT I
6100 GO TO 2195
6110 INPUT 'DO YOU WANT A LIST OF RESIDUALS';BS
6120 PRINT 'Y = ';EXP(B);' TIMES X^';M
6130 GO TO 2250
7010 PRINT 'YOU HAVE CHOSEN TO FIT LN(Y) = M/X + B.'
7020 INPUT 'TYPE IN THE NUMBER OF DATA POINTS.';N
7030 PRINT 'TYPE IN THE DATA POINTS WITH X(I) FIRST, THEN Y(I).'
7040 Q% = 6
7050 FOR I = 1 TO N
7060 PRINT 'TYPE IN X(';I;'), Y(';I;')'
7070 INPUT X(I), Y(I)
7080 X(I) = 1/X(I) \ Y(I) = LOG(Y(I))
7090 NEXT I
7100 GO TO 2195
7110 INPUT 'DO YOU WANT A LIST OF RESIDUALS';BS
7120 IF B >= 0 THEN 7130 ELSE 7150
7130 PRINT 'LN(Y) = ';M;' TIMES 1/X   + ';B
7140 GO TO 2250
7150 PRINT 'LN(Y) = ';M;' TIMES 1/X   ';B
7160 GO TO 2250
8010 PRINT 'YOU HAVE CHOSEN TO FIT Y = B EXP(AX).'
8020 INPUT 'TYPE IN THE NUMBER OF DATA POINTS.';N
8030 PRINT 'TYPE IN THE DATA POINTS WITH X(I) FIRST, THEN Y(I).'
8040 Q% = 7
8050 FOR I = 1 TO N
8060 PRINT 'TYPE IN X(';I;'),   Y(';I;')'
8070 INPUT X(I), Y(I)
8080 Y(I) = LOG(Y(I))
8090 NEXT I
8100 GO TO 2195
8110 INPUT 'DO YOU WANT A LIST OF RESIDUALS';BS
8120 PRINT 'Y = ';B;' TIMES EXP(';M;'X)'
8130 GO TO 2250
10010 PRINT 'YOU HAVE CHOSEN TO SPECIFY SOME FUNCTION OF YOUR'
10020 PRINT 'INPUT VALUES AS DEPENDENT VARIABLE, AND SOME OTHER'
10030 PRINT 'FUNCTION AS YOUR INDEPENDENT VARIABLE IN A LINEAR FIT.'
10040 INPUT 'TYPE IN THE NUMBER OF DATA POINTS.',N
10042 PRINT 'TYPE IN BASIC STATEMENTS TO FORM THE DESIRED FUNCTIONS.'
10043 PRINT 'CALL THE OLD INDEPENDENT VARIABLE X1(I) AND THE NEW'
10044 PRINT 'INDEPENDENT VARIABLE X(I).  CALL THE OLD DEPENDENT VARIABLE'
10045 PRINT 'Y1(I) AND THE NEW DEPENDENT VARIABLE Y(I). USE STATEMENT'
10046 PRINT 'NUMBERS BETWEEN 10100 AND 10200.  THEN TYPE CONT.'
10049 STOP
10050 PRINT 'TYPE IN THE DATA POINTS WITH X(I) FIRST, THEN Y(I).'
10060 Q% = 9
```

```
10070 FOR I = 1 TO N
10080 PRINT 'TYPE IN X(';I;'),   Y(';I;')'
10090 INPUT X1(I), Y1(I)
10210 NEXT I
10220 GO TO 2195
10240 INPUT 'DO YOU WANT A LIST OF RESIDUALS'; B$
10250 PRINT 'REMEMBER THAT YOU SPECIFIED TWO FUNCTIONS.   YOUR OUTPUT'
10260 PRINT 'IS IN TERMS OF A LINEAR FIT WITH WHATEVER THESE FUNCTIONS ARE.'
10270 GO TO 2250
32000 PRINT 'THERE ARE NO MORE OPTIONS. TYPE RUN TO TRY AGAIN.'
32767 END
```

Common Logarithms*

N	0	1	2	3	4	5	6	7	8	9	1 2 3	4 5 6	7 8 9
10	0000	0043	0086	0128	0170	0212	0253	0294	0334	0374	4 8 12	17 21 25	29 33 37
11	0414	0453	0492	0531	0569	0607	0645	0682	0719	0755	4 8 11	15 19 23	26 30 34
12	0792	0828	0864	0899	0934	0969	1004	1038	1072	1106	3 7 10	14 17 21	24 28 31
13	1139	1173	1206	1239	1271	1303	1335	1367	1399	1430	3 6 10	13 16 19	23 26 29
14	1461	1492	1523	1553	1584	1614	1644	1673	1703	1732	3 6 9	12 15 18	21 24 27
15	1761	1790	1818	1847	1875	1903	1931	1959	1987	2014	3 6 8	11 14 17	20 22 25
16	2041	2068	2095	2122	2148	2175	2201	2227	2253	2279	3 5 8	11 13 16	18 21 24
17	2304	2330	2355	2380	2405	2430	2455	2480	2504	2529	2 5 7	10 12 15	17 20 22
18	2553	2577	2601	2625	2648	2672	2695	2718	2742	2765	2 5 7	9 12 14	16 19 21
19	2788	2810	2833	2856	2878	2900	2923	2945	2967	2989	2 4 7	9 11 13	16 18 20
20	3010	3032	3054	3075	3096	3118	3139	3160	3181	3201	2 4 6	8 11 13	15 17 19
21	3222	3243	3263	3284	3304	3324	3345	3365	3385	3404	2 4 6	8 10 12	14 16 18
22	3424	3444	3464	3483	3502	3522	3541	3560	3579	3598	2 4 6	8 10 12	14 16 17
23	3617	3636	3655	3674	3692	3711	3729	3747	3766	3784	2 4 6	7 9 11	13 15 17
24	3802	3820	3838	3856	3874	3892	3909	3927	3945	3962	2 4 5	7 9 11	12 14 16
25	3979	3997	4014	4031	4048	4065	4082	4099	4116	4133	2 4 5	7 9 10	12 14 16
26	4150	4166	4183	4200	4216	4232	4249	4265	4281	4298	2 3 5	7 8 10	11 13 15
27	4314	4330	4346	4362	4378	4393	4409	4425	4440	4456	2 3 5	6 8 9	11 12 14
28	4472	4487	4502	4518	4533	4548	4564	4579	4594	4609	2 3 5	6 8 9	11 12 14
29	4624	4639	4654	4669	4683	4698	4713	4728	4742	4757	1 3 4	6 7 9	10 12 13
N	0	1	2	3	4	5	6	7	8	9	1 2 3	4 5 6	7 8 9

* The proportional parts are stated in full for every tenth at the right-hand side. The logarithm of any number of four significant figures can be read directly by adding the proportional part corresponding to the fourth figure to the tabular number corresponding to the first three figures. There may be an error of 1 in the last place.

APPENDIX

N	0	1	2	3	4	5	6	7	8	9	1	2	2	4	5	6	7	8	9
30	4771	4786	4800	4814	4829	4843	4857	4871	4886	4900	1	3	4	6	7	9	10	11	13
31	4914	4928	4942	4955	4969	4983	4997	5011	5024	5038	1	3	4	5	7	8	10	11	12
32	5051	5065	5079	5092	5105	5119	5132	5145	5159	5172	1	3	4	5	7	8	9	11	12
33	5185	5198	5211	5224	5237	5250	5263	5276	5289	5302	1	3	4	5	7	8	9	11	12
34	5315	5328	5340	5353	5366	5378	5391	5403	5416	5428	1	2	4	5	6	8	9	10	11
35	5441	5453	5465	5478	5490	5502	5514	5527	5539	5551	1	2	4	5	6	7	9	10	11
36	5563	5575	5587	5599	5611	5623	5635	5647	5658	5670	1	2	4	5	6	7	8	10	11
37	5682	5694	5705	5717	5729	5740	5752	5763	5775	5786	1	2	4	5	6	7	8	9	11
38	5798	5809	5821	5832	5843	5855	5866	5877	5888	5899	1	2	3	5	6	7	8	9	10
39	5911	5922	5933	5944	5955	5966	5977	5988	5999	6010	1	2	3	4	5	7	8	9	10
40	6021	6031	6042	6053	6064	6075	6085	6096	6107	6117	1	2	3	4	5	6	8	9	10
41	6128	6138	6149	6160	6170	6180	6191	6201	6212	6222	1	2	3	4	5	6	7	8	9
42	6232	6243	6253	6263	6274	6284	6294	6304	6314	6325	1	2	3	4	5	6	7	8	9
43	6335	6345	6355	6365	6375	6385	6395	6405	6415	6425	1	2	3	4	5	6	7	8	9
44	6435	6444	6454	6464	6474	6484	6493	6503	6513	6522	1	2	3	4	5	6	7	8	9
45	6532	6542	6551	6561	6571	6580	6590	6599	6609	6618	1	2	3	4	5	6	7	8	9
46	6628	6637	6646	6656	6665	6675	6684	6693	6702	6712	1	2	3	4	5	6	7	7	8
47	6721	6730	6739	6749	6758	6767	6776	6785	6794	6803	1	2	3	4	5	6	7	7	8
48	6812	6821	6830	6839	6848	6857	6866	6875	6884	6893	1	2	3	4	5	6	7	7	8
49	6902	6911	6920	6928	6937	6946	6955	6964	6972	6981	1	2	3	4	4	5	6	7	8
50	6990	6998	7007	7016	7024	7033	7042	7050	7059	7067	1	2	3	3	4	5	6	7	8
51	7076	7084	7093	7101	7110	7118	7126	7135	7143	7152	1	2	3	3	4	5	6	7	8
52	7160	7168	7177	7185	7193	7202	7210	7218	7226	7235	1	2	3	3	4	5	6	7	7
53	7243	7251	7259	7267	7275	7284	7292	7300	7308	7316	1	2	2	3	4	5	6	6	7
54	7324	7332	7340	7348	7356	7364	7372	7380	7388	7396	1	2	2	3	4	5	6	6	7
55	7404	7412	7419	7427	7435	7443	7451	7459	7466	7474	1	2	2	3	4	5	5	6	7
56	7482	7490	7497	7505	7513	7520	7528	7536	7543	7551	1	2	2	3	4	5	5	6	7
57	7559	7566	7574	7582	7589	7597	7604	7612	7619	7627	1	1	2	3	4	5	5	6	7
58	7634	7642	7649	7657	7664	7672	7679	7686	7694	7701	1	1	2	3	4	4	5	6	7
59	7709	7716	7723	7731	7738	7745	7752	7760	7767	7774	1	1	2	3	4	4	5	6	7
60	7782	7789	7796	7803	7810	7818	7825	7832	7839	7846	1	1	2	3	4	4	5	6	6
61	7853	7860	7868	7875	7882	7889	7896	7903	7910	7917	1	1	2	3	3	4	5	6	6
62	7924	7931	7938	7945	7952	7959	7966	7973	7980	7987	1	1	2	3	3	4	5	5	6
63	7993	8000	8007	8014	8021	8028	8035	8041	8048	8055	1	1	2	3	3	4	5	5	6
64	8062	8069	8075	8082	8089	8096	8102	8109	8116	8122	1	1	2	3	3	4	5	5	6
65	8129	8136	8142	8149	8156	8162	8169	8176	8182	8189	1	1	2	3	3	4	5	5	6
66	8195	8202	8209	8215	8222	8228	8235	8241	8248	8254	1	1	2	3	3	4	5	5	6
67	8261	8267	8274	8280	8287	8293	8299	8306	8312	8319	1	1	2	3	3	4	5	5	6
68	8325	8331	8338	8344	8351	8357	8363	8370	8376	8382	1	1	2	3	3	4	4	5	6
69	8388	8395	8401	8407	8414	8420	8426	8432	8439	8445	1	1	2	3	3	4	4	5	6
N	0	1	2	3	4	5	6	7	8	9	1	2	3	4	5	6	7	8	9

N	0	1	2	3	4	5	6	7	8	9	1	2	3	4	5	6	7	8	9
70	8451	8457	8463	8470	8476	8482	8488	8494	8500	8506	1	1	2	3	3	4	4	5	6
71	8513	8519	8525	8531	8537	8543	8549	8555	8561	8567	1	1	2	3	3	4	4	5	6
72	8573	8579	8585	8591	8597	8603	8609	8615	8621	8627	1	1	2	3	3	4	4	5	6
73	8633	8639	8645	8651	8657	8663	8669	8675	8681	8686	1	1	2	2	3	4	4	5	5
74	8692	8698	8704	8710	8716	8722	8727	8733	8739	8745	1	1	2	2	3	4	4	5	5
75	8751	8756	8762	8768	8774	8779	8785	8791	8797	8802	1	1	2	2	3	3	4	5	5
76	8808	8814	8820	8825	8831	8837	8842	8848	8854	8859	1	1	2	2	3	3	4	4	5
77	8865	8871	8876	8882	8887	8893	8899	8904	8910	8915	1	1	2	2	3	3	4	4	5
78	8921	8927	8932	8938	8943	8949	8954	8960	8965	8971	1	1	2	2	3	3	4	4	5
79	8976	8982	8987	8993	8998	9004	9009	9015	9020	9025	1	1	2	2	3	3	4	4	5
80	9031	9036	9042	9047	9053	9058	9063	9069	9074	9079	1	1	2	2	3	3	4	4	5
81	9085	9090	9096	9101	9106	9112	9117	9122	9128	9133	1	1	2	2	3	3	4	4	5
82	9138	9143	9149	9154	9159	9165	9170	9175	9180	9186	1	1	2	2	3	3	4	4	5
83	9191	9196	9201	9206	9212	9217	9222	9227	9232	9238	1	1	2	2	3	3	4	4	5
84	9243	9248	9253	9258	9263	9269	9274	9279	9284	9289	1	1	2	2	3	3	4	4	5
85	9294	9299	9304	9309	9315	9320	9325	9330	9335	9340	1	1	2	2	3	3	4	4	5
.86	9345	9350	9355	9360	9365	9370	9375	9380	9385	9390	1	1	2	2	3	3	4	4	5
87	9395	9400	9405	9410	9415	9420	9425	9430	9435	9440	1	1	2	2	3	3	4	4	5
88	9445	9450	9455	9460	9465	9469	9474	9479	9484	9489	0	1	1	2	2	3	3	4	4
89	9494	9499	9504	9509	9513	9518	9523	9528	9533	9538	0	1	1	2	2	3	3	4	4
90	9542	9547	9552	9557	9562	9566	9571	9576	9581	9586	0	1	1	2	2	3	3	4	4
91	9590	9595	9600	9605	9609	9614	9619	9624	9628	9633	0	1	1	2	2	3	3	4	4
92	9638	9643	9647	9652	9657	9661	9666	9671	9675	9680	0	1	1	2	2	3	3	4	4
93	9685	9689	9694	9699	9703	9708	9713	9717	9722	9727	0	1	1	2	2	3	3	4	4
94	9731	9736	9741	9745	9750	9754	9759	9763	9768	9773	0	1	1	2	2	3	3	4	4
95	9777	9782	9786	9791	9795	9800	9805	9809	9814	9818	0	1	1	2	2	3	3	4	4
96	9823	9827	9832	9836	9841	9845	9850	9854	9859	9863	0	1	1	2	2	3	3	4	4
97	9868	9872	9877	9881	9886	9890	9894	9899	9903	9908	0	1	1	2	2	3	3	4	4
98	9912	9917	9921	9926	9930	9934	9939	9943	9948	9952	0	1	1	2	2	3	3	3	4
99	9956	9961	9965	9969	9974	9978	9983	9987	9991	9996	0	1	1	2	2	3	3	3	4
N	0	1	2	3	4	5	6	7	8	9	1	2	3	4	5	6	7	8	9

Answers to Selected Problems

1.2b 1.1×10^6

1.2d   34.55

1.3b 6.7342×10^7

1.3d 6.432×10^3

1.4b 25 m s^{-1}

1.4d $2.18 \times 10^{-18} \text{ J}$

1.6b $1609.344000 \cdots$

1.8b 459.67^0 R

2.3 $\cot(\alpha) \approx \cos(\alpha) \approx \dfrac{\pi}{2} - \alpha$

2.6b   4.606

2.6d   1.3024

2.7b 1.28×10^8

2.10b $\rho = 11.18$
           $\phi = 2.0344 \text{ radians} = 116.565^0$

2.11b $(-1, -1), \sqrt{2}$

2.14b $(12.5, 21.65 \cdots, 17.5)$

2.15c   1.30025 radians

2.19 $-\mathbf{k}$ (10 volts)

2.24b $(3 - 7i)^3 - (7i)^2$

2.26 $R = 0.25877, I = -0.18322, \phi = -0.6161 \text{ radians}$

2.30 $r = c$ sphere

2.33 $(-5, 5, 10\sqrt{2})$

2.35   0

2.37 $-2\mathbf{i} - 4\mathbf{j} - \mathbf{k}$

2.39 $0.3288 \text{ kg m}^{-1} \text{ s}^{-1}$

2.42 $1.534e^{1.89839}, 1.534e^{3.99279}, 1.534e^{6.08718}$

3.3   1.5121

3.5b   differentiable everywhere except $x = \pm\pi/2, \pm3\pi/2, \pm5\pi/2$, etc.

3.8b Q/T^2

3.8d $ax^2b\cos(bx) + 2ax\sin(bx)$

3.9b $d^2y/dx^2 = ab^2e^{bx}, d^3y/dx^3 = ab^3e^{bx}$

3.11b $v_{mp} = (2kT/m)^{1/2}$ 421 m s^{-1}

3.14b 0

3.17c $dy/dx = a/(3x), d^2y/dx^2 = -a/(3x^2)$

3.17e $dy/dx = 2/(c - x)^3, d^2y/dx^2 = 6/(c - x)^4$

3.17g $$\frac{dy}{dx} = -\frac{1}{x(1 + x)^2} - \frac{1}{x^2(1 + x)}$$

$$\frac{d^2y}{dx^2} = \frac{2}{x(1 + x)^3} + \frac{2}{x^2(1 + x)^2} + \frac{2}{x^3(1 + x)}$$

3.18b $-k$

3.18d ab^2

3.19   3.00% error, compare with 3.0301% error

3.23   minimum at $x = x_0/2$, maximum value $= G_0 + RTx_0\ln(x_0)$

3.24b 0.866 km along road

3.27 $h = 1.08385$ m

3.28b 4.5

3.28d 0

3.28f 0

4.2 $2e^{-5x} + 18$

4.4b   24

4.9b $1.0017\cdots$

4.10b diverges

4.10d 1

4.12   5787, 76.07

4.14   7.5, 7.746

4.16   199.8 m s^{-1}

4.19 $\pi^2 - 4$

4.21 $42\ln(x + 2) + C, -36\ln(x + 1) + C$

4.22   2340, $2333\frac{1}{3}$

4.26b $x^2/4 - (x/4)\sin(2x) - (1/8)\cos(2x)$

4.27b $4\ln(2) - 3/2$

4.27d $(1/4)\sin^2(\pi^4)$

4.28b diverges

4.31b 0

5.6b   converges by ratio test

5.6d   converges by alternating series test

5.11 $\ln(x) = \ln(2) + (1/2)(x-2) - (1/8)(x-2)^2 + (1/24)(x-2)^3$
$$-(1/64)(x-2)^4 + \cdots + (-1)^{n-1}\frac{1}{n2^n}(x-2)^n + \cdots$$

5.12b all finite values of x

5.13 $x_2 \le 0.02$

5.16c $a_0 = L, a_n = \dfrac{2L}{n^2\pi^2}(-1)^n, b_n = 0$

5.20 $\tan(x) = 1 + 2(x - \pi/4) + 2(x - \pi/4)^2 + 2(x - \pi/4)^3 + \cdots$

5.22 $x \le 0.14855$

5.24 $a_n = 0, b_n = \dfrac{2aT}{n^2\pi^2}[1 - (-1)^n](-1)^{(n-1)/2}$

5.27 $\tan(x) = x + x^3/3 + 2x^5/15 + 17x^7/315 + \cdots -\pi/2 < x < \pi/2$

6.1 $\Delta E \approx 2588$ J mol^{-1}

6.3b $(\partial S/\partial T)_{E,n} = (\partial S/\partial T)_{E,V} + (\partial S/\partial V)_{E,T}(\partial V/\partial T)_{E,n}$

6.3d $-(x/u^2)\sin(x/u) + (2y^2/u^3)e^{-y^2/u^2} - (4y/u^2)$
$$= -(x/u^2)\sin(x/u) + (-2we^{-w^2} + 4)(-y/u^2)$$

6.11b $e^4 - 1$

6.12 path 1: x_1
path 2: $x_1 + x_1 y_1$

6.13 12

6.14 17.5

6.16 $\pi h a^2/3$

6.18 $\mathbf{i}3ax^2 + \mathbf{j}e^{bz} + \mathbf{k}bye^{bz}$

6.19 $\mathbf{F} = \dfrac{-Gm_s m_e}{(x^2 + y^2 + z^2)^{3/2}}(\mathbf{i}x + \mathbf{j}y + \mathbf{k}z), |\mathbf{F}| = \dfrac{Gm_s m_e}{r^2}$

6.20 3

6.22 $[(x^4 - 2x) + (y^4 - 2y) + (z^4 - 2z)]e^{x^2}e^{y^2}e^{z^2}$

6.24 $\mathbf{e}_\rho(\partial f/\partial \rho) + \mathbf{e}_\phi \dfrac{1}{\rho}(\partial f/\partial \phi) + \mathbf{e}_z(\partial f/\partial z)$

$[\mathbf{e}_\rho(-2\rho/a^2)\sin(\phi) + \mathbf{e}_\phi(1/\rho)\cos(\phi) + \mathbf{e}_z(-2z/a^2)\sin(\phi)]e^{-(\rho^2 + z^2)/a^2}$

6.27a $(-1, 0)$

6.27b $(-1/2, 1/2)$

6.30 $dE = (2E/3nR)\,dS + (5E/3n)\,dn - (2E/3V)\,dV$ with E given as before

6.31d $(\partial f/\partial x)_y = -3(x + y)^{-4}, (\partial f/\partial y)_x = -3(x + y)^{-4}$

6.32b $(\partial^2 f/\partial x^2)_y = -(1/y^2)\cos(x/y)$
$(\partial^2 f/\partial y^2)_x = -(x^2/y^4)\cos(x/y) - (2x/y^3)\sin(x/y)$
$(\partial^2 f/\partial y\,\partial x) = (2/y^3)\cos(x/y) + (x/y^4)\sin(x/y)$

6.34b exact

6.34d exact

6.34f exact

6.36 inexact, 8, not necessarily

6.37 $f(x, y) = \ln(x + y) - \ln(2)$

6.38 $I = \pi a\rho^6/3 + \pi b\rho^4/2$

6.40 $(3, -4), f(3, -4) = -25$

6.41 $(9/2, -5/2), f(9/2, -5/2) = -20.5$

7.1 $z(t) = a_0 b(t + be^{-t/b}) - a_0 b^2$

$v(30s) = 49.876 \text{ m s}^{-1}$

$z(30s) = 1250.62 \text{ m}$

$\lim_{t \to \infty} v(t) = 50 \text{ m s}^{-1}$

7.5 $8.65 \times 10^{13} \text{ s}^{-1}$

7.12 $y = -2x - C/x$

7.14 193.6 N

7.18 $x = (mv_0/\zeta)(1 - e^{-\zeta t/m})$

7.20 $x = v_0 t - F_0 t^2/(2m), v = v_0 - F_0 t/m, \text{ if } 0 < t < mv_0/F_0$

$x = mv_0^2/(2F_0), v = 0, \text{ if } t > mv_0/F_0$

7.21 $n = Ce^{-Bt} + A/B$ where $C = \text{constant}$

$n(5 \text{ h}) = 0.048771 \text{ mol liter}^{-1}$

7.23 $y = Ae^x + Be^{-x} + Ce^{2x} + e^x(x + x^2)/4$

7.24b $y = Ce^{-x}$

7.24c $y = [C + \sin(x)]/(x^2 - 1)$

7.25 $y = xe^{-\sin(x)}$

7.26 $y_p = (18/5)e^{3x} - (1/2)\sin(x)$

7.27 22800 years

8.1 $f = Ce^{iax}, a = \text{eigenvalue}$

8.3 $-2 - 4x\dfrac{d}{dx}$

8.5b $\hat{B}^2 = x^2 \hat{D}_x^2 + x\hat{D}_x, \hat{B}^2 bx^4 = 16bx^4$

8.7a $(4, 1, 6)$

8.7b $(1, -2, 3)$

8.8a $(-2, 1, -3)$

8.13 $\begin{bmatrix} -1 & 10 & -3 \\ 5 & 14 & 3 \\ -5 & 3 & 1 \end{bmatrix}$

$\begin{bmatrix} 12 & 6 & 10 \\ 7 & 9 & 6 \\ 0 & -1 & -7 \end{bmatrix}$

8.16b $\begin{bmatrix} -2 & 1 \\ 3/2 & -1/2 \end{bmatrix}$

8.18   21

8.20 $47, -47, 47, 0$

8.23b $\begin{bmatrix} -1 & 0 & 0 \\ 0 & -1 & 0 \\ 0 & 0 & 1 \end{bmatrix}$

8.27 $\hat{E}: 3, \hat{C}_3: 0, \hat{C}_3^2: 0, \hat{\sigma}_a: 1, \hat{\sigma}_b: 1, \hat{\sigma}_c: 1$

8.29c $2(\hat{D}_x f)\hat{D}_x + (\hat{D}_x^2 f)$

8.30 $a = \sqrt{km}/(2\hbar)$

8.32 $0, \hbar^2\pi^2/a^2$

8.35b $(-1/2 - \sqrt{3}/2, \sqrt{3}/2 - 1/2, 1)$

8.35d $(1, 1, 1)$

8.36b $\begin{bmatrix} 1 & 0 & 0 \\ 0 & -1 & 0 \\ 0 & 0 & -1 \end{bmatrix}$

8.36d $\begin{bmatrix} -1 & 0 & 0 \\ 0 & 1/2 & -\sqrt{3}/2 \\ 0 & \sqrt{3}/2 & 1/2 \end{bmatrix}$

8.37b $x \cos(x/y)$

8.37d $(y, -x, -z)$

8.38b $\begin{bmatrix} 38 & 26 & -20 & -61 \\ -12 & -56 & 69 & 50 \\ 19 & -13 & 8 & -24 \\ 46 & -15 & 17 & -103 \end{bmatrix}$

8.38d $\begin{bmatrix} 15 & 45 & -27 & 13 \\ 9 & 36 & 1 & 11 \end{bmatrix}$

8.41b $\begin{bmatrix} -0.107143 & 0.214286 & 0.035714 \\ 0.195055 & -0.159341 & -0.013736 \\ 0.082418 & -0.010989 & -0.104396 \end{bmatrix}$

8.41d $\begin{bmatrix} -1/4 & 1/2 & 1/4 \\ -1/4 & 3/2 & -3/4 \\ 1/2 & -1 & 1/2 \end{bmatrix}$

8.42 $\begin{bmatrix} -1 & 0 \\ 4 & 3 \end{bmatrix}$

8.43a $\hat{E} \leftrightarrow \begin{bmatrix} 1 & 0 & 0 \\ 0 & 1 & 0 \\ 0 & 0 & 1 \end{bmatrix}$ $\hat{C}_2 \leftrightarrow \begin{bmatrix} -1 & 0 & 0 \\ 0 & -1 & 0 \\ 0 & 0 & 1 \end{bmatrix}$

$\hat{\sigma}_b \leftrightarrow \begin{bmatrix} 1 & 0 & 0 \\ 0 & -1 & 0 \\ 0 & 0 & 1 \end{bmatrix}$ $\hat{\sigma}_a \leftrightarrow \begin{bmatrix} -1 & 0 & 0 \\ 0 & 1 & 0 \\ 0 & 0 & 1 \end{bmatrix}$

9.2b   pH $= 11.119$

9.4 1575 K

9.5 4.49341

9.7 1.7549

9.11 $x = 60/7$, $y = 50/7$

9.12   no nontrivial solution

9.13 $x = 3$, $y = 1$

9.14 $x_2 = 3$, $x_3 = 5$

9.15   Equations are linearly dependent. $x_1 = 1$, $x_2 = 2$, $x_3 = 3$

9.16 $x_1 = 1/2$, $x_2 = 0$, $x_3 = 3/2$

9.18b
$$\mathbf{X} = \begin{bmatrix} 1/2 \\ 1/\sqrt{2} \\ 1/2 \end{bmatrix}$$

9.19a $x = 1, 2$

9.19c $x = -1 \pm i$

9.22   pH $= 2.955$

9.23 5.140×10^4 K

9.25 3.39925

9.27 $x_1 = 6$, $x_2 = 2$, $x_3 = 1$

9.28a   linearly dependent

9.28c $x_1 = 2$, $x_2 = 1$

9.29   Two sets of roots, both equal: $x = 1$, $y = 1$

9.30b $x_1 = 1$, $x_2 = 1$, $x_3 = 1$, $x_4 = 1$

9.31   Only one eigenvector with nonzero eigenvalue $b = 3$,
$$\mathbf{X} = \begin{bmatrix} 1/\sqrt{3} \\ 1/\sqrt{3} \\ 1/\sqrt{3} \end{bmatrix}$$

10.2 $\bar{x} = 2.8757$ m, $s = 0.0083$ m

10.5 108.00 ± 2.97^0

10.6 131.69 ± 0.21 s

10.7 128.25 ± 2.15 g mol^{-1}

10.13 44.6 kJ mol^{-1}

10.14   first order, $k = 0.0537$ h^{-1}

10.16 2.6×10^{-4} min^{-1}

10.17 2.76×10^{-3} h^{-1}

10.22a 11.002 in, 0.023 in
        9.499 in, 0.019 in

10.22b 11.002 in, 0.014 in
        0.014 in, 0.016 in

10.22c 93.506 in^2, 0.205 in^2

10.22d 93.507 in^2, 0.150 in^2

10.22e $93.506 \ in^2, 0.107 \ in^2$

10.23 $24.6 \ kJ \ mol^{-1}$

10.24a $1.53 \ kJ \ mol^{-1}$

10.25 $1.862 \ J \ K^{-1} \ mol^{-1} \pm .020 \ J \ K^{-1} \ mol^{-1}$

10.26 $3\sigma^4$

10.27 second order $k = 2.06 \times 10^{-2} \ atm^{-1} \ min^{-1} \pm 0.13 \times 10^{-2} \ atm^{-1} \ min^{-1}$

$P_0 = 0.938 \ atm$

10.30 $0.6263 \pm 0.0036 \ mol$

10.31 $1430.16 \pm 1.58 \ liter \ mol^{-1} \ cm^{-1}$

11.1a $74.956 \cdots$

11.3b $(5.48915 \cdots) \times 10^9$

11.4a $R = (V * 3/(4 * P))\uparrow.3333333$

11.23a $-1/3 \pm 0.745356i$

11.23b $0.357143 \pm 0.123718i$

11.26a $Z = X * Y/(A * C)$

11.27c no errors

11.27e $Y = EXP(X)$

$Y = EXP(Y)$

11.26g $Y = LOG10(COS(X))$ or $Y = 2.302585*LOG(COS(X))$

11.27b $Z = SQR(17*Y\uparrow2 + 3*X\uparrow2)/(14*A*B)$

11.27d 100 INPUT N

110 S = 0

120 F = 1

130 FOR I = 1 TO N

140 S = S + A(I)

150 F = F*I

160 NEXT I

170 S = S/F

11.27f $H = 1/(A*X\uparrow2 + B*X+C)$

Index

Absolute value (magnitude) of a number, 14
Acceleration, 75, 179
  due to gravity, 75, 357
Accuracy, 266
Activity coefficients, 243
Adjacent side of a right triangle, 15
Adjoint (hermitian conjugate) of a matrix, 225
Algorithm, 9, 340
Amplitude of a wave, 199
Analytic function, 117
Analytical methods of integration, 95ff
Angle, 15ff
  negative, 16
Antiderivative, 74ff
Area, element of, 158ff
Argand diagram, 39
Argand plane (complex plane), 39
Argument (phase) of a complex number, 40
Argument of a function, 16
Arithmetic assignment statement in BASIC, 313
Arrays in BASIC, 325ff
Assignment statement in BASIC
  arithmetic, 313
  string, 329
Associate (hermitian conjugate) of a matrix, 225
Associative property, 13, 38, 208, 220, 229
Average, 270
  time, 94
Axes, coordinate, 25
  real and imaginary, 39

Bar-graph approximation in numberical integra-
  tion, 100
Base of logarithms, 20
BASIC computer language, 310ff

Basis functions, 110, 123
Batch processing in BASIC, 338ff, 344ff
Beating, 191
Bisection, method of, for numerical solution of
  algebraic equations, 248
Block-diagonal matrix, 234
Boltzmann probability distribution, 246
Boundary conditions, 184, 199
Branching in BASIC, 319ff

Calling a subroutine in BASIC, 348
Card reader, 338
Carriage control symbols in BASIC, 337
Cartesian coordinates, 26, 31
Central limit theorem of statistics, 273
Central processing unit (CPU), 310
Chain rule, 61, 145, 359
Character of an operator in a symmetry group,
  235
Characteristic equation, 183
Clapeyron equation, 283
Coefficients
  of a linear differential equation, 181
  of a series, 110
Cofactor (minor) of a determinant, 226
Column vector, 219
Columns of a matrix, 218
Combination, linear, 182
Combination of errors, 277ff
Commutative property, 13, 38, 208, 220, 229
Commutator, 208
COMPILE statement in BASIC, 343
Compiler program, 311, 340
Complementary equation, 190
Complementary function, 190

397

398